WHITTINGTON'S DICTIONARY OF PLASTICS

SECOND EDITION

WHITTINGTON'S
DICTIONARY OF PLASTICS

by

Lloyd R. Whittington

Sponsored by the

SOCIETY OF PLASTICS ENGINEERS, INC.

a **TECHNOMIC**® publication

TECHNOMIC Publishing Co., Inc.
265 Post Road West, Westport, CT. 06880

PREFACE

During the ten years that have passed since the first edition of this work was published, many changes have occurred in the plastics industry that affect the terms used by those persons who are engaged in its raw material preparation, processing methods, applications, sales of its products, and testing procedures.

During this period numerous new polymers, processing methods and applications have been developed.

The growing influence of toxicity, ecological impacts, flammability and other safety aspects have also introduced the plastics technician to terms formerly of concern to laboratory scientists but now are of importance in all aspects of plastics manufacture and use.

International standards setting organizations have cooperated in developing a universal systems of weights, measures and standard terms to be used throughout the world. The resulting publications which have been accepted in a large degree by most countries and organizations goes far beyond conversion to the old "cgs" metric system with which most U.S. technicians are familiar; in fact much of this old metric system has been abandoned in favor of the "SI," abbreviation for the French name "Le System International d'Unites." Although not officially enforced in the United States, on December 23, 1975 President Ford signed a bill termed the "Metric Conversion Act" which established a board to coordinate the "voluntary conversion to the SI metric system in the U.S." Subsequently, the National Bureau of Standards published "Guidelines for the Metric System of Weights and Measures" for use in the U.S. It recognizes that certain units which are not part of the SI are used so widely that it is impractical to abandon them. In voluntary compliance with the Act, several technical organizations such as the American Society for Testing and Materials and the American Chemical Society have stipulated that papers and definitions submitted to them for publication must give SI units.

In addition to the broadened scope of this dictionary required by the above developments, resulting in new terms not present in the first edition, the author has attempted to broaden many definitions in the first edition to make them more comprehensive. While still called a dictionary, it has been attempted to explain materials and processes in this edition in a manner somewhat midway between the bare, concise definitions typical of the conventional dictionary and full treatises such as appear in encyclopedia.

As in the case of the first edition, it has been attempted wherever possible to employ simple language understandable to students and workers without advanced academic educations.

By permission of the American Society for Testing and Materials, a selected group of ASTM definitions and abstracts of standards, appearing in the 1976 Annual Book of ASTM Standards©, has been reproduced in full or in part.

L. R. Whittington

Ashland, Ohio
June, 1978

Å. Abbreviation for ANGSTROM UNIT, which see.

AB. A prefix attached to the names of the practical electric units to indicate the corresponding unit in the old cgs system (emu), e.g. abampere, abvolt.

ABFA. Abbreviation for AZOBISFORMAMIDE, which see.

Abherent. (abhesive) A coating or film applied to one surface to prevent or reduce its adhesion to another surface brought into intimate contact with it. Abherents applied to plastic films are often called *anti-blocking agents.* Those applied to molds, calendar rolls and the like are sometimes called *release agents* or PARTING AGENT, which see.

Abietic Acid. $C_{19}H_{31}CH_2OH$. A monocarboxylic acid derived from rosin. Its derivatives used as plasticizers include HYDROABIETYL ALCOHOL, HYDROGENATED METHYL ABIETATE and METHYL ABIETATE, which see.

Ablation. Derived from the Latin word *ablatio,* meaning a carrying away, this term was first used in medicine to denote the excision of harmful growths or organs from the body, then by astronomers and astrophysicists to describe the erosion and disintegration of meteors entering the atmosphere, and is now used in the plastics industry for the layer-by-layer decomposition of a plastic mass when heated suddenly to a very high temperature.

Ablative Plastics. A term applied to any polymer or resin which decomposes layer-by-layer when its surface is heated, leaving a heat-resisting layer of charred material which eventually breaks away to expose virgin material. Ablative plastics are used on nose cones of projectiles and space vehicles.

ABL Bottle. An internal pressure test vessel about 18″ diameter by 24″ long, used to determine the quality and strength of filament wound reinforcement material in the vessel.

Abrasion Cycle. The number of repetitive abrading motions to which a specimen is subjected in an abrasion resistance test.

Abrasion Resistance. The ability of a material to withstand mechanical action such as rubbing, scraping, or erosion, that tends progressively to remove material from its surface. ASTM tests for abrasion resistance are D 1044-73, "Resistance of Transparent Plastics to Surface Abrasion;" and D 1242-56, "Resistance to Abrasion of Plastic Materials." Both are located in Part 35 of the 1976 Book of ASTM Standards.

Abrasive Finishing. A method of removing flash, gate marks and rough edges from plastic articles by means of abrasive belts or wheels. The process is usually employed on large rigid or semi-rigid products with intricate surfaces which cannot be treated by tumbling or other more efficient deflashing methods.

ABS. Abbreviation for acrylonitrile-butadiene-styrene. See ABS RESINS.

Absolute Specific Gravity. The ratio of the weight of a given volume of a substance to that of an equal volume of water at the same temperature, as determined by an apparatus which provides correction for the effects of air buoyancy. See also SPECIFIC GRAVITY.

Absolute Viscosity. (dynamic viscosity) The tangential force or unit area of either of two parallel planes at unit distance apart when the space between the planes is filled with the fluid in question and one of the planes moves with unit differential velocity in its own plane. The C.G.S. unit for absolute viscosity is the poise (dyne-sec/sq. cm.). The centipoise (0.01 poise) is also used. See also VISCOSITY.

Absorptiometry. An analytical technique utilizing the absorption of electromagnetic radiation by a specimen as a property related to the composition and/or the quantity of a given material in the specimen. When the electromagnetic radiation is in the ultraviolet, the visible, and the near infrared portions of the spectrum, the technique is often called *spectrophotometry.* When the absorbing medium is in the gaseous state, the absorption spectrum will consist of dark lines or bands, being the reverse of the emission spectrum of the absorbing surface. When the absorbing medium is in the solid or liquid state the spectrum of the transmitted light shows broad dark regions which are not resolvable into lines and have no sharp or distinct edges.

Absorption. (1) The penetration of a substance into the mass of another substance by molecular or chemical action. See, for example, WATER ABSORPTION. (2) The process whereby energy is dissipated within a specimen placed in a field of radiation energy. Since processes other than absorption occur, e.g. scattering, the total amount of energy removed from the source of energy will be greater than the amount actually dissipated within the sample.

ABS Resins. A family of thermoplastics based on acrylonitrile, butadiene and styrene combined by a variety of methods involving polymerization, graft copolymerization, physical mixtures and combinations thereof. Over 70 standard grades of ABS resins are available, plus some special grades modified to yield unusual properties. The standard grades are rigid, hard and tough but not brittle, and possess good impact strength, heat resistance, low temperature properties, chemical resistance and electrical properties. ABS compounds in pellet form can be extruded, blow molded, calendared and injection molded. ABS powders are available for rotational molding. ABS powders are used as modifiers for other resins, for example PVC. Typical applications for ABS resins are household appliances, automotive parts, business machine and telephone components, pipe and pipe fittings, packaging and shoe heels.

Abvolt. The cgs electromagnetic unit of potential difference and electromotive force. It is the potential difference that must exist between two points in order that one erg of work be done when one abcoulomb of charge is moved from one point to the other. One abvolt is one $\times \ 10^{-8}$ volt.

Accelerated Test. A test procedure in which conditions are intensified to reduce the time required to obtain a deteriorating effect similar to one resulting from normal service conditions.

Accelerator. (promoter) A substance which hastens a reaction, usually by acting in conjunction with a CATALYST or a CURING AGENT, which see. Accelerators are sometimes used in the polymerization of thermoplastics, but are used most widely in curing systems for thermosets and natural and synthetic rubbers. Sometimes called *cocatalyst.*

Accumulator. (1) In blow molding, an auxiliary ram extruder used to provide fast parison delivery. The accumulator cylinder is filled with plasticated melt from the main extruder between parison deliveries or "shots," and stores this melt until the plunger is required to deliver the next parison. (2) A device for conserving energy in hydraulic systems of molding equipment.

Acetal Copolymer Resins. A family of highly crystalline thermoplastics prepared by copolymerizing trioxane with small amounts of a comonomer which randomly distributes carbon-carbon bonds in the polymer chain. These bonds, as well as hydroxyethyl terminal units, give the acetal copolymers a high degree of thermal stability and resistance to strong alkaline environments. The acetal copolymer resins marketed by Celanese Plastics Co. under the registered trademark Celcon range in melt index from 1.0 to 27.0 and yield products which maintain high levels of performance in long-term applications. For example, Celcon pipe fittings are used in hot water distribution systems and appliances.

Acetaldehyde. CH_3CHO. (ethanal, ethyl aldehyde, acetic aldehyde) A colorless, flammable liquid made by the hydration of acetylene, the oxidation or dehydrogenation of ethyl alcohol, and the oxidation of saturated hydrocarbons or ethylene. Among its many applications as a highly reactive intermediate are the production of thermosetting resins by reaction with cresol or phenol, with polyvinyl alcohol to form polyvinyl acetal resins.

Acetal Resins. (polyformaldehyde, polyoxymethylene resins) Thermoplastics produced by the addition polymerization of aldehydes through the carbonyl function, yielding unbranched polyoxymethylene chains of great length. Examples are Du Pont's "Delrin" and Celanese Corporation's "Celcon" (an acetyl copolymer based on trioxane). The acetal resins are among the strongest and most stiff of all thermoplastics, and are also characterized by good fatigue life, resilience, low moisture sensitivity, high solvent and chemical resistance, and good electrical properties. They may be processed by conventional injection molding and extrusion techniques, and fabricated by welding methods used for other plastics.

Acetate. (1) A derivative of acetic acid. (2) A generic name for cellulose acetate plastics, particularly for fibers thereof. Where at least 92% of the hydroxyl groups are acetylated, the term *triacetate* may be used as the generic name of the fiber. (3) A compound containing the acetate group, CH_3COO-.

Acetate Fibers. Fibers made by partially acetylating cellulose. The term *triacetate fibers* is applied to almost completely acetylated cellulose. At one time, acetate fibers were third in place among synthetic fibers, surpassed only by rayon and nylon. Their major application was for ladies' apparel and home furnishing fabrics. The triacetates were developed later and became more widely used due to their similarity to acrylics, nylons and polyesters. According to the FTC an acetate fiber must be one in which not less than 92% of the hydroxyl groups are acetylated in order to be called a triacetate fiber. At one time acetate fibers were called rayons, which is no longer permissible because rayons are of a completely different chemical structure.

Acetic Acid. CH_3COOH. (ethanoic acid, methanecarboxylic acid, vinegar acid) A colorless liquid with the familiar taste and odor of vinegar, of which it is the chief constituent in dilute form. Acetic acid was originally derived by souring wine and beer, but is most often synthesized today by oxidation of acetaldehyde in the presence of a catalyst. Among the uses of acetic acid in the plastics industry are the manufacture of cellulosic plastics such as CA, CAB and CAP; vinyl acetate; and acetate esters for plasticizing thermoplastics.

Acetic Aldehyde. See ACETALDEHYDE.

Acetic Ether. See ETHYL ACETATE.

Acetone. CH_3COCH_3. (dimethyl ketone, 2-propanone) The simplest and most important member of the ketone family of solvents. All of the cellulosic plastics, PVC, PVAc, PMMA, epoxies, and some thermosetting resins are soluble in acetone. It is also an intermediate in the production of Bisphenol A, antioxidants, and resins such as methyl methacrylate, cellulose acetate, and epoxies.

Acetone Extraction. In molded phenolic products, the amount of acetone-soluble material that can be extracted from the material is an indication of the degree of cure. A test for determining such material is given in ASTM D 494-46 found in part 35 of the 1976 ASTM Book of Standards.

Acetone Resin. A synthetic resin produced by the reaction of acetone with materials such as phenol or formaldehyde.

Acetylation. The introduction of an acetyl group into the molecule of an organic compound having an −OH or −NH_2 group, by treatment with acetic anhydride or acetyl chloride.

Acetyl Cyclohexane Sulfonyl Peroxide. A polymerization initiator, often used in conjunction with a dicarbonate such as di-sec-butyl peroxydicarbonate in vinyl chloride polymerization. These newer initiators have largely been replaced by benzoyl and lauroyl peroxides which were the principle initiators in the early years of PVC production.

Acetylene. C_2H_2. (ethyne, ethine) A colorless gas derived by reacting water with calcium carbide, or by cracking petroleum hydrocarbons. In the plastics industry it is an important intermediate in the production of vinyl chloride, neoprene, acrylonitrile, and trichloroethylene. See also POLYACETYLENES.

Acetylene Black. Carbon black produced by incomplete combustion of acetylene. It is used as a filler in plastics, imparting electrical conductivity. See also CARBON BLACK.

Acetylene Polymers. See POLYACETYLENES.

n-Acetyl Ethanolamine. $CH_3CONHC_2H_4OH$. (hydroxyethyl acetamide) A plasticizer for polyvinyl alcohol and cellulosic plastics.

Acetyl Peroxide. $(CH_3CO)_2O_2$. (diacetyl peroxide) A resin catalyst.

4-Acetyl Resorcinol. $C_6H_3(OH)_2COCH_3$. (2,4-dihydroxy-acetophenone) A light stabilizer for plastics.

Acetyl Ricinoleates. A generic term for a number of important plasticizers.

Acetyl Triallyl Citrate. $CH_3COOC_3H_4(COOCH_2CH:CH_2)_3$. A cross linking agent for polyesters, and a polymerizable monomer. Easily polymerized with peroxide catalysts, it forms a clear, hard thermosetting resin.

Acetyl Tributyl Citrate. $CH_3COOC_3H_4(COOC_4H_9)_3$. A plasticizer commonly used in vinyl plastics, derived from the esterification and acetylation of citric acid. It has been FDA approved for food contact use.

Acetyl Triethyl Citrate. $CH_3COOC_3H_4(COOC_2H_5)_3$. A plasticizer derived from the esterification and acetylation of citric acid, used in cellulose nitrate, cellulose acetate, and certain vinyls such as polyvinyl acetate. It has been FDA approved for food contact use.

Acetyl Tri-2-Ethylhexyl Citrate. $C_{32}H_{58}O_8$. A plasticizer for vinyls, with limited compatibility for cellulose nitrate and ethyl cellulose.

Acetyl Value. The number of milligrams of KOH necessary to neutralize the acetic acid liberated by hydrolysis of one gram of an acetylated compound.

Acid-Acceptor. A compound which acts as a stabilizer by chemically combining with acid which may be initially present in minute quantities in a plastic, or which may be formed by the decomposition of the resin. See also STABILIZER.

Acid Number. See ACID VALUE.

Acidolysis. The process of reacting an acid with an ester. See also ESTER EXCHANGE.

Acid Resistance. The ability of a plastics material to withstand attack by an acid. Most plastics have a high degree of acid resistance. Tests for determining resistance of plastics to some acids are given in ASTM D 543-67, published in Part 35 of the 1976 Book of ASTM Standards.

Acid Value. The measure of free acid content of a substance. It is expressed as the number of milligrams of KOH neutralized by the free acid present in one gram of the substance. This value, also called *acid number,* is sometimes used in connection with the end-group method of determining the molecular weight of polyesters. It is also used in evaluating plasticizers, in which acid values should be as low as possible.

Acrolein. CH_2CHCHO. (acrylic aldehyde, propenal, allyl aldehyde, acraldehyde) A liquid derived from the oxidation of allyl alcohol or propylene, used as an intermediate in the production of polyester resins and polyurethanes.

Acrylamide. $CH_2CHCONH_2$. A crystalline solid at room temperature, capable of polymerization or copolymerization.

2-Acrylamido-2-Methylpropanesulfonic Acid. (AMPS) A solid aliphatic sulfonic acid monomer produced by Lubrizol Corp. Its homopolymers are water-soluble and hydrolytically stable. It can be incorporated into other polymers by crosslinking.

Acrylate Resins. See ACRYLIC RESINS.

Acrylic Acid. $CH_2CHCOOH$. (acroleic acid, ethylene-carboxylic acid, vinylformic acid, propenoic acid) A colorless, unsaturated acid which polymerizes readily. The homopolymer is not often used except as a textile sizing agent, but esters of acrylic acid are widely used in the production of acrylic resins.

Acrylic Aldehyde. See ACROLEIN.

Acrylic Esters. (acryl esters) Esters of acrylic or methacrylic acid or of structural derivatives thereof. Polymers derived from these esters vary from soft, elastic, film-forming materials to hard plastics. They are readily polymerized as homopolymers or copolymers with many other monomers, contributing improved resistance to heat, light and weathering. Some members of the acrylic ester family, e.g. butylene dimethacrylate and trimethylolpropane trimethacrylate, function as reactive plasticizers in PVC and elastomers. They serve as plasticizers during processing, then polymerize during cure to impart hardness to the finished article. See also ACRYLIC RESINS.

Acrylic Fiber. Generic name for a manufactured fiber in which the fiber-forming material is any long chain synthetic polymer composed of at least 85% by weight of acrylonitrile units $-CH_2CH(CN)-$. (Federal Trade Commission).

Acrylic Foam. The acrylic foam that has gained widespread acceptance for lining drapes and coating fabrics is made by mixing an emulsified acrylic with compressed air in the ratio of one part of emulsion with four or five parts of air, spreading the foam on a substrate, then drying in an oven. The emulsion may contain fillers and pigments to provide opacity, and a foaming aid such as ammonium stearate. When the coated fabric must have abrasion resistance for washing and cleaning, the acrylic foam can be crushed between rollers to partially collapse the cell structure.

Acrylic Resins. Polymers of acrylic or methacrylic esters, sometimes modified with non-acrylic monomers such as the ABS group. The acrylates may be methyl, ethyl, butyl or 2-ethylhexyl. Usual methacrylates are the methyl, ethyl, butyl, laural and stearyl. The resins may be in the form of molding powders or casting syrups, and are noted for their exceptional clarity and optional properties. Acrylics are widely used in lighting fixtures because they are slow burning or may be made self-extinguishing, and do not produce harmful smoke or gases in the presence of flame.

Acrylic Rubber. (AR) A synthetic rubber at least partially made from acrylonitrile, or from ethyl acrylate copolymerized with many of the monomers or block polymers of the synthetic rubber family.

Acrylonitrile. (propenenitrile, vinyl cyanide) A monomer with the structure $(CH_2:CHCN)$. It is most useful in copolymers. Its copolymer with butadiene is nitrile rubber, and several copolymers with styrene exist that are tougher than polystyrene. It is also used as a synthetic fiber and as a chemical intermediate.

Acrylonitrile-Butadiene Copolymers (NBR). A family of copolymers ranging from about 18 to 50% of acrylonitrile, and sometimes including small amounts of a third monomer. The family includes the German materials Perbunan and Buna-N, and the nitrile rubbers. The outstanding property of this so-called nitrile rubber family is excellent resistance to oils, fats and hydrocarbons such as motor fuels, making them useful for motor gaskets, abrasion linings, conveyor belts and hoses for oils and fuels.

Acrylonitrile-Butadiene-Styrene Resins. See ABS RESINS, NITRILE BARRIER RESINS.

Acrylonitrile-Styrene Copolymers. A series of copolymers which have the transparency of polystyrene, but with improved stress cracking and solvent resistance.

Activation. (1) The process of inducing radioactivity in a specimen by bombardment with neutrons or other types of radiation. (2) The process of rendering a thermoplastic surface more receptive to printing inks, paints or adhesives by means of chemical treatments, corona discharge or flame treatment.

Activator. (1) An agent added to the accelerator in natural or synthetic resins to enhance the action of the accelerator in the vulcanization process. (2) A chemical additive used to initiate the chemical reaction in a specific mixture.

Actuators. Devices that control the movement or mechanical action of a machine indirectly rather than directly or by hand. They can perform linear or rotary motions, and are usually motivated by means of pneumatic or hydraulic cylinders.

Adaptor. (die adaptor) In an extruder, the portion of the die assembly that attaches the die to the extruder and provides a flow channel for the molten plastic between the extruder and the die.

Adapter Plate. In injection molding, the plate holding the mold to the press or platen.

Adapter Ring. An annular retaining part for extrusion and injection apparatus.

ADC. Abbreviation for ALLYL DIGLYCOL CARBONATE, which see.

Addition Polymerization. A polymerization reaction in which molecules of the monomer join together to form a polymeric product in which the molecular formula of the repeating unit is identical with that of the monomer. The molecular weight of the polymer so formed is thus the total of the molecular weights of all of the combined monomer units.

Additive. Any substance which is added to another substance. In the plastics industry the term is most often employed for materials added in minor amounts to basic resins or compounds to alter properties.

Adducts. (inclusion complexes) Crystalline mixtures, not true compounds, in which molecules of one of the components are contained within the crystal lattice framework of the other component. The complexes are stable at room temperatures but the entrapped component can escape when the mixture is melted or dissolved.

Adherend. (n) A body which is held to another by an adhesive.

Adherometer. An instrument which measures the strength of an adhesive bond.

Adhesion. The state in which two surfaces are held together by interfacial forces which may consist of valence forces or interlocking action or both. One method for testing the strength of adhesive bonds is ASTM D 952-75, found in Part 35 of the 1976 Book of ASTM Standards. See ADHESION, MECHANICAL; ADHESION, SPECIFIC.

Adhesion, Mechanical. Adhesion between surfaces in which the adhesive holds the parts together by interlocking action. See ADHESION, SPECIFIC.

Adhesion Promoter. A coating which is applied to a substrate before it is coated with a plastic, serving to improve the adhesion of the coating to the substrate. Typical adhesion promoters are based on silanes and silicones with hydrolyzable groups on one end of their molecules which react with moisture to yield silanol groups, which in turn react with or adsorb to inorganic surfaces to enable strong bonds to be made. At the other ends of the molecules are reactive, but non-hydrolyzable groups that are compatible with resins or elastomers in adhesive formulations. Adhesion promoters are added to the adhesive as water or ethanol solutions.

Adhesion, Specific. Adhesion between surfaces which are held together by valence forces of the same type as those which give rise to cohesion. See ADHESION, MECHANICAL.

Adhesive-Assembly. The process of joining two or more plastic parts other than flat sheets (for which the term *laminating* is used) by means of an adhesive. A related term is SOLVENT WELDING, which see.

Adhesive Film. A thin, dry film of resin, usually a thermoset, used as an interleaf in the production of laminates such as plywood. Heat and pressure applied in the laminating process cause the film to bond both layers together.

Adhesives. Materials capable of joining one surface to another. Adhesives are used in the plastics industry to join a plastic article to another article of (a) the same plastic, (b) a different plastic, or (c) a non-plastic material. Adhesives based on plastics are also used in other industries to join non-plastic materials, for example plywood, glass, cloth and metals. Adhesives used in all of these applications can be classified into five types. A *monomeric cement* is a monomer of at least one of the polymers to be joined, and is catalyzed so that a bond is produced by polymerization. A *solvent cement* is one which dissolves the plastics being joined, forming strong intermolecular bonds, then evaporates. *Bonded adhesives* are solvent solutions of resins, sometimes containing plasticizers, which dry at room temperature. *Elastomeric adhesives* contain natural or synthetic rubbers either dissolved in solvents or suspended in water or other liquid, and are cured at room or elevated temperatures. *Reactive adhesives* are those containing partially polymerized resins, e.g. expoxies, polyesters or phenolics, which cure with the aid of catalysts to form a bond. See also ELECTROMAGNETIC ADHESIVE, HOT MELTS.

Adiabatic. (adj) Denoting a process in which no heat is deliberately added or removed. The term is used somewhat incorrectly to describe the method of extrusion in which heat is developed from mechanical action of the screw to an extent sufficient to plastify the compound.

Adiabatic Extrusion. See EXTRUSION, AUTOTHERMAL.

Adipate Plasticizers. For plasticizers derived from adipic acid, see benzyloctyl bis(2,2,4-trimethyl-1,3-pentanediol) monoisobutyrate- dibutoxyethoxy ethyl- dibutoxyethyl- dibutyl-dicapryl- didecyl- diethoxyethyl- diethyl- di(2-ethylhexyl)- di-n-hexyldi-isobutyl- di-iso-decyl- di-isooctyl- dimethoxyethyl- di(methylcyclohexyl)- dinonyl- ditetrahydrofurfuryl-octyl decyl- and polypropylene adipates.

Adipic Acid. $COOH(CH_2)_4COOH$. (hexanedioic acid, 1,4-butanedicarboxylic acid) A dibasic acid used in the production of polyamides, alkyd resins and urethane foams. Esters of adipic acid are used as plasticizers and lubricants. A list of plasticizers derived from adipic acid is given under ADIPATE PLASTICIZERS.

Adiponitrile. $NC(CH_2)_4CN$. An intermediate used in the manufacture of nylon 6/6.

Adipodinitrile Carbonate. (ADNC) See 5,5-TETRAMETHYLENE DI(1,3,4-DIOXAZOL-2-ONE).

ADNC. Abbreviation for adipodinitrile carbonate. See 5,5'-TETRAMETHYLENE DI(1,3,4-DIOXAZOL-2-ONE).

Adsorption. The adhesion of the molecules of gases, dissolved substances, or liquids in more or less concentrated form to the surfaces of solids or liquids with which they are in contact; a concentration of a substance at a surface or interface of another substance.

Aeolotrophy. See ANISOTROPY.

Aerosol. A suspension of liquid or solid particles in a gas under pressure. In the packaging industry, the term is defined as a self-contained sprayable product in which the propellant force is supplied by a liquefied gas, for example "Freon."

Affine Deformation. A type of deformation in which each element in the volume distorts in the same way as does the volume as a whole.

Affinity. With respect to an adhesive, affinity is an attraction or polar similarity between the adhesive and an adherend.

After-Bake. A technique used with phenolic or amino resins to increase the output of a molding press by ejecting moldings before they are fully cured, subsequently baking them. Also, after-baking may be used on fully cured parts to improve their electrical properties and heat resistance.

Ageing (aging). The process of, or the results of, exposure of plastics to natural or artificial environmental conditions for a prolonged period of time. See also ARTIFICIAL AGEING, ARTIFICIAL WEATHERING.

Aggregate. In the reinforced plastics industry, a hard fragmented material used with an epoxy binder as a flooring or surfacing medium, or in epoxy tooling.

AI. Abbreviation for amide-imide polymers. See POLYAMIDE-IMIDE RESINS.

Air-Assist Vacuum Forming. A modification of the process of THERMOFORMING, which see, in which partial preforming of the sheet is effected by air flow or air pressure before vacuum pull-down.

Air Brush. A small spray gun designed for manually spraying small areas and relatively fine lines, for example in the decoration of toys and doll faces.

Air Bubble Viscometer. An instrument used to measure the viscosity of oils, varnishes and resin solutions by matching the rate of rise of an air bubble in the sample liquid with the rate of rise in one of a series of standard liquids. The standard liquids are numbered to designate the viscosity index. The Gardner-Holt Bubble Viscometer utilizes this principle.

Air Gap. (1) In extrusion coating, the distance from the die opening to the nip formed by the pressure roll and the chill roll. (2) In the radio frequency heating of plastics, the space between the electrode and the surface of the material.

Air-Knife Coating. A knife coating technique especially suitable for thin coatings such as adhesives, wherein a high pressure jet of air is forced through orifices in the knife to meter and control the thickness of the coating. See also SPREAD COATING.

Airless Blast Deflashing. The process of deflashing molded parts by bombarding them with tiny nonabrading pellets which break off flash by impact. See also BLAST FINISHING.

Airless Spraying. A spraying process used for decorating plastics or other materials with plastics coatings, in which the coating material is forced through a small orifice at such a high pressure that no air is needed to atomize the material. The process has also been used in the reinforced plastics field for the SPRAY-UP process, which see.

Air Locks. Surface depressions on a molded part, caused by air trapped between the mold surface and the plastic material.

Air Ring. (air cooling ring) In the process of blowing tubular film, a circular manifold with an annular opening concentric with and just above the extrusion die opening, which blows a uniform stream of cooling air on the plastic tube. The air is sometimes refrigerated.

Air-Slip Forming. A vacuum forming process in which a male mold is enclosed in a box which forms an air cushion as the mold advances toward the hot plastic sheet, thus keeping the mold from contacting the sheet until the end of its travel. At this point, vacuum is applied to destroy the air cushion and pull the sheet against the mold. See also THERMO-FORMING.

Air Vent. A small outlet, usually a groove, to prevent entrapment of gases as molding material enters a mold.

Albedo. The fraction of electromagnetic radiation reflected by a surface.

Alcohol. A generic term for organic compounds having the general structure ROH. In the simplest alcohols, R is a C_nH_{2n+1} group, for example CH_3OH (methanol) or C_2H_5OH (ethanol). In more complex alcohols R may be other alkyl, acyclic or alkaryl groups.

Alcohols are classified according to the number of OH (hydroxyl) groups they contain — monohydric, dihydric, trihydric or polyhydric. Dihydric alcohols are also called glycols; trihydric alcohols are also known as glycerol, glycerin or glycyl alcohol; and the term polyol is used for polyhydric alcohol. Alcohols have many important applications in the plastics industry. They are used directly as solvents and diluents. Many esters of alcohols with organic acids are plasticizers. As intermediates, alcohols are used in the production of resins such as acrylics, alkyds, aminos, polyurethanes and epoxies.

Alcohol, Absolute. Ethyl alcohol which has been rendered anhydrous by drying, containing in excess of 99.9% alcohol.

Alcohol, Denatured. Ethyl alcohol which has been adulterated with toxic material so as to render it unfit for internal consumption but remaining suitable for use as an industrial solvent or reactant.

Alcoholysis. The process of reacting an ester with an alcohol. See also ESTER EXCHANGE.

Aldehyde. A generic term used to describe organic compounds containing a carbonyl group attached to a terminal carbon atom. They may be represented by the general formula R-CHO. For formaldehyde, the simplest and most widely used member in plastics manufacture, "R" is hydrogen. For all other aldehydes, "R" represents a hydrocarbon radical.

Alfin Catalysts. Catalysts obtained from alkali alcoholates derived from a secondary alcohol, used for polymerizing olefins.

Alginates. Derivatives of alginic acid used as emulsifying agents, thickeners and films.

Aliphatic. (adj.) The class of organic compounds such as petroleum and propane whose molecules do not have their carbon atoms arranged in a ring structure.

Alizarin. 1,2-dihydroxyanthraquinone. The basis for many pigments.

Alkali. Any of the hydroxides and carbonates of the alkali metals (potassium, sodium and lithium) and the radical ammonium. The term is also used more generally for any strong base in aqueous solution capable of forming salts.

Alkali Cellulose. See REGENERATED CELLULOSE.

Alkali Resistance. The ability of a plastic material to resist the effects of an alkali. Several alkalis are listed as reagents to be used in testing plastics for chemical resistance in ASTM D 543-67.

Alkane-Imide Resin. A thermoplastic introduced by Raychem Corp. under the trademark Polyimidal. The polymer retains high strength up to 392°F. and melts at 575°F. It has good electrical properties, low water absorption, and high solvent resistance. Molding and extrusion grades are available.

Alkanes. The generic term for saturated hydrocarbons, that is, compounds containing carbon and hydrogen only and all of the hydrogen atoms they can possibly contain. Alkanes can be represented by the general formula C_nH_{2n+2}. The first member of the alkane series is methane, CH_4.

Alkenes. Same as OLEFINS, which see.

Alkyd. The term derived from the names of the combining functional groups in alcohols and acids – "a" from alcohols and "cid" from acids.

Alkyd Molding Compounds. Compounds based on alkyd resins containing fillers, pigments, lubricants and other additives. Alkyd molding compounds are chemically similar to polyester resins, but the term alkyd is usually applied to those polyester formulations which use lesser quantities of monomers of the high viscosity or dry type, resulting in free-flowing granular and nodular types. The compounds are used for applications requiring good electrical properties and long-term dimensional stability such as automotive distributor caps, rotors and coil caps. Alkyds can be compression molded under low pressures, cure rapidly and present no venting problems because no volatiles are liberated during cure.

Alkyd Resins. Unsaturated polyesters produced by reacting an organic alcohol with an organic acid are dissolved in and reacted with unsaturated monomers such as styrene, diallyl phthalate, diacetone acrylamide or vinyl toluene to form alkyd resins. A peroxide catalyst is used to initiate crosslinking between the polyester resin and the monomers, which results in a cured thermoset system. These resins are used as coatings and finishes, and in the production of alkyd molding compounds. These compounds may be free-flowing powders, putty-like masses in bulk or rope form, and may be glass reinforced.

Alkyl. A general term for a monovalent aliphatic hydrocarbon radical, which may be represented as having been derived from an alkane by dropping one hydrogen from the formula. Corresponding aromatic radicals are known as *aryl*. Examples of alkyl groups are C_2H_5 – (ethyl), $CH_3CH_2CH_2$ – (propyl), and $(CH_3)_2CH$– (isopropyl).

Alkyl Aluminum Compounds. A family or organo-aluminum compounds widely used as catalysts in the polymerization of olefins. Members include trialkyl compounds such as triethyl, tripropyl and triisobutyl aluminums; alkyl aluminum hydrides such as diisobutyl aluminum hydride and diethyl aluminum hydride; and alkyl aluminum halides such as diethyl aluminum chloride.

Alkyl Aryl Phosphate. $(C_8H_{17}O)(C_6H_5O)_2PO$. A plasticizer for cellulose acetate butyrate, ethyl cellulose, polystyrene and vinyl resins.

Alkyl Aryl Phthalate. Plasticizer for cellulosic plastics, polymethyl methacrylate, polystyrene and vinyl plastics.

Alkylene Oxide Polymers. A term sometimes used for EPOXIDES, which see.

Alkyl Epoxy Stearate. A plasticizer for PVC and some other resins, with good low temperature properties. Also acts as a stabilizer.

Alkyl Polysulfide. See POLYSULFIDES.

bis(Alkylthio) Cadmium. A stabilizer for PVC.

Allobar. A form of an element having different atomic weights and thus differing in isotropic composition from the naturally occurring form of the element.

Allomerism. A similarity of crystalline form with a difference in chemical composition. See POLYALLOMERS.

Allophanamide. See BIURET.

Allotrophy. The ability of a substance to exist in two or more solid, liquid or gaseous forms due to differences in the arrangement of atoms or molecules.

Alloy. A term sometimes used in the plastics industry to denote blends of polymers or copolymers with other polymers or elastomers. An example is a blend of styrene-acrylonitrile copolymer with butadiene-acrylonitrile rubber. The term *polyblend* is sometimes used for such mixtures.

Allyl. (n) The unsaturated radical C_3H_5, which upon liberation forms diallyl, C_6H_{10}, a pungent volatile liquid.

Allyl Alcohol. CH_2:$CHCH_2OH$. (propenyl alcohol, AA, 2-propen-1-ol). A colorless liquid with a characteristic pungent odor, obtained from the hydrolysis of allyl chloride (from propylene) with dilute caustic, or by the dehydration of propylene alcohol. It is a basic material for all allyl resins, and its esters are used as plasticizers.

Allyl Aldehyde. See ACROLEIN.

Allyl Chloride. CH_2CHCH_2Cl. (3-chloropropene, alpha-chloropropylene, AC, chloroallylene) Used in the preparation of allyl alcohol and various thermosetting resins.

Allyl Cyanide. CH_2:$CHCH_2CN$ (3-butenenitrile, vinyl-acetonitrile). Used as a crosslinking agent.

Allyl Diglycol Carbonate. (ADC) A colorless, water-clear monomer which can be polymerized and cast into a variety of transparent, optical grade products. The thermosetting polymer has the highest scratch and abrasion resistance of all transparent plastics. It can be copolymerized with other unsaturated monomers such as vinyl acetate or methyl methacrylate to produce polymers with a wide variety of properties.

Allyl Diglycol Carbonate Resins. Thermosetting resins with outstanding optical clarity, good abrasion resistance and mechanical properties. The resins are made by polymerizing the monomer of the same name with catalysts such as benzoyl peroxide or, preferably, diisopropyl peroxy dicarbonate. The monomer may also be copolymerized with a variety of monomers and unsaturated compounds such as vinyl acetate, methylmethacrylate and maleic anhydride to yield copolymers with a wide range of properties.

Allyl Esters. Esters of allyl alcohol, used in the production of plasticizers and resins.

Allyl Resins. Resins formed by the addition polymerization of compounds containing the group $CH_2:CH-CH_2$, such as esters of allyl alcohol and dibasic acids. They are commercially available as monomers, as partially polymerized prepolymers, or as molding compounds. The most important member of the family is diallyl phthalate (DAP). Others are diallyl isophthalate (DAIP), diallyl maleate (DAM) and diallyl chlorendate (DAC). The monomers and partial polymers may be cured with peroxide catalysts to thermosetting resins of good high temperature performance, solvent and chemical resistance. The molding compounds may be reinforced with glass fibers or other reinforcements, and are easily molded by compression and transfer molding techniques.

Alpha. A prefix, usually abbreviated as the Greek letter α, denoting the location of a substituting group of atoms in the main group of a compound.

Alpha Cellulose. A colorless filler obtained by treating wood pulp with alkalis, used in light colored thermosetting resins such as urea formaldehyde and melamine formaldehyde. The material is sometimes treated with resinous agents to coat the individual particles and reduce water absorption of the finished parts.

Alpha Olefins. A sub-group of the Olefin group of unsaturated hydrocarbons of the general formula C_nH_{2n} in which the number of carbon atoms ranges from 5 to 20.

Alpha Paper. Paper made from purified wood cellulose pulp, used as surfacing sheets of decorative laminated plastics.

Alpha Particle. (alpha ray) One of the particles emitted in radioactive decay. It is identical with the nucleus of the helium atom and consists, therefore, of two protons plus two neutrons bound together. A moving alpha particle loses energy rapidly in traversing through matter, for example only a few centimeters of air before coming to rest.

Alternating Copolymer. A copolymer in which the two different monomeric units alternate along the chain in an $-A-B-A-B-$ manner.

Alternating Strain Amplitude. Analogous to ALTERNATING STRESS AMPLITUDE, which see.

Alternating Stress Amplitude. A test parameter of a dynamic fatigue test; one half of the algebraic difference between the maximum and minimum stress in one cycle.

Alternative Stress. A stress varying between two maximum values which are equal but with opposite signs, according to a law determined in terms of the time.

Alumina Trihydrate. (aluminum hydroxide, aluminum hydrate, hydrated alumina, hydrated aluminum oxide) $Al_2O_3 \cdot 3H_2O$ or $Al(OH)_3$. A white crystalline powder used as a fire and smoke retarding filler in glass-reinforced polyesters and other cross-linked thermosetting or thermoplastic resins. When heated, alumina trihydrate releases water which lowers the surface temperature of the plastic surface and helps to provide a barrier to combustion.

Aluminum Alkyls. (Aluminum trialkyls) Catalysts used in the Ziegler process of polymerization. Examples are triethylaluminum and triisobutylaluminum.

Aluminum Chelates. Chemically modified aluminum secondary butoxide, used as curing agents for epoxy, phenolic and alkyd resins.

Aluminum Distearate. $Al(OH)(C_{18}H_{35}O_2)_2$ A white powder used as a lubricant for plastics.

Aluminum Hydroxystearate. $Al(OH)[OOC(CH_2)_{10}CHOH(CH_2)_5CH_3]_2$ A white powder used as a plastics lubricant.

Aluminum Isopropylate. $Al(OC_3H_7)_3$ (Aluminum isopropoxide) A white solid, used as a cross-linking agent.

Aluminum Monostearate. $Al(OH)_2[OOC(CH_2)_{16}CH_3]$ A white to yellowish white powder, used as a stabilizer.

Aluminum Oleate. $Al(C_{18}H_{33}O_2)_3$ A plastics lubricant.

Aluminum Palmitate. $Al(OH)_2(C_{16}H_{31}O_2)$ A plastics lubricant.

Aluminum Silicates. Minerals with varying proportions of Al_2O_3 and SiO_2, occurring naturally in clays. They are used as pigments and fillers in plastics.

Ambient. (adj) Completely surrounding; indicative of the surrounding environmental conditions such as temperature, pressure, atmosphere, etc.

Ambient Temperature. The temperature of the medium surrounding an object. The term is often used to denote prevailing room temperature.

Amide. An organic compound containing the $-CONH_2$ group. It is closely related to the organic acids with the COOH grouping, and a common example is urea: $CO(NH_2)_2$.

Amide-Imide Resins. See POLYAMIDE-IMIDE RESINS.

Amine. A compound derived from ammonia by substitution of one or more hydrogen atoms by hydrocarbon radicals.

Amine Resins. See AMINO RESINS.

Amino-. (Amido) A prefix denoting the presence of an $-NH_2$ or NH group.

Amino Acids. Organic acids containing an amino group, obtained by the hydrolysis of a protein or by synthesis. They contain carboxyl and amino groups linked to the same carbon atom.

n-beta-(Aminoethyl)-gamma-Aminopropyl Trimethoxy Silane. A silane coupling agent used in reinforced epoxy, phenolic, melamine and polypropylene resins.

Aminoplasts. A coined term denoting thermosetting resins made by the polycondensation of formaldehyde with a nitrogen compound and a higher aliphatic alcohol. The two general types are *urea-formaldehyde* and *triazine-formaldehyde.* Melamine is the triazine most often used.

gamma-Aminopropyl Triethoxy Silane. $NH_2CH_2CH_2CH_2Si(OC_2H_5)_3$ A silane coupling agent used in reinforced epoxy, phenolic, melamine and many thermoplastic resins.

Amino Resins. (polyalkylene amides) A generic term designating a group of nitrogen-rich polymers containing nitrogen in the amino form, NH_2. The starting amino-bearing material is usually reacted with formaldehyde to form a reactive monomer, which is condensation polymerized to a thermosetting resin. Included are urea, melamine, copolymers of both with formaldehyde; and, of limited use, thiourea, aniline, dicyanodiamide, toluenesulfonamide, benzoguanamine, ethylene urea, and acrylamide. Not included, because properties warrant separate classification, are polyamides of the nylon type, polyurethanes, polyacrylamide, and acrylamide copolymers. The most important members of the amino resin family are melamine-formaldehyde and urea-formaldehyde resins. The basic resins are clear water-white syrups or white powdered materials which can be dispersed in water to form colorless syrups. They cure at elevated temperatures with appropriate catalysts. Molding powders are made by adding fillers to the uncured syrups, forming a consistency suitable for compression and transfer molding.

AMMA. Abbreviation for copolymers of acrylonitrile and methyl methacrylate.

Amorphous. Devoid of crystallinity or stratification. Most plastics are amorphous at processing temperatures, many retaining this state under all normal conditions.

Ampere. The unit of measure of the quantity of electrical current flowing through a conductor. For practical purposes, an ampere is the transfer of one coulomb per second. The *International Ampere* is the constant electric current which, when passed through a silver nitrate solution of specified strength, deposits silver at the rate of .001118 grams per second. The Ampere and its symbol "A" are retained in the SI system of International metric units. However, the SI definition is the constant current which, if maintained in two straight parallel conductors of infinite length, of negligible circular sections, and placed 1 meter apart in a vacuum, will produce between these conductors a force equal to 2×10^{-7} newtons per meter of length.

Amphoteric. Pertaining to a material which has the capacity of behaving either like an acid or a base. Polymerization emulsifiers having both anionic and cationic groups are known as *amphoteric emulsifiers.*

Amyl. (n) The radical C_5H_{11}, also known as *pentyl.* The amyl radical occurs in many isomeric forms, and the term amyl usually refers to any mixture of the isomers.

Amyl Acetate. $CH_3COOC_5H_{11}$ (banana oil, pear oil, amylacetic ester) A commercial solvent for several resins, including the cellulosics, vinyls, acrylics, polystyrene, alkyd and phenolics. Its primary constituent is isoamyl acetate, but other isomers such as normal- and secondary-amyl acetates are present in amounts determined by the grade and derivation.

Amyl Formate. $HCOOC_5H_{11}$ Solvent for resins and cellulose derivatives.

Amyl Oleate. $C_{17}H_{33}COOC_5H_{11}$ Solvent and plasticizer for cellulosic and vinyl resins.

para-tert-Amyl Phenol. $(CH_3)_2C_2H_5CC_6H_4OH$ A white crystalline material made by alkylating phenol with amyl chlorides or amylenes, then separating by distillation. Resins made by reacting para-tert-amyl phenol with formaldehyde or paraformaldehyde are used in varnishes for wood, wire coating and coil insulation. They are also used as a plasticizer-stabilizer in hot melt adhesives based on ethyl cellulose.

Amyl Salicylate. See ISOAMYL SALICYLATE.

Anaerobic Adhesives. Adhesives that cure only in the absence of air after being confined between assembled parts. An example is dimethacrylate adhesive, used for bonding assembly parts, locking screws and bolts, retaining gears and other shaft-mounted parts, and sealing threads and flanges.

Anatase. (octahedrite) A naturally crystallized form of TITANIUM DIOXIDE, which see.

Anchorage. Part of an insert that is molded inside of a plastic and held fast by shrinkage of the plastic.

Andrade Creep. A term applied to the stress-strain behavior of a material exhibiting a type of creep compliance which is linear in the cube-root of time. A variety of polymers exhibit this form of creep.

Anelasticity. The dependence of elastic strain on both stress and time, resulting in a lag of strain behind stress. In materials subjected to cyclic stress, the anelastic effect causes DAMPING, which see.

Angel's Hair. Fiber-like strands of material pulled away from thermoplastic films, especially polypropylene, in heat-sealing and cutting operations which employ hot knives or wires. The Angel's hair accumulates on the cutting mechanism, eventually affecting performance and requiring removal.

Angle Head. An extruder head positioned at an angle with the axis of the extruder screw.

Angle Press. A hydraulic molding press equipped with horizontal and vertical rams, used in the production of complex moldings containing deep undercuts.

Angstrom Unit. Originally abbreviated as ($\mathring{A}$), and defined as one hundred millionth (10^{-8}) of a centimeter, this term was later defined in terms of the wave length of the red line of cadmium (6438.4694 $\mathring{A}$). In the new SI terminology, the Angstrom is defined as 1 meter $\times 10^{-10}$, or in SI form as 1.000 000*E-10 meters.

Angular Welding. See FRICTION WELDING.

Anhydride. (1) A compound from which water has been extracted. (2) An oxide of a metal (basic anhydride) or of a nonmetal (acid anhydride) which forms a base or an acid, respectively, when united with water.

Aniline. (phenylamine, aminobenzene) $C_6H_5NH_2$ A colorless, oily liquid made by reacting chlorobenzene with aqueous ammonia in the presence of a catalyst. It is used in the production of aniline formaldehyde resins and certain catalysts and antioxidants.

Aniline Formaldehyde Resins. Members of the aminoplastics family, made by condensing formaldehyde and aniline in an acid solution. The resins are thermoplastic, and are used in the production of molded and laminated insulating materials with high dielectric strength and good chemical resistance. See also AMINO RESINS.

Aniline Inks. Fast-drying inks used tor printing on cellophane, polyethylene, etc. Aniline inks were first made from solutions of coal tar dies in organic solvents, hence the name. However, modern inks generally employ pigments rather than dyes.

Aniline Resins. See ANILINE FORMALDEHYDE RESINS and AMINO RESINS.

Animal Black. (animal char, animal charcoal) A form of charcoal derived from animal bones, used as a pigment. See also CARBON BLACK.

Anion. An atom, molecule or radical which has gained an electron to become negatively charged.

Anion Exchange Resins. See ION EXCHANGE RESINS.

Anionic. Pertaining to any negatively charged atom, radical or molecule; or to any compound or mixture containing negatively charged groups.

Anionic Polymerization. See IONIC POLYMERIZATION.

Anisotropy. The tendency of a material to react differently to stresses applied in different directions.

Annealing. The process of relieving stresses in molded plastic articles by heating to a predetermined temperature, maintaining this temperature for a predetermined length of time, and slowly cooling the articles. The process is used to relieve stresses in articles to be painted which might craze due to solvent attack.

ANSI. Abbreviation for American National Standards Institute, located at 1430 Broadway, New York, N.Y. 10018.

Anthophyllite. A type of ASBESTOS, which see, the major source of which is in Finland. Anthophyllite is a naturally occurring magnesium-iron silicate, used widely in polypropylene because of its superior heat stability.

Antiblocking Agents. Agents which are incorporated in plastic compounds to reduce the adhesion of surfaces of products made from the compounds to each other or to other surfaces, and which function by producing a slight roughening of the surface. The term is not generally used for coatings, dusts or sprays applied to surfaces after products have been formed, for the same purpose, or for a SLIP AGENT, which see. Antiblocking agents are usually finely divided, solid infusible materials, usually minerals but sometimes waxes. They function by forming small protruding "asperites" on surfaces which maintain small air spaces which reduce the coefficient of friction and prevent surfaces from sticking together.

Anti-Foaming Agent. An additive which reduces the surface tension of a solution or emulsion, thus inhibiting or modifying the formation of a foam. Commonly used agents are insoluble oils, dimethyl polysiloxanes and other silicones, certain alcohols, stearates and glycols. The agents are necessary in many polymerization reactions to prevent foaming altogether, and are also used to delay foaming in the production of cellular plastics.

Antifogging Agents. Additives which prevent or reduce the condensation of water on a plastic film in the form of small droplets which resemble fog. Such additives function as mild wetting agents which exude to the surface and lower the surface tension of water, thereby causing it to spread into a continuous phase. Antifogging agents are particularly useful in PVC films used for wrapping meats and other moist foods. Examples of antifogging agents are alkylphenol ethoxylates, complex polyol mono-esters, polyoxyethylene esters of oleic acid, polyoxyethylene sorbitan esters of oleic acid, and sorbitan esters of fatty acids.

Anti-Gelling Agent. An additive which prevents a solution from forming a gel.

Antimicrobial Agents. See BIOCIDES.

Antimony Trioxide. Sb_2O_3 (antimony white, flowers of antimony, antimony bloom, antimony oxide) A very fine white powder made by vaporizing antimony metal or ore in an oxidizing atmosphere, then cooling and collecting the fumes. It is used as a flame retardant in plastics, particularly PVC.

Antioxidant. A substance incorporated in a material for the purpose of inhibiting oxidation at normal or elevated temperatures. Antioxidants are used primarily with natural and synthetic rubbers, petroleum-based resins, and other such polymers which oxidize readily due to structural unsaturation. However, some thermoplastics, especially polypropylene, ABS, rubber-modified polystyrene, acrylic and vinyl resins also require protection by antioxidants for some applications. There are two main classes of antioxidants: (1) Those that inhibit oxidation through reaction with chain-propagating radicals (radical or chain terminators); such as hindered phenols or secondary aryl amines which intercept either the R· or RO_2· free radicals. These also are referred to as primary antioxidants or free radical scavengers. (2) Those that decompose peroxide into nonradical and stable products such as phosphites and various types of sulfur compounds such as esters of thiodipropionic acid. These also are referred to as secondary antioxidants, synergists, or peroxide decomposers.

Antiozonant. A substance added to elastomers to retard or prevent deterioration caused by exposure to air containing ozone.

Anti-Sag Agents. See THICKENING AGENTS.

Antistatic Agents. Chemicals which impart a slight to moderate degree of electrical conductivity to plastics compounds, thus preventing the accumulation of electrostatic charges on finished articles. They may be incorporated in the materials before molding, or applied to their surfaces after molding. They function either by being inherently conductive or by absorbing moisture from the air. Examples of antistatic additives are long-chain aliphatic amines and amides, phosphate esters, quaternary ammonium salts, polyethylene glycols, polyethylene glycol esters, and ethoxylated long-chain aliphatic amines. See also STATIC ELIMINATORS, SOOT-CHAMBER TEST.

Antistats. See ANTISTATIC AGENTS.

Ant Oil. See FURFURAL.

Apparent Density. The weight per unit volume of a material including voids inherent in the material as tested. Note: The term BULK DENSITY is commonly used for material such as molding powder. See also BULK FACTOR, DENSITY.

Apparent Specific Gravity. The specific gravity of a porous solid when the volume used in the calculations is considered to exclude the permeable voids. See also SPECIFIC GRAVITY.

Apparent Viscosity. See VISCOSITY.

Aprotic Solvent. An organic solvent that does not offer or accept protons to or from a substance dissolved in it. Benzene, C_6C_6, is an example of such a solvent.

Aprotic Substance. A substance which can act neither as an acid or a base.

AR. Abbreviation for ACRYLIC RUBBER, which see. (British Standards Institution)

Aragonite. See CALCIUM CARBONATE.

Aramides. A term used for aromatic nylon fibers which are used in high performance, low specific gravity composites, and in radial tires in competition with steel.

Aramid Fibers. A family of high-strength aromatic amide reinforcing fibers produced by Du Pont under the trademark Kevlar-49. The fibers are available in the same forms as glass reinforcements, and produce composites with high modulus, fatigue resistance, low thermal expansion coefficient, and good electric properties. The fibers are used for reinforcing auto tires.

Arc Resistance. The ability of a plastic material to resist the action of a high voltage electrical arc, usually stated in terms of time required to render the material electrically conductive. Failure of the specimen may be caused by heating to incandescence, burning, tracking or carbonization of the surface. A test method for arc resistance is described in ASTM D 495-73.

Arc Tracking. See TRACKING.

Aromatic Compounds. A class of organic compounds containing an unsaturated ring of carbon atoms. Included are benzene, naphthalene, anthracene and their derivatives. The term *aromatic* stems from the fact that many of these compounds have an agreeable odor.

Aromatic Hydrocarbon. A hydrocarbon with a molecular structure involving one or more rings of six carbon atoms, and having properties similar to benzene, which is the simplest of the aromatic hydrocarbons. Other members of the family include many solvents for plastics.

Artificial Ageing. The accelerated testing of plastics to determine their changes in properties such as dimensional stability, water resistance, resistance to chemicals and solvents, light stability and fatigue resistance.

Artificial Weathering. The process of exposing plastics to continuous or repeated environmental conditions developed by laboratory methods designed to simulate conditions encountered in actual outdoor exposure. Such conditions include temperature, humidity, moisture, light in the ultraviolet range, and direct water spray. The laboratory conditions are usually intensified to a degree greater than those normally encountered in actual outdoor conditions in order to decrease the time necessary to achieve significant results. ASTM tests pertaining to artificial weathering of plastics are D 756-56; D 794-68; D 1435-75; D 1499-64; D 1501-71; D 1920-69; D 2565-70; and G 23-69. All are published in Part 35 of the 1976 Book of ASTM Standards.

ASA. Abbreviation for terpolymers of acrylonitrile, styrene and acrylates.

Asbestos. The commercial term for a family of fibrous mineral silicates comprising some 30 known varieties, only six of which are of commercial importance. These are classified under two general types, the *serpentine* type and the *amphibole* type. The serpentine type contains CHRYSOTILE, which see, and is the most widely used. It is used as a reinforcement in thermosetting resins and laminates; and, in finer form, as a filler in polyethylene, polypropylene, nylons and vinyls. In the vinyl field its main use is in vinyl-asbestos flooring. The amphibole type includes the minerals anthophyllite, tremolite, actinolite, amosite, and crocidolite (blue asbestos containing iron). The amphiboles, especially crocidolite, are used in plastics when some unique property such as good chemical resistance, more neutral pH, or low water absorption is required. Asbestos fibers, with lengths up to about 3/4" and diameters in the vicinity of 0.02 micron, possess tensile strengths between 200,000 and 400,000 psi. Their primary advantages as reinforcements are high resistances to heat and chemicals. Health hazards are associated with exposure to asbestos fibers. It is essential that appropriate precautions be taken when processing asbestos, and that exposure limits for those exposed to asbestos fibers be held to levels set by OSHA, to become 2 fibers in an 8-hour time-weighted average after July 1, 1976.

Ascaridole. $C_{10}H_{16}O$ (1,4-peroxide-para-menthene-2) A naturally-occurring peroxide with uses as a polymerization initiator.

Ash Content. The solid residue remaining after a substance has been incinerated or heated to a temperature sufficient to drive off all combustible or volatile substances.

Ashing. A finishing process used to produce a satin-like finish on plastic articles, or to remove cold spots or teardrops from irregular surfaces which cannot be reached by wet sanding. The part is applied to a loose muslin disc loaded with wet ground pumice, rotating at a speed such as 4000 linear feet per minute.

Aspect Ratio. In fiber technology, the ratio of length to diameter of a fiber. In flake-like materials such as mica, the ratio between equivalent diameter and thickness of the flake.

Assembly of Plastics. Plastic parts may be joined to other articles by many methods. Self-tapping screws are made with special thread configurations to suit specific resins. Threaded inserts to receive mounting screws may be molded in or installed by press-fitting or by means of self-tapping external threads. Press fitting may be employed to join plastics to similar or dissimilar materials. Snap fitting joints are made by molding or machining an undercut in one part, and providing a lip to engage this undercut in a mating component. Other methods are BUTT FUSION, CEMENTING, FABRICATE, HEAT SEALING, HOT PLATE WELDING, LASER, STAKING, THERMOBANDE WELDING, ULTRASONIC INSERTING, ULTRASONIC STAKING, and WELDING, which see.

A-Stage. (n) An early stage in the preparation of certain thermosetting resins in which the material is still soluble in certain liquids and fusible. Sometimes referred to as *resol*. See also B-STAGE and C-STAGE. (ASTM D 883-75a).

ASTM. Abbreviation for American Society for Testing and Materials, said to be the largest non-governmental standards-writing body in the world. Its publications can be obtained from ASTM, 1916 Race Street, Philadelphia, PA 19103.

Asymmetric. The opposite of symmetrical. Of such form that no point, line or plane exists about which opposite portions are exactly similar.

Asymmetry. A molecular arrangement in which a particular carbon atom is joined to four different groups.

Atactic. (adj) Pertaining to an arrangement which is more or less random.

Atactic Block. A regular block that has a random distribution of equal numbers of the possible configurational base units. (IUPAC).

Atactic Polymers. Polymers with molecules in which substituent groups or atoms are arranged at random above and below the backbone chain of atoms, when the latter are arranged so as to be all in the same plane. The opposite of STEREOSPECIFIC POLYMERS, which see. Note: The IUPAC and ISO definition for atactic polymer is a regular polymer whose molecules have a random distribution of equal numbers of the possible configurational base units.

ATE. Abbreviation for aluminum triethyl (triethyl aluminum), a polymerization catalyst for olefins.

Attenuation. The process for making slim and slender, for example the formation of fibers from molten glass.

atto (a). The SI-approved prefix for a multiplication factor to replace 10^{-18}.

AU. Abbreviation for the polyester type of polyurethane rubber.

Autoacceleration. In some vinyl polymerization reactions, as the reaction approaches completion and the viscosity of the reaction medium rises, there is an increase in the rate of increase of molecular weight of the polymer chains that have not yet been terminated. This increase in molecular weight as conversion progresses is called autoacceleration, or the *Trommsdorff Effect,* or *gel effect.*

Autoadhesion. (Tackiness) The ability of two contiguous surfaces of the same material, when pressed together, to form a strong bond which prevents their separation at the place of contact.

Autocatalytic Degradation. A type of degradation in which the breakdown products produced in the initial phase of degradation accelerate the rate at which subsequent degradation proceeds.

Autoclave. A strong pressure vessel with a quick-opening door and means for heating and applying pressure to its contents. Autoclaves are widely used for bonding and curing reinforced plastic laminates such as polyesters, polyurethanes, epoxies and phenolics.

Autoclave Molding. See BAG MOLDING.

Autogeneous Extrusion. Equivalent to autothermal extrusion. See EXTRUSION, AUTOTHERMAL.

Automatic Mold. A mold, e.g. for compression, transfer or injection molding, that is equipped to perform all operations of the molding cycle, including ejection of the molded parts, in a completely automatic manner without human assistance.

Autooxidation. After polyolefins have been exposed to an oxidizing agent such as molecular oxygen for the purpose of rendering them receptive to inks or adhesives, the oxidation may continue for a time after exposure to the oxidizing agent has been terminated. Such self-sustaining oxidation is called autooxidation.

Autothermal Extrusion. See EXTRUSION, AUTOTHERMAL.

Average Molecular Weight. (viscosity method) The molecular weight of polymeric materials determined by viscosity of the polymer in solution as a specific temperature. This gives an average molecular weight of the molecular chains in the polymer independent of the specific chain length. The value falls between weight average and number average molecular weight.

Axial Winding. A method of FILAMENT WINDING, which see, in which the filaments are wound in a direction parallel to the axis of rotation.

Azelaic Acid. $CO_2 H(CH_2)_7 CO_2 H$ (nonanedioic acid, 1,7-heptanedicarboxylic acid) A yellowish to white crystalline powder, derived from a fatty acid such as oleic acid by oxidation with ozone. It is an intermediate used in the production of plasticizers, polyamides and alkyd resins. For plasticizers derived from azelaic acid see DICYCLOHEXYL-, DI-2-ETHYLBUTYL-, DI-2-ETHYLHEXYL-4-THIO-, DI-n-HEXYL-, DIISOBUTYL-, DI(2-ETHYLHEXYL)-, and DIISOQCTYL azelates.

Azeotropic Copolymer. A copolymer in which the relative numbers of the different kinds of units (MERS) are the same as in the mixture of monomers from which it was obtained.

2(1-Aziridinyl)ethyl Methacrylate. A vinyl monomer introduced in 1969 under the trademark "Sipomer IM" by Alcolac Chemical Corp. The monomer combines a reactive vinyl group with an aziridinyl functional group. It can be polymerized alone or with other vinyl monomers to yield polymers with pendant aziridinyl groups which promote adhesion of coatings to the polymer. As little as 0.5 to 2.0% of Sipomer IM in the monomer system is effective for achieving good coating adhesion.

Azobisformamide. $H_2N-CO-N:N-CO-NH_2$. (ABFA, azodicarbonamide) An aliphatic azo compound widely used as a chemical blowing agent in PVC, polystyrene, polyolefins, many other plastics, and in natural and synthetic rubbers. It is non-toxic, odorless, non-staining, and unlike other organic blowing agents, is self-extinguishing and does not support combustion. Since ABFA in the pure state decomposes at temperatures above $420°F$, when used with heat-sensitive plastics such as PVC it is used in the presence of an activator which lowers its decomposition temperature into the normal processing range. Such activators are compounds of cadmium, lead and zinc, which also act as heat stabilizers either directly or synergistically with other stabilizers.

Azobis(isobutyronitrile). A blowing agent developed in Germany for use in rubber and PVC. It is non-staining and yields white PVC foam of fine uniform cell structure. However, its decomposition product, tetramethyl succinonitrile, is toxic and must be eliminated from the expanded product. For this reason the material is not used commercially in the U.S. as a blowing agent, but it is used as a polymerization initiator.

Azodicarbonamide. See AZOBISFORMAMIDE.

Azo Dyes. An important family of dyes containing the -N=N- group, produced from amino compounds by the processes of diazotiazation and coupling. By varying the composition it is possible to produce acid, basic, direct, or mordant dyes. The family is subdivided into monoazo, diazo, trisazo and tetrazo types depending on the number of -N=N- groups in the molecule.

Azo Group. The structural grouping -N=N-.

Back Draft. (back taper, counterdraft) A slight undercut or tapered area in a mold tending to prevent removal of the molded part. See also UNDERCUT.

Backing Plate. In injection molding, a plate used as a support for the cavity blocks, guide pins, bushing, etc. Sometimes called *support plate.*

Back Pressure. In extrusion, the resistance of the molten plastic material to forward flow. In molding, the viscosity resistance of the material to continued flow when the mold is closing. Back pressure increases the temperature of the melt, and contributes to better mixing of colors and homogeneity of the material.

Back-Pressure Relief Port. An opening in an extrusion die for escape of excess material.

Back Taper. See BACK DRAFT.

Bactericide. An agent capable of destroying bacteria. See also BIOCIDES.

Bacteriostat. An agent which when incorporated in a plastics compound will prevent the growth of bacteria on surfaces of formed articles. See also BIOCIDES.

Baffle. A plug or other device inserted in a flow channel to restrict the flow of material or to divert it to a desired path.

Bagasse. (megass) A tough fiber derived from crushed sugar cane remaining after the sugar juice has been extracted. It is used as a reinforcement in laminates and molding powders.

Bag Molding. A method of forming and curing reinforced plastic laminates employing a flexible bag or mattress to apply pressure uniformly over one surface of the laminate. A preform comprising a fibrous sheet impregnated with an A- or B-stage resin is placed over or in a rigid mold forming one surface of the article. The bag is applied to the other surface, then pressure is applied by vacuum, an autoclave, press, or by inflating the bag. Heat may be applied by steam in the autoclave, or through the rigid mold portion. When an autoclave is used the process is sometimes called *Autoclave Molding.*

Bakelite. A trademark for phenolic resins, derived from that of Dr. Leo H. Baekeland, a Belgian who developed phenolic resins in the early 1900's. The trademark was once used by the Bakelite Corporation, and is now owned by Union Carbide.

Balanced Construction. In plywood, a laminate with an odd number of plies, symmetrical on both sides of its center line.

Balanced Design. In reinforced plastics, a winding pattern so designed that the stresses in all filaments are equal.

Balanced Runner. In injection molding, a runner system designed to place all cavities at the same distance from the sprue.

Balata, Natural. A substance identical in composition and properties to GUTTA-PERCHA, which see, obtained from trees in South America.

Balata, Synthetic. A stereospecific synthetic rubber. The trans isomer of polyisoprene.

Ball and Ring Test. A method of determining the softening point of thermoplastics. A specimen is cast or molded in a ring of metal 5/8" inside diameter × 3/32" thick × 1/4" deep. This ring is placed above a metal plate in a fluid heating bath, and a steel ball 3/8" diameter weighing 3.5 grams is placed in the center of the specimen. The softening point is considered to be the temperature of the fluid when the ball penetrates the specimen and touches the lower plate.

Ball Mill. A cylindrical or conical shell rotating about a horizontal axis, partially filled with a grinding medium such as natural flint pebbles, ceramic pellets or metallic balls. The material to be ground is added to just or slightly more than fill the voids between the pellets. The shell is rotated at a speed which will cause the pellets to cascade, thus reducing particle sizes by impact. It has been proposed that in the plastics industry the term ball mill be reserved for metallic grinding media, and the term pebble mill for non-metallic grinding media. This differentiation is not observed in other industries. See also JAR MILL.

Ball Rebound Test. A method for measuring the energy response of polymeric materials by dropping a ball, e.g. a 1/8″ dia. steel ball, on a specimen from a fixed height and determining the rebound height. The difference between the two heights indicates the energy absorbed. By conducting the tests over a range of temperatures, useful data can be obtained with regard to first and second order transition points, average molecular weight distribution, and effects of additives and plasticizers.

Ball-Up. A term used in the adhesives industry to describe the tendency of an adhesive to stick to itself.

Ball Viscometer. A type of viscosity measuring apparatus employing solid balls of specified weight and diameter as the shearing mechanism.

Balsams. See OLEORESINS.

Banana Liquid. A solution of nitrocellulose in amyl acetate or similar solvent.

Banana Oil. See AMYL ACETATE.

Banbury Mixer. A mixing apparatus originally used for rubber, now also used for mixing plastics such as cellulosic, vinyl, polyethylene and pyroxylin compounds. It consists of two contra-rotating spiral-shaped blades encased in segments of cylindrical housings, intersecting so as to leave a ridge between the blades. The blades may be cored for circulating heating or cooling media.

Band Heaters. See HEATER BANDS.

Bank. In calendering, a reservoir of material at the opening between the rolls on the material feeding side.

Bar. A unit of pressure or stress, equal to 10^6 dynes per cm^2, or 0.987 atmospheres. This term is discouraged by the new SI system, in which 1 bar is to be stated as 10^5 Pa. (100 kPa). The term *bar* has been used most widely in the fluid and gas pressure fields.

Barcol Hardness. A hardness value obtained by measuring the resistance to penetration of a sharp steel point under a spring load with an instrument called the Barcol Impressor. Direct readings are obtained on a scale ranging from 0 to 100. This test is often used to measure the degree of a cure of a thermosetting plastic.

Barite. See BARIUM SULPHATE.

Barium-Cadmium Stabilizers. A family of stabilizers based on salts of the title metals with organic acids, often in combination with a zinc salt of such acids, phosphites, and epoxides. These stabilizers are the most widely used for vinyl compounds. They provide moderate to good heat stability at low cost, but cannot be used in compounds for food or drinking water contact.

Barium Ferrite. A magnetic material which cannot be incorporated directly into a thermoplastic, but when encapsulated in Nylon-12 can be used in a composite which acts as a magnet. After being injection molded, the parts are magnetized by standard pulse discharge equipment.

Barium Hydroxide, Monohydrate. $Ba(OH)_2 \cdot H_2O$ (barium monohydrate) A white powder used in the production of phenol formaldehyde resins and barium soaps.

Barium Peroxide. BaO_2 or $BaO_2 \cdot 8H_2O$ An oxidizing catalyst used in some polymerization reactions.

Barium Ricinoleate. $Ba(CO_2 C_7 H_4 CH=CH \cdot CH_2 CHOH \cdot C_5 H_{10} CH_3)_2$ A heat stabilizer imparting good clarity, used most often in vinyl plastisols and organosols.

Barium Stearate. $Ba(C_{17} H_{35} COO)_2$ A heat stabilizer, used particularly when sulfur staining is to be avoided. Also used as a lubricant where high temperatures are involved.

Barium Sulphate. $BaSO_4$ (blanc fixe, heavy spar, barytes, permanent white, terra ponderosa) A white powder obtained from the mineral barite or synthesized chemically. One of the synthetic varieties, *blanc fixe,* is made by reacting aqueous solutions containing sulphate ions and barium ions. As a filler in plastics and rubber, barium sulphate imparts opacity to X-rays but only a low order of optical opacity. Thus, it is useful as a filler when it is desired to increase specific gravity without adversely affecting the tinctorial power of pigments.

Bar Mold. A mold in which the cavities are arranged in rows on separate bars, which may be removed individually to facilitate stripping.

Barrel. (n) (1) The tubular portion of an extruder. See EXTRUDER BARREL. (2) A container, agitated by rotation or vibration, used for tumbling moldings to remove flash and sharp edges.

Barrel. (v) (barreling) See TUMBLING.

Barrier Plastics. Thermoplastics with very low permeability to gases. Most of the barrier plastics on the market are NITRILE BARRIER RESINS, which see. Several others, however, are based on various copolymers, some of which have gas permeabilities greater than nitrile resins but are easier to process. The largest application expected for barrier resins is in bottles for carbonated beverages.

Barrier Sheet. An inner layer of a laminate, placed between the core and an outer layer.

Barus Effect. In extrusion, the swelling of an extrudate to a dimension larger than the corresponding dimension of the die. It is caused by changes in stress which occur in viscoelastic polymer melts as they emerge from a die.

Barye. The old cgs unit of force applied to, or distributed over, a surface, expressed as dynes per square centimenter. In the new SI, the barye is expressed as $1.000\ 000*E01$ pascals.

Barytes. A filler material made from the naturally occurring form of BARIUM SULPHATE, which see.

Basebox. In the metal coating industry, a unit of area comprising 31,360 square inches.

Basic Lead Carbonate. $2PbCO_3 Pb(OH)_2$ A very effective heat stabilizer, used where toxicity is of no concern as in electrical insulating compounds. Its use today is limited because of its tendency to form blisters in processing and cause spew when exposed to outdoor weathering.

Bast Fibers. A family of vegetable fibers taken from the inner bark of plants, which run the length of the stem, are surrounded by enveloping tissue and are cemented together by pectic gums. Included in the family are jute, flax, sunn, hemp and ramie, some of which are used in reinforced plastics.

Baumé. (Bé) A system of specific gravity units devised by the French chemist Antoine Baumé for the graduation of hydrometers. The relations to sp. gr. (at $60/60°$ F) **are:**

$$°Bé = 145 - \frac{145}{sp.\ gr.}\quad \text{for liquids heavier than water, and}$$

$$°Bé = \frac{140}{sp.\ gr. - 130}\quad \text{for liquids lighter than water.}$$

BBP. Abbreviation for BUTYL BENZYL PHTHALATE, which see.

Bé. Abbreviation for BAUMÉ, which see.

Beaders. Devices for rolling beads on the edges of thermoplastic sheets or cylinders.

Bead Polymer. A polymer in the form of small globules. See SUSPENSION POLYMERIZATION.

Bead Polymerization. A type of polymerization identical to SUSPENSION POLYMERIZATION, which see, except that the monomer is dispersed as relatively large droplets in water or other suitable inert diluent by vigorous agitation.

Bearing Strength. A term used in the plastic industry to denote the ability of sheets to sustain edgewise loads that are applied by pins, rods or rivets used to assemble the sheets to other articles.

Becquerel. The SI unit of activity of a radionuclide having one spontaneous nuclear transition per second.

Bending Strength. See FLEXURAL STRENGTH.

Bentonite. A type of clay, used as a filler, resulting from the weathering of volcanic ash and consisting essentially of montmorillonite, a hydrous silicate of alumina. The name is derived from Fort Benton, Wyoming, where it was discovered. The material has the unique quality of absorbing large quantities of water.

Benzaldehyde. C_6H_5CHO (benzoic aldehyde, oil of bitter almonds, benzoyl hydride, benzene carbonal) A solvent, particularly for polyester resins and cellulosic plastics.

Benzene. C_6H_6 (benzol) A solvent and intermediate in the production of phenolics, epoxies, styrene and nylon. Benzene is alkylated with ethylene to give ethylbenzene, which is then dehydrogenated to form styrene. Hydrogenation of benzene yields cyclohexane, a solvent and raw material for preparing adipic acid, from which nylon is derived. As a solvent, benzene will dissolve ethyl cellulose, polyvinyl acetate, PMMA, polystyrene, coumarone-indene resins and certain alkyds.

Benzene Ring. The six carbon atoms, shown arranged at the angles of a hexagon, each with a hydrogen atom attached in the case of benzene, or with one or more hydrogens replaced by other atoms or radicals in the case of all organic compounds derived from benzene.

Benzenesulphonbutylamide. $C_6H_5SO_2NHC_4H_9$ A plasticizer for cellulosics and polyvinyl acetate.

Benzene Sulfonyl Hydrazide. 4,4′-oxybis(benzene sulfonyl hydrazide) OBSG. A blowing agent marketed by Uniroyal under the tradename Celogen BSH. The product is a white crystalline solid that melts and begins to decompose near 105°C. It produces a white unicellular foam when incorporated in PVC plastisol but has a strong residual odor that does not occur when used in rubber. It is also used in epoxy and phenolic foams, and serves as a cross-linking agent in rubber compositions and rubber-resin blends.

Benzidine Orange, Benzidine Yellow. Metal-free diazo pigments based on dichlorobenzidine. They are highly transparent, bright in color and low in cost due to high strength, but tend to bleed and fade upon exposure to light.

Benzine. (ligroin) A saturated petroleum fraction, used as a solvent. The term benzine is outmoded due to confusion with benzene, and *Ligroin* is preferred.

Benzofuran. See COUMARONE.

Benzofuran Resin. See COUMARONE-INDENE RESINS.

Benzoguanamine. $C_6H_5C_3N_3(NH_2)_2$ (2,4-diamino-6-phenyl-s-triazine) A crystalline material which reacts with formaldehyde to give thermosetting resins with heat resistance, alkali resistance and gloss properties generally superior to those of melamine-formaldehyde resins. Benzoguanamine resins are used for protective coatings, paper additives and finishes, laminating agents, textile finishes and adhesives.

Benzoic Acid. C_6H_5COOH (carboxybenzene, benzene carboxylic acid, phenylformic acid) A white crystalline substance occurring naturally in benzoin gum and some berries, also synthesized from phthalic acid or toluene. It is used in the manufacture of plasticizers such as 2-ethylhexyl-p-oxy-benzoate, diethylene glycol dibenzoate, dipropylene glycol dibenzoate, ethylene glycol dibenzoate, triethylene glycol dibenzoate, polyethylene glycol (200)- and (600)-dibenzoate, and benzophenone.

Benzoic Ether. See ETHYL BENZOATE.

Benzophenones. Class name for a family of UV Stabilizers based on substituted 2-hydroxy-benzophenones. Typical members are 2-hydroxy-4-methoxybenzophenone, 2,hydroxy-4-octyloxybenzophenone, 4-dodecyloxy-2-hydroxybenzophenone, 2,2'-dihydroxy-4-methoxybenzophenone, and 2,2'-dihydroxy-4-4'-dimethoxybenzophenone. They function both as direct UV absorbers and, in the case of polyolefins, as energy transfer agents and radical scavengers.

p-Benzoquinone. $C_6H_4O_3$ (quinone, 1,4-benzoquinone, chinone) A yellow crystalline material used, along with many of its derivatives, as an inhibitor in unsaturated polyester resins to prevent premature gelation during storage.

Benzotriazoles. A family of UV Stabilizers. All are derivatives of 2-(2'-hydroxyphenyl)benzotriazole and function primarily as UV absorbers. Typical examples are 2-(2'-hydroxy-5'-methylphenyl)benzotriazole and the corresponding 5'-t-octylphenyl analog. As a class the benzotriazoles offer strong intensity and broad UV absorption with a fairly sharp wavelength cutoff close to the visible region. The higher alkyl derivatives are less volatile and therefore more suitable for higher temperature processing conditions.

Benzoyl Peroxide. $(C_6H_5CO)_2O_2$ A catalyst used in the polymerization of styrene, vinyl and acrylic resins; and for curing polyester and silicone resins. It can be dispersed in diluents or plasticizers to minimize hazards usually associated with the dry product.

Benzyl. The compound radical $C_6H_5CH_2$, which exists only in combination.

Benzyl Acetate. $C_6H_5CH_2OOCCH_3$ (phenylmethyl acetate) A colorless liquid with a pleasant odor, used as a solvent for cellulose acetate, cellulose nitrate, etc.

Benzyl Alcohol. $C_6H_5CH_2OH$ (alpha-hydroxy-toluene, phenylcarbinol) A water-white liquid used as a solvent for cellulosics and some other resins.

Benzyl Benzoate. $C_6H_5CH_2OOCC_6H_5$ A water-white liquid which readily freezes to a solid, used as a plasticizer.

Benzyl Butyrate. $C_3H_7COOCH_2C_6H_5$ A liquid with a heavy fruity odor, used as a plasticizer.

Benzyl Cellulose. A member of the cellulosic family of plastics, a benzyl ether of cellulose. It is used in lacquers or may be formulated for making films and compounds for molding and extrusion.

Benzyl Formate. $C_6H_5CH_2OOCH$ A solvent for cellulosics.

Benzyloctyl Adipate. $C_6H_5CH_2COO(CH_2)_4COOC_8H_{17}$ A plasticizer for polystyrene, vinyl and cellulosic resins.

Benzyltrimethylammonium Chloride. $C_6H_5CH_2N(CH_3)_3 \cdot Cl$ A quaternary ammonium salt, used as a solvent for cellulosics and a catalyst for phenolic resins.

Berlin Blue. A term used generally for any of the varieties of iron blue pigments.

Berlin Red. A pigment consisting essentially of red iron oxide.

Beta-. A prefix, usually abbreviated as the Greek letter β, denoting the location of a substituting group of atoms in the main group of a compound, or a type of radiation. See also BETA PARTICLE.

Beta Gage. (Beta-ray gage) A device for measuring the thickness of plastics films, sheets or extruded shapes, consisting of a Beta-ray emitting source and a detecting element. When material is passed between these elements, some of the rays are absorbed, the percent absorbed being a measure of the thickness of the material. Signals from the detecting element can be used to control automatic equipment for regulating the thickness. The radioactive source of beta rays is usually Krypton 85 or Strontium 90. Also used for particular applications are the radioisotopes Cesium 137, Promethium 147 and Ruthenium 106. See also THICKNESS GAGING.

Beta Particle. A particle created at the instant of emission from a radioactive atomic nucleus, having a mass about 1/1837 of that of the proton. A negatively charged beta particle is identical to an ordinary electron. A positively charged beta particle (positron) differs from an electron by having equal but opposite electrical properties. A stream of beta particles is called a *beta ray*. Such rays are used in equipment for measuring and controlling the thickness of plastic films and extrudates.

Betatron. An accelerator used to impart high velocities to electrons (beta particles). The propellant is an electromagnetic field. Five to six Mev will produce X-rays equivalent to the gamma radiation of 12 to 20 grams of radium.

BHT. Abbreviation for butylated hydroxytoluene. See DI-tert-BUTYL-para-CRESOL.

Biaxial Orientation. The process of stretching a hot plastic film or other article in two directions under conditions that result in molecular reorientation. See also ORIENTATION.

Biaxial Winding. A term used in the reinforced plastics industry to denote a type of winding in which the helical band is laid in sequence, side by side, with no crossover of fibers.

Bicarburetted Hydrogen. See ETHYLENE.

Bierbaum Scratch Hardness. See SCRATCH HARDNESS.

Billow Forming. A variation of the THERMOFORMING (which see) process, in which the hot plastic sheet is clamped in a frame and billowed upwards against a male plug or die as the plug or die descends into the frame. The process is suitable for thin-walled containers with a high draw ratio.

Bin Activators. Devices that promote the steady flow of granular or powdered plastic materials from storage bins or hoppers. Among the many types of equipment are vibrators or mallets acting upon the outside of the container, prodding devices or air jets acting directly on the material, inverted cone baffles with vibrating means located at the bottom of the hopper, and other "live bottom" devices such as scrapers, rolls and chains.

Binder. An adhesive material used for holding particles of dry material together. For example, resinous adhesives used in foundry sands are called binders. The term is also sometimes used for the continuous phase, a thermoplastic or thermosetting resin, in reinforced plastics.

Bingham Body. A substance that behaves somewhat like a Newtonian fluid in that there is a linear relationship between rate of shear and shearing force, but also has a yield value. See also NEWTONIAN FLUID.

Biocides. Agents incorporated in or applied to surfaces of plastics to destroy bacteria, fungi, marine organisms and like living matter. Some plastics, for example acetals, acrylics, epoxies, phenoxies, ABS, nylons, polycarbonate, polyesters, fluorocarbons and polystyrene, are normally resistant to attack by bacteria or fungi. Others, e.g. alkyds, phenolics, low-density polyethylene, urethanes and flexible vinyls can under some circumstances be affected by growth of these organisms on their surfaces. Even though the resins themselves might be resistant, additives such as plasticizers, stabilizers, fillers and lubricants can serve as food for fungi and bacteria. Examples of biocides are organotins, brominated salicylanilides, mercaptans, quaternary ammonium compounds, mercury compounds, and compounds of copper and arsenics.

Biodegradation. The degradation of a plastic by living organisms such as bacteria, fungi and yeasts. Most of the commonly used plastics are essentially non-biodegradable, exhibiting limited susceptability to assimilation by micro-organisms. One exception is POLYCAPROLACTAM, which see. However, the growing emphasis on environmental aspects of discarded plastics has caused researchers to seek for ways of attaining biodegradation after a predetermined period of time. One method under development is to add an UV light sensitizer that will cause photodegradation after a period of exposure to light, which will be followed by biodegradation. Another method utilizes starch which will cause the polymer to swell and break up after prolonged exposure to the elements, after which bacteria will take over. A third method is the formation of weak links in the polymer chain, temporarily held together by a degradable stabilizer. See also PHOTODEGRADATION.

Bipolymer. A polymer derived from two species of monomer. (IUPAC) The more commonly used term is *copolymer*.

Birefringence. (double refraction) The difference between any two refractive indices. When the refractive indices measured along three mutually perpendicular axes are identical, the polymeric material is said to be optically isotropic. Orientation by drawing may alter the refractive index parallel to the direction of draw so that it is no longer identical to that perpendicular to this direction, in which event the material is said to display birefringence. Crystalline polymers which are normally birefringent may become optically isotropic at their melting points. Such optical studies provide useful information regarding the shape of molecules, degrees of orientation and other polymer characteristics.

Bis(4-t-Butylcyclohexyl) Peroxy Dicarbonate. A catalyst of the organic peroxide family, used in reinforced plastics and vinyl polymerization. Unlike other percarbonates, it does not require refrigeration for storage or handling. The material is sold as Noury's trademark Percadox 16.

Biscuit. See PREFORM.

Bis(ethoxyethoxyethyl) Phthalate. $C_6 H_4 (COOC_2 H_4 OC_2 H_4 OC_2 H_5)$ A good primary plasticizer for polyvinyl acetate, nitrocellulose, cellulose acetate and many other polymers. It has limited compatibility in PVC.

Bis(beta-Hydroxyethyl)-gamma-Aminopropyltriethoxy Silane. A silane coupling agent used in reinforced epoxy resins, and also in many reinforced thermoplastics such as PVC, polycarbonates, nylon, polypropylene and polysulfones.

Bisphenol A. $(CH_3)_2 C(C_6 H_4 OH)_2$ (para, para'isopropylidenediphenol). An intermediate used in the production of epoxy, polycarbonate and phenolic resins. The name was coined after the condensation reaction by which it may be formed – two (bis) molecules of phenol with one of acetone (A).

Bis(Tri-n-Butyltin) Oxide. A liquid derived by the hydrolysis of tributyl tin chloride, used as an agent to control the growth of most fungi, bacteria and marine organisms in plastics used in boat construction and in urethane foams.

Bis(Tri-n-Butyltin) Sulfosalicylate. An antimicrobial agent used in flexible PVC film and urethanes.

Bis (2,2,4-Trimethyl-1,3-Pentanediol Monoisobutyrate) Adipate. A plasticizer for cellulosic resins and polystyrene.

Bite. The ability of an adhesive to penetrate surfaces and thereby produce an adhesive bond.

Bitter Almond Oil, Synthetic. See BENZALDEHYDE.

Biuret. (allophanamide, carbamylurea) $NH_2 CONHCONH_2 \cdot H_2 O$ A white crystalline material derived from urea by heat or by reacting an isocyanate with urea. It is used primarily in analytical chemistry, but the biuret group is formed during some polymerization reactions such as primary bonds in urethane elastomers.

Bivinyl. See BUTADIENE.

Blanc Fixe. A synthetic form of barium sulphate prepared by reacting aqueous solutions containing barium ions with aqueous solutions containing sulphate ions, and precipitating the reactant. It is used as a special-purpose filler to impart X-ray opacity and high specific gravity.

Blanking. (die cutting) The cutting of flat sheet stock to shape by striking it sharply with a punch while it is supported on a mating die. Punch presses are often used for the operation.

Blanking Die. A metal die used in the blanking process.

Blast Finishing. The process of removing flash from molded objects, and/or dulling their surfaces, by impinging media such as steel balls, crushed apricot pits, walnut shells or plastic pellets upon them with sufficient force to fracture the flash. When the material being deflashed is not sufficiently brittle at room temperatures, the articles can be chilled to a temperature at which they are sufficiently brittle. The majority of blast finishing machines comprise wheels rotating at high speeds, fed at their centers with the media, which is thrown out at high velocity against the article.

Bleed. (n) An escape passage at the parting line of a mold, similar to a vent but deeper, serving to allow material to escape or bleed out. See also BLEEDING.

Bleeding. The diffusion of color from a plastic article into a surrounding surface or part, caused by inherent solubility of the pigment in one or more ingredients of the composition. The terms *migration, crocking, blooming* and *bronzing* are sometimes used loosely to describe the same phenomenon.

Bleedout. In filament winding, the excess liquid resin that migrates to the surface of a winding.

Blending Resin. (extender resin) With respect to vinyl plastisols and organosols, a blending resin is one of larger particle size and lower cost than the dispersion resins normally used, and which can be used as a partial replacement for the primary resin. Blending resins are sometimes used to alter properties as well as reduce costs.

Blister. (n) 1. An imperfection on the surface of a plastic article caused by a pocket of air or gas beneath the surface. 2. A thermoformed canopy or pocket roughly hemispherical in shape, for example an aircraft cockpit cover or a shape used in blister packaging.

Blister Packaging. A method of packaging articles in thermoformed "blisters" or pouches shaped to more-or-less fit the contours of the article. The preformed blisters, usually slightly oversized to provide ample room, are made of thermoplastics such as vinyls, polystyrene, or cellulosic plastics. They are placed inverted in fixtures, loaded with the articles, then cards coated with an adhesive are applied and sealed to the flanges of the blisters by means of heat and pressure.

Block. A portion of a polymer molecule comprising many constitutional units that has at least one constitutional or configurational feature not present in the adjacent portions. (IUPAC)

Block Copolymer. A copolymer with chains composed of shorter homo-polymeric chains which are linked together. These "blocks" can be either regularly alternating or random. Such copolymers usually have higher impact strengths than either of the homopolymers or physical mixtures of the two homopolymers.

Blocking. An undesirable adhesion between layers of plastic such as that which may develop under pressure during storage or use. Blocking can be prevented by use of agents added to the plastic compound or applied to the surfaces of finished articles. Such agents are called *anti-blocking agents.*

Block Polymer. A polymer whose molecules consist of blocks connected linearly. The blocks are connected directly or through a constitutional unit that is not part of the blocks. In the polymer molecule $A_k - B_l - A_m - B_n$, the individual blocks are regular and of the same species. When blocks are of different monomer species, the term *block copolymer* is used.

Block Press. (1) A press used for the agglomeration of laminate squares under heat. The squares are cut from a laminated sheet are are superimposed so that they are perpendicularly crossed in order to reduce the anisotropy caused by laminating. (2) A press used to mold very large blocks (up to 8 cubic feet) of polystyrene foam.

Blood Red. A pigment consisting essentially of red iron oxide.

Bloom. An undesirable cloudy effect or whitish powdery deposit on the surface of a plastic article caused by the exudation of a compounding ingredient such as a lubricant, stabilizer, pigment, plasticizer, etc. The term is also used to describe a discoloration of a metal mold.

Blowing Agent. (foaming agent) Any substance which alone or in combination with other substances is capable of producing a cellular structure in a plastic or rubber mass. Thus, the term includes compressed gases which expand when pressure is released, soluble solids that leave pores when leached out, liquids that develop cells when they change to gases, and chemical agents that decompose or react under the influence of heat to form a gas. Liquid foaming agents include certain aliphatic and halogenated hydrocarbons, low boiling alcohols, ethers, ketones and aromatic hydrocarbons. The chemical blowing agents range from simple salts such as ammonium or sodium bicarbonate to complex nitrogen releasing agents, of which azobisformamide (ABFA) is an important example.

Blow Molding. The process of forming hollow articles by expanding a hot plastic element against the internal surfaces of a mold. In its most common form, the process comprises extruding a tube (parison) downward between the opened halves of a metal mold, closing the mold to pinch off and seal the parison at top and bottom, injecting air through a needle inserted through the parison wall, cooling the mass in contact with mold, opening the mold and removing the formed article. Many variations of the process exist. In the earliest use of the process, two sheets of cellulose nitrate were used instead of a parison. This method is still in use today. The parison is sometimes formed by injection molding, and sometimes is extruded in advance, cut into lengths and reheated when needed. A new variation of the process announced in 1970, called Cold Parison Blow Molding (the Corpolast Process), employs a pre-formed parison made by extruding a tube and forming one of its ends into a closed hemisphere so that the preform resembles a test tube. These preformed parisons are heated by infra-red radiation and then blow molded in the usual manner. A more recent variation of the blow molding process called *Stretch Blow Molding* achieves biaxial orientation of the polymer by stretching the parison longitudinally as well as radially within a narrow temperature range wherein the stretching produces molecular orientation. The advantages of biaxial orientation are better clarity, reduced creep, higher impact strength, improved gas and water vapor barriers, and lower weight.

Blown Film. See FILM BLOWING.

Blow-Up Ratio. (1) The ratio between the diameter of a blow molding parison and the maximum diameter of the cavity in which it is to be blown. (2) In tubular extrusion blowing of film, the ratio of the extrusion die diameter and the diameter of the blown film.

Blue Asbestos. (crocidolite) An iron-rich form of ASBESTOS, which see, fibers of which are used in reinforced plastics when good chemical resistance is essential.

Blueing. A mold blemish in the form of a blue oxide film on the polished surface of a mold resulting from the use of abnormally high mold temperatures.

Blueing Off. A mold making term for the process of checking the accuracy of mating of two surfaces by applying a thin coating of Prussian Blue on one surface, pressing the coated surface against the other surface, and observing the areas of intimate contact where the blue color has been transferred.

Blushing. The formation of a whitish discoloration on a freshly applied solution coating or lacquer which occurs when fast evaporation of a solvent cools the film below the dew point of the surrounding atmosphere, causing moisture to condense on the wet surface. It is encountered most frequently in periods of high humidity, and can sometimes be avoided by using slower drying and solvents in the formulation. The term "blushing" is sometimes used for the tendency of a plastic article to turn white or chalky in areas that are highly stressed, but this use of the term is not approved by ASTM. See also GATE BLUSH.

Body. (1) A term used loosely in the paint and adhesives industries to denote all-over consistency, that is a combination of viscosity, specific gravity, pastiness, tackiness, etc. (2) An aspect of fabric quality, related to hand and drape.

Body Putty. A paste-like mixture of resin, often a polyester, and a filler such as talc, used to repair metal surfaces such as auto bodies.

Bolster. A spacer or filler in a mold.

Bolus Alba. See KAOLIN.

Bon-Arylamide Reds. A group of metal-free monazo pigments based on substituted 2-hydroxy-3-naphthoic acid.

Bonded Adhesives. See ADHESIVES.

Bonded Fabric. A web of fibers held together by an adhesive medium which does not form a continuous film.

Bonding Resins. A term used for all resins used for bonding aggregates such as foundry sand, grinding wheels, abrasive papers, asbestos brake linings, and concrete masses. Also sometimes used for resinous adhesives for plywood, etc.

Bond Strength. (1) With regard to plastics laminates, a measure of the interlaminar or intralaminar strength. (2) The degree of attraction existing between atoms within a molecule.

BON (B-O-N) Pigments. A family of brilliant reds and maroons widely used in plastics and rubbers, resistant to bleeding, migration and crocking. The initials BON stand for beta-oxynaphoic acid, the basic raw material.

BON Red Pigments. A class of organic azo pigments made by coupling beta-hydroxynaphthoic acid to various amines and forming the barium, calcium, strontium or manganese salts thereof. The colors range from yellowish red to deep maroon.

Booster Ram. A hydraulic ram used as an auxiliary to the main ram of a molding press.

BOP. Abbreviation for butyl octyl phthalate. See BUTYL ETHYLHEXYL PHTHALATE.

Boric Acid Esters. Flame retardants for plastics, etc., and plasticizers. Examples are the trimethyl, tri-n-butyl, tricyclohexyl tridodecyl, and tri-p-cresyl borates.

Bornyl Acetate. $C_{10}H_{17}OOCCH_3$ A solvent and plasticizer for nitrocellulose.

Boron Fibers. Filaments produced by a chemical vapor plating process of depositing boron on a tungsten wire core, or, more recently, on a glass filament core. The core filament is preheated and drawn through a reactor containing a vaporized boron compound such as boron trichloride (in the case of tungsten wire) or diborane (in the case of the newer glass filament core). The filaments thus produced have nominal diameters ranging from .004 to .008 inches. They are characterized by low density, high tensile strength and high modulus of elasticity, but are extremely stiff, e.g. five times stiffer than glass fibers. This stiffness makes boron filaments difficult to weave, braid or twist, but they can be formed into resin impregnated tapes for hand lay-up and filament winding processes. The high cost of boron filaments has limited their use to experimental aircraft and aerospace applications.

Boron Hydride. See DIBORANE.

Boss. A protuberance provided on an article to add strength, facilitate alignment during assembly or for attaching the article to another part.

Boyer-Beaman Rule. A statement of the relationship between the glass transition temperature T_g and the melting temperature T_m of a polymer. The ratio of T_g to T_m (with T expressed in degrees Kelvin) usually lies between 0.5 and 0.7. For symmetrical polymers such as polyethylene the ratio is close to 0.5. For unsymmetrical polymers such as polystyrene and polyisoprene it is approximately 0.7.

BR. Abbreviation for butadiene rubber. (British Standards Institution) See POLYBUTADIENE.

Brabender Plastograph. (Brabender Plasti-Corder) An instrument which continuously measures the torque exerted in shearing a polymer or compound specimen over a wide range of shear rates and temperatures, including those conditions anticipated in actual plant practice. The instrument records torque, time and temperature on a graph called a plastogram, from which much information can be obtained with regard to processability of an experimental compound. It shows the effects of additives and fillers, measures and records lubricity, plasticity, scorch, cure, shear and heat stability and polymer consistency. The Brabender Plastograph is made in Germany. A similar instrument is made in the U.S. under the registered trademark Brabender Plasti-Corder.

Branched Polymer. A polymer in which the molecules have been formed by BRANCHING, which see. The opposite of a *linear polymer.*

Branching. The growth of a new polymer chain from an active site on an established chain, in a direction different from that of the original chain. Branching occurs as a result of chain transfer processes or from the polymerization of difunctional monomers, and is an important factor in polymer properties.

Breakdown Voltage. The voltage required, under specific conditions, to cause failure of an insulating material. See also DIELECTRIC STRENGTH.

Breaker Plate. A perforated plate located at the rear end of an extruder head or die adapter, serving to support the screen pack if one is used, otherwise functioning as a screen to remove foreign particles and also to create back pressure to help stabilize the flow of material through the extruder. Breaker plates are also used without screens in the nozzles of injection molding machines for improving the distribution of color particles.

Breaking Extension. See ELONGATION.

Breathable Film. A film which is at least slightly permeable to gases due to the presence of open cells throughout its mass or to perforations.

Breathing. (1) The passage of air through a plastic film due to a degree of porosity. (2) In injection molding, the momentary opening and closing of a mold during the early stages of the cycle to permit the escape of air or gas from the heated compound.

Brightening Agents. (optical brighteners, fluorescent bleaches, optical whitening agents) Chemical agents used primarily in fibers, but also to some extent in molded and extruded products, to overcome yellow casts and to enhance clarity or brightness. In contrast to *blueing agents* which act by removing yellow light, the optical brighteners absorb the invisible ultra violet rays and convert their energy into visible blue-violet light. Thus, they cannot be used in compounds that also contain U.V. absorbing agents. Optical brighteners are used in PVC sheet and film, fluorescent lighting fixtures, vinyl flooring, nylon fishing line, polyethylene bottles etc. A few examples of optical brighteners are coumarins, naphthotriazolyl-stilbenes, benzoxazolyl, benzimidazolyl, naphthylimide, and diaminostilbene disulfonates.

Brinell Hardness. The hardness of a material as determined by pressing a hardened steel ball of 10 mm diameter into the specimen under a constant load, expressed as the load in kilograms divided by the area in square mm of the spherical impression formed by the ball. For non-ferrous materials, the load is 500 kg applied for 30 seconds.

British Thermal Unit. (Btu) Prior to the introduction of the new SI system, Btu meant simply the quantity of energy required to raise the temperature of one pound mass of water

1°F, averaged from 32 to 212°F. It was equal to 1054.350 Joules. Slight differences exist for the actual starting temperature of the water. In the ASTM Standard for Metric Practice there are 31 types of Btu's listed, equated variously to Joules, Watts per meter kilvin, Watts, joules per meter^2, watts per meter^2, watts per meter^2 kelvin, joules per kilogram, and joules per kilogram kelvin. The only type in this list of 31 kinds of Btu's that seems to correspond to the old-fashioned Btu ("Btu, thermochemical/lb. °F, c, heat capacity) is to be translated to 4.184.000 E+03 joules per kilogram kelvin. (J/Kg·K).

Brittleness Temperature. The temperature at which plastics and elastomers rupture by impact under specified conditions. One method, described by ASTM D746, consists of determining the temperature at which 50% of a group of specimens fail by a single impact. The brittleness temperature is related to that of the GLASS TRANSITION, which see, in the case of plastics with glass transition temperatures below room temperature, such as flexible PVC.

Broadgoods. A term used in the fabric industry for woven materials, including glass fabrics, that are over eighteen inches in width.

Bronze Pigments. Simulated bronze or gold colored pigments made by staining aluminum flakes with yellow or brown colorants.

Bronzing. A term sometimes used for BLEEDING, which see, but more specifically referring to the appearance of an iridescent metallic luster caused by a film of dry pigment on a glossy surface.

Brookfield Viscometer. The Brookfield Synchro-Lectric Viscometer is the most widely used instrument for measuring the viscosity of plastisols and other liquids of a thixotropic nature. The instruments measures the shearing stress on a spindle rotating at a definite, constant speed while immersed in the sample. The degree of spindle lag is indicated on a rotating dial. This reading multiplied by a conversion factor based on spindle size and rotational speed, gives a value for viscosity in centipoises. By taking measurements at different rotational speeds an indication of the degree of thixotrophy of the sample obtained.

BSI. Abbreviation for British Standards Institution.

B-Stage. An intermediate stage in the reaction of certain thermosetting resins in which the material swells when in contact with certain liquids and softens when heated, but may not entirely dissolve or fuse. The resin in an uncured thermosetting molding compound is usually in this stage. Sometimes referred to as *resistol*. See also A-STAGE and C-STAGE. (ASTM D883-75a).

Btu. See BRITISH THERMAL UNIT.

BTX. Abbreviation for the group of solvents comprising benzene, toluene and xylene.

Bubble Forming. A thermoforming process in which the plastic sheet is clamped in a frame suspended above a mold, heated, blown into a blister shape by air, then molded to shape by means of a descending plug applied to the blister forcing it downward into the mold. See also THERMOFORMING.

Bubbler. A device inserted into a mold force, cavity or core which allows water to flow deep inside the hole into which it is inserted and to discharge through the open end of the hole. Uniform cooling of the molds and of isolated mold sections can be achieved in this manner.

Buckling. A crimping of the fibers in a composite material, often occurring in glass reinforced thermosets due to resin shrinkage during cure.

Bulk Density. The density of a molding material in loose form (granular, nodular, etc.) expressed as a ratio of weight to volume (e.g., g/cm^3 or lb/ft^3).

Bulk Factor. The ratio of the volume of any given mass of loose plastic material to the volume of the same mass of the material after molding or forming. The bulk factor is also equal to the ratio of the density after molding or forming to the apparent density of the material as received.

Bulking Value. The volume of a weight unit of material, usually expressed in gallons per pound.

Bulk Modulus. The modulus of volume elasticity, M_b, which is equal to

$$\frac{p_2 - p_1}{\dfrac{v_1 - v_2}{v_1}}$$

wherein p_1, p_2 : v_1, and v_2 are the initial and final pressure and volume respectively. See also MODULUS OF ELASTICITY.

Bulk Molding Compound. See PREMIX.

Bulk Polymerization. (mass polymerization) The polymerization of a monomer in the absence of any medium other than a catalyst or accelerator. The monomers are usually liquids, but the term also applies to the polymerization of gases and solids in the absence of solvents or any other dispersing medium. Polystyrene, PMMA, low density polyethylene, and styrene-acrylonitrile copolymers are examples of polymers most frequently produced by bulk polymerization. Acrylic monomers may be simultaneously polymerized and formed into products by conducting the polymerization in molds such as those for rods and sheets. Other bulk polymerizations are conducted in heated kettles, usually equipped with agitators and means for controlling the atmosphere.

Bulk Specific Gravity. The specific gravity of a porous solid when the volume of the solid as used in the calculation includes both the permeable and impermeable voids. See also SPECIFIC GRAVITY.

Buna-N. See ACRYLONITRILE-BUTADIENE COPOLYMERS.

Buna-S. A synthetic elastomer produced by the copolymerization of butadiene and styrene. Also called GR-S, and STYRENE-BUTADIENE RUBBERS, which see.

Burned. Showing evidence of excessive heating during processing or use of a plastic, as evidenced by blistering, discoloration, distortion or destruction of the surface.

Burning Rate. See FLAMMABILITY, OXYGEN INDEX FLAMMABILITY TEST, SELF EXTINGUISHING.

Bushing. In an extruder, the outer ring of any type of a circular tubing or pipe die which forms the outer surface of the tube or pipe.

Butadiene. CH_2:CH·CH:CH_2 (erythrene;butadiene-1,3; vinylethylene; bivinyl; divinyl B) A gas, insoluble in water but soluble in alcohol and ether, obtained from cracking of petroleum, from coal tar benzene, or from acetylene. It is widely used in the formation of copolymers with styrene, acrylonitrile, vinyl chloride and other monomers, imparting flexibility to the subsequent moldings.

Butadiene-Acrylonitrile Copolymers. (NBR) See ACRYLONITRILE-BUTADIENE CO-POLYMERS.

Butadiene Rubber. (BR) See POLYBUTADIENE.

Butadiene-Styrene Thermoplastics. See STYRENE-BUTADIENE THERMOPLASTICS.

Butaldehyde. See BUTYRALDEHYDE.

Butanal. See BUTYRALDEHYDE.

1,4-Butanedicarboxylic Acid. See ADIPIC ACID.

1,3-Butanediol. See 1,3-BUTYLENE GLYCOL.

1,4-Butanediol. See 1,4-BUTYLENE GLYCOL.

1,2,4-Butanetriol. $HOCH_2CHOHCH_2CH_2OH$ A nearly colorless liquid, used as an intermediate for alkyd resins and a plasticizer for cellulosics.

Butanoic Acid. See BUTYRIC ACID.

n-Butanol. See n-BUTYL ALCOHOL.

2-Butene-1,4-Diol. $HOCH_2CH$:$CHCH_2OH$ A nearly colorless odorless liquid, used as an intermediate for alkyd resins, plasticizers, nylon, and cross-linking agent for resins.

Butenes. See BUTYLENES.

2-Butoxyethanol. See ETHYLENE GLYCOL MONOBUTYL ETHER.

2-Butoxyethyl Pelargonate. $CH_3(CH_2)_7COOC_2H_4OC_4H_9$ A plasticizer for polystyrene, vinyl chloride polymers and copolymers, and cellulosic plastics.

Butoxyethyl Stearate. $CH_3(CH_2)_{16}COOC_2H_4OC_4H_9$ A high boiling ester type plasticizer for nitrocellulose, polystyrene, ethyl cellulose and polyvinyl acetate.

Butt-Fusion. A method of joining pipe, sheet or other forms of a thermoplastic resin wherein the ends of the two pieces to be joined are heated to the molten state and then rapidly pressed together to form a homogeneous bond.

Butyl. (1) The radical C_4H_9, occurring only in combination. (2) Abbreviation used by British Standards Institution for BUTYL RUBBER, which see.

Butyl Acetate. $CH_3COOC_4H_9$ A solvent of moderate strength for ethyl cellulose, cellulose nitrate, vinyls, PMMA, polystyrene, coumarone-indene resins and certain alkyds and phenolics.

sec-Butyl Acetate. $CH_3COOCH(CH_3)(C_2H_5)$ (2-butanol acetate) A solvent for nitrocellulose, ethyl cellulose, PVC, acrylics, polystyrene, phenolics and alkyd resins.

Butyl Acetoxystearate. $CH_3(CH_2)_5CH(CH_3COO)(CH_2)_{10}COOC_4H_9$ A plasticizer similar to butyl acetyl ricinoleate, except that the double bond is saturated. It is compatible with cellulosic and vinyl resins.

Butyl Acetyl Ricinoleate. $CH_3(CH_2)_5CH(CH_3CO_2)CH_2(CH_2)_7CO_2C_4H_9$ A yellow, oily liquid derived from castor oil, butyl alcohol and acetic anhydride, used as a plasticizer. It is compatible with cellulosics and vinyls.

n-Butyl Acrylate. $CH_2:CHCOOC_4H_9$ A colorless liquid which polymerizes readily on heating.

n-Butyl Alcohol. $CH_3(CH_2)_2CH_2OH$ A medium-boiling point alcohol used as a solvent for cellulosic, phenolic and urea-formaldehyde resins. It is also used as a diluent-reactant in the manufacture of urea-formaldehyde and phenol-formaldehyde resins, and as an intermediate in the production of butyl acetate, dibutyl phthalate and dibutyl sebacate.

n-Butyl Aldehyde. See BUTYRALDEHYDE.

Butylated Resins. Resins containing the butyl radical, C_4H_9.

Butylated Hydroxytoluene. (di-tert-butyl-para-cresol, BHT) A white crystalline solid, the most widely used antioxidant for plastics such as ABS and LDPE. It is FDA-approved as an antioxidant for food and food packaging materials.

Butyl Benzenesulphonamide. $C_6H_5SO_2NHC_4H_9$ (N-n-butyl benzenesulphonamide) A plasticizer for some synthetic resins and an intermediate for resin manufacture.

Butyl Benzoate. $C_6H_5COOC_4H_9$ (n-butyl benzoate) A plasticizer and solvent for cellulosics.

Butyl Benzyl Phthalate. $C_6H_4(COO)_2C_4H_9C_7H_7$ (BBP) A clear, oily liquid used as a plasticizer for cellulosic and vinyl resins. It imparts good stain resistance, low volatility at calendering and extruding temperatures, low oil extraction and good heat and light stability.

Butyl Benzyl Sebacate. $C_4H_9OOC(CH_2)_8COOC_7H_7$ An ester-type plasticizer with a light straw color. It combines the desirable properties of dibenzyl sebacate and dibutyl sebacate.

Butyl Borate. See TRIBUTYL BORATE.

Butyl Cyclohexyl Phthalate. $C_6H_4(COOC_4H_9)(COOC_6H_{11})$ A plasticizer for PVC, other vinyls, cellulosic plastics and polystyrene.

Butyl Decyl Phthalate. A plasticizer for polystyrene, PVC and vinyl chloride-acetate co-polymers.

Butyl Diglycol Carbonate. $(C_{14}H_{26}O_7)$ (diethylene glycol bis[n-butylcarbonate]) A color-less liquid of low volatility, used as a plasticizer and solvent for many resins.

1,3-Butylene Diamethacrylate. A polymerizable monomer sold under the trade name "Monomer X-970," used in PVC and rubber systems to obtain rigid or semi-rigid products from materials that are normally flexible. The monomer acts as a plasticizer at room temperatures, and cross-links at processing temperatures.

1,3-Butylene Glycol. (1,3-butanediol) $CH_3CHOHCH_2CH_2OH$ A colorless liquid made by catalytic hydrogenation of acetaldol. Its most important use is as an intermediate in the manufacture of polyester plasticizers.

1,4-Butylene Glycol. (1,4-butanediol, tetramethylene glycol) $HOCH_2CH_2CH_2CH_2OH$ A stable, hygroscopic colorless liquid, used in the production of polyesters by reaction with dibasic acids, and in the production of polyurethanes by reaction with diisocyanates.

Butylenes. The class of four C_4 monounsaturated hydrocarbons including the following:

IUPAC NAMES	Alternative Names
1-butene	alpha butylene
cis-2-butene	cis-beta-butylene
trans-2-butene	trans-beta-butylene
methylpropene	isobutylene

The term *butenes* refers to the first three members above as a group. The butylenes are used as monomers for rubbery homopolymers and copolymers with sytrene, acrylics, other olefins and vinyls. They are also used in adhesives for many plastics, plasticizers and flame retardants.

1,3-Butylene Glycol Adipate Polyester. (Saniticizer® 334F) A polymeric plasticizer for PVC.

Butyl Epoxy Stearate. A plasticizer for PVC, imparting low temperature flexibility.

Butyl Ethylhexyl Phthalate. (butyl octyl phthalate) A mixed ester of butanol and 2-ethyl-hexanol, widely used as a primary plasticizer for PVC compounds and plastisols in which it performs like DOP in most respects. It is also compatible with vinyl chloride-acetate copolymers, cellulose nitrate, ethyl cellulose, polystyrene, chlorinated rubber and, at lower concentrations, PMMA.

Butyl Formate. $HCOOC_4H_9$ A solvent for several resins, including nitrocellulose and cellulose acetate.

tert-Butyl Hydroperoxide. $(CH_3)_3COOH$ A highly reactive peroxy compound used as a polymerization catalyst.

Butyl Isodecyl Phthalate. $C_4H_9OCOC_6H_4COOC_{10}H_{21}$ (decyl butyl phthalate) A plasticizer for PVC and polystyrene.

Butyl Isohexyl Phthalate. $(COOC_4H_9)C_6H_4(COOC_6H_{13})$ A plasticizer for cellulosics, acrylic resins, polystyrene, PVC and other vinyl resins.

Butyl Lactate. $CH_3CHOHCOOC_4H_9$ A solvent for nitrocellulose, ethyl cellulose and many synthetic resins.

Butyl Laurate. $C_{11}H_{23}COOC_4H_9$ A plasticizer for cellulosic plastics, polystyrene and vinyl resins.

n-Butyl Methacrylate. $H_2C:CCH_3COOC_4H_9$ A polymerizable monomer used in the production of acrylic resins and potting compounds.

n-Butyl Myristate. $CH_3(CH_2)_{12}COOC_4H_9$ A butyl ester of myristic acid. An oily liquid used as a plasticizer for cellulosic plastics.

Butyl Octadecanoate. See BUTYL STEARATE.

Butyl Octyl Phthalate. See BUTYL ETHYLHEXYL PHTHALATE.

Butyl Oleate. $CH_3(CH_2)_7CH:CH(CH_2)_7COOC_4H_9$ A solvent, plasticizer and lubricant, used mainly with chloroprene and other synthetic rubbers, chlorinated rubber and ethyl cellulose. It is also used as a mold lubricant.

n-Butyl Palmitate. $C_{15}H_{31}COCC_4H_9$ A plasticizer for polystyrene and cellulosic plastics.

tert-Butyl Perbenzoate. $C_6H_5CO \cdot O_2 \cdot C(CH_3)_3$ A catalyst for the polymerization of acrylic and styrene monomers, and the curing of polyesters. Also used in the compounding of silicones and polyethylene. t-butyl perbenzoate has long been the workhorse in sheet molding compounds, because it is stable enough for all practical purposes but is slow-reacting, requiring activation temperatures of 250 to 260° F unless augmented by a less stable peroxide.

tert-Butyl Permaleic Acid. $(CH_3)_3CCO_2COCH:CHCOOH$ A polymerization catalyst.

t-Butyl Peroxy Neodecanoate. A polymerization initiator for vinyl chloride.

t-Butyl Peroxypentanoate. A member of the peroxyester catalyst family.

t-Butyl Perphthalic Acid. $(CH_3)_3CO_2COC_6H_4COOH$ A polymerization catalyst.

para-tert-Butyl Phenol. $(CH_3)_3CC_6H_4OH$ A white crystalline solid used as a plasticizer for cellulose acetate.

p-tert-Butylphenyl Salicylate. A plasticizer approved by FDA for food contact use, also used as a light absorbing agent.

n-Butylphosphoric Acid. $C_4 H_9 H_2 PO_4$ A reddish amber liquid used as a catalyst, for example in urea resin production.

Butyl Phthalyl Butyl Glycollate. $C_4 H_9 OCOC_6 H_4 COOCH_2 COOC_4 H_9$ A plasticizer with good light stability, used mainly with PVC and polystyrene, but compatible with most other thermoplastics. It has been FDA approved for food contact use.

n-Butyl Propionate. $C_2 H_5 CO_2 C_4 H_9$ A colorless liquid with an apple-like odor, used as a solvent for nitrocellulose.

Butyl Ricinoleate. $C_{17} H_{32} (OH)COOC_4 H_9$ A plasticizer for vinyl resins and cellulose acetate butyrate, derived from castor oil and butyl alcohol.

Butyl Rubber. (GR-1) A synthetic elastomer produced by copolymerizing isobutylene with a small amount (about 2%) of isoprene or butadiene. It has good resistance to heat, oxygen and ozone, and to the permeation of gases. Thus, it is widely used in inner tubes. Other polymers of isobutylene alone range from oils to tacky waxes.

Butyl Stearate. $C_{17} H_{35} COOC_4 H_9$ (butyl octodecanoate) A mold lubricant and plasticizer, compatible with natural and synthetic rubbers, chlorinated rubber and ethyl cellulose. It can be used in vinyls in very low concentrations as a non-toxic secondary plasticizer and lubricant. In the production of polystyrene, butyl stearate is added to the emulsion polymerization system to impart good flow properties to the resin.

Butyraldehyde. $CH_3 (CH_2)_2 CHO$ (butaldehyde, n-butanal, n-butyl aldehyde, butyric aldehyde) An aldehyde sometimes used in place of formaldehyde in the production of resins. Butyraldehyde reacts with polyvinyl alcohol to form polyvinyl butyrate.

Butyrate. (1) The salt or ester of BUTYRIC ACID, which see. (2) The common name for CELLULOSE ACETATE BUTYRATE (CAB), which see.

Butyric Acid. $CH_3 CH_2 CH_2 COOH$ (n-butyric acid, butanoic acid, ethylacetic acid, propylformic acid) A liquid used in the production of CELLULOSE ACETATE BUTYRATE, which see. Derivatives of butyric acid are used in the production of plasticizers for cellulosic plastics.

Butyrolactone. $CH_2 CH_2 CH_2 COO$ (gamma-butyrolactone) A hygroscopic, colorless liquid obtained by the dehydrogenation of 1,4-butanediol. It is a solvent for cellulosics, epoxy resins and vinyl copolymers.

CA. Abbreviation for CELLULOSE ACETATE, which see.

CAB. Abbreviation for CELLULOSE ACETATE BUTYRATE, which see.

Cadmium Ethylhexoate. A metallic soap used as a stabilizer for vinyls, especially to avoid plate-out in calendering compounds.

Cadmium Pigments. Inorganic pigments based on cadmium sulphide and cadmium sulpho-selenides, used widely in PVC, polystyrene and polyolefins. Included are cadmium maroon, -orange, -red and -yellow. The cadmium pigments have good resistance to heat (up to 500°C) and to alkalis, and are non-bleeding. Light stability is good in solid colors, but may be poor when used for tints with white pigments. Resistance to acids is poor.

Cadmium Ricinoleate. $Cd[CH_3(CH_2)_5CHOHCH_2CH:(CH_2)_7CO_2]_2$ A white powder derived from castor oil, used as a heat stabilizer for vinyl chloride polymers and copolymers.

Cadmium Stearate. $Cd(C_{17}H_{35}COO)_2$ A heat- and light-stabilizer, used when good clarity is desired.

Calcined Clays. See CLAYS.

Calcium Acetate. $Ca(C_2H_3O_2)_2 \cdot H_2O$ (vinegar salts, gray acetate, lime acetate, brown acetate) A stabilizer.

Calcium Carbonate. $CaCO_3$ (aragonite, calcite, chalk, limestone, lithographic stone, marble, marl, travertine, whiting) Grades of calcium carbonate suitable as fillers for plastics are obtained from naturally-occurring deposits as well as by chemical precipitation. The natural types are prepared by dry grinding, yielding particles usually over 20 microns, used in stiff products such as floor tiles; or by wet grinding, yielding particles under 16 microns, used in flexible products. The chemically precipitated types range from .05 to 11 microns in size, and are most often used in plastisols and highly flexible products. Both the wet ground and precipitated types are available with coatings such as resins, fatty acids and calcium stearate. These coated grades have low oil absorption, of particular value in compounding plastisols. The calcium stearate coatings provide improved electrical properties, heat stability and lubricity during processing, beneficial in extrusion compounds.

Calcium Glycerophosphate. $CaC_3H_7O_2PO_4$ (calcium glycerinophosphate) A white, crystalline powder, odorless and nearly tasteless, used as a stabilizer for plastics.

Calcium Oxide. CaO (lime, quicklime, burnt lime) A white powder with affinity for water, with which it combines to form calcium hydroxide. It has been used to remove traces of water in vinyl plastisols.

Calcium Phosphate, Dibasic. $CaHPO_4$ or $CaHPO_4 \cdot 2H_2O$ (dicalcium ortho-phosphate, bicalcic phosphate, secondary calcium phosphate) A stabilizer.

Calcium Phosphate, Monobasic. $CaH_4(PO_4)_2 \cdot H_2O$ (calcium biphosphate; acid calcium phosphate, calcium phosphate, primary; monocalcium phosphate) A stabilizer.

Calcium Phosphate, Tribasic. $Ca_3(PO_4)_2$ (Calcium orthophosphate, tricalcium phosphate, precipitated calcium phosphate, tricalcium orthophosphate, tricalcic phosphate, tertiary calcium phosphate) A stabilizer.

Calcium Ricinoleate. $Ca[CH_3(CH_2)_5CHOHCH_2CHCH(CH_2)_7CO_2]_2$ A white powder derived from castor oil, used as a non-toxic stabilizer for PVC.

Calcium Silicate. $CaSiO_3$ (Wollastonite) A naturally occurring mineral found in metamorphic rocks, used as a reinforcing filler in polyester molding compounds, low density polyethylene and other thermosetting resins. It imparts smooth molded surfaces and low water absorption.

Calcium Stearate. $Ca(C_{17}H_{35}COO)_2$ A non-toxic stabilizer and lubricant. It is not often used alone because of its early color development, but is used in combination with zinc and magnesium derivatives and epoxides in the production of non-toxic stabilizers.

Calcium Sulphate. (anhydrite) $CaSO_4$ A filler and white pigment. The hydrated forms are known as gypsum and terra alba.

Calcium Thiocyanate. $Ca(SCN)_2 \cdot 3H_2O$ (Calcium sulphocyanate, calcium rhodanate) A solvent for acrylic and cellulosic resins.

Calcium-Zinc Stabilizers. A family of stabilizers based on compounds and mixtures of compounds of calcium and zinc. Their effectiveness is limited, but they are among the few that have been approved by the FDA for materials to be contacted by foods.

Calender. (n) The machine performing the operation of CALENDERING, which see.

Calender Coating. The process of coating substrates such as paper or fabric by passing both the substrate and a plastic film through calender rolls.

Calendering. The process of forming thermoplastics sheeting or film by passing the material through a series of heated rollers. The gap between the last pair of heated rollers determines the thickness of the sheet. Subsequent cold rollers cool the sheet. The plastic compound is usually premixed and plasticated on separate equipment, then fed continuously into the nip of the first pair of calender rolls.

Calorie. In the cgs system, a calory is the amount of energy required to raise one gram mass of water $1°C$. In the new SI in which the joule is the standard unit of mechanical, electrical and thermal energy, the calorie at $20°C$ is equal to 4.181 90 E+00 joules.

Calorimeter. A device for measuring the heat liberated during thermal reactions.

Camphor. $C_{10}H_{16}O$ (2-camphanone; 2-keto-1,7,7-trimethylnorcamphane; 1,7,7-trimethyl-2-oxobicyclo [2.2.1] heptane) A colorless to white crystalline material, derived originally by distilling the leaves, twigs and stems of the camphor tree, and since the 1930's from a constituent of turpentine known as pinene. Its importance to the plastics industry dates from 1865, when Parker discovered that it could be used to plasticize cellulose nitrate to form celluloid.

Camphoric Acid. $C_{10}H_{16}O_4$ A plasticizer for cellulose nitrate, derived by oxidizing camphor with nitric acid.

CAN. Abbreviation for cellulose acetate nitrate.

Candela. (cd) The new SI unit of luminous intensity, defined as the luminous intensity, in the perpendicular direction, of a surface of 1/600,000 square meter of a black body at the temperature of freezing platinum under a pressure of 101,325 newtons per square meter. In older literature, the same abbreviation (cd) is used for *candlepower* and *candle* which have different values.

Cantilever Beam Stiffness. A method of determining the stiffness of plastics by measuring the force and angle of bend of a cantilever beam made of the specimen material. The ASTM test is D 747.

Caoutchouc. An early name for natural rubber, still in use in the French language.

CAP. Abbreviation for CELLULOSE ACETATE PROPIONATE, which see.

Capillary Rheometer. An instrument for measuring the flow properties of polymer melts, comprising a capillary tube of specified diameter and length, means for applying desired pressures to force the molten polymer through the capillary, means for maintaining desired temperatures of the apparatus, and means for measuring differential pressures and flow rates. The data obtained from capillary viscometers is usually presented as graphs of shear stress against shear rate at constant temperature.

Capillary Viscometer. This term is frequently used for two types of capillary instruments — one used for concentrated solutions or polymer melts described under CAPILLARY RHEO-METER, and the other used for measuring dilute solution viscosities. The most widely used of the latter types employ a glass capillary tube and means for timing the flow of a measured volume of solution through the tube under the force of gravity. This time is then compared with the time taken for the same volume of pure solvent, or of another liquid of known viscosity, to flow through the same capillary.

Capric Acid. $CH_3(CH_2)_8COOH$ (decanoic acid, decoic acid, decylic acid) A plasticizer and an intermediate for resins.

Caprolactam. $CH_2(CH_2)_4NHCO$ (epsilon-caprolactam, aminocaproic lactam) A cyclic amide type compound containing a ring of 6 carbon atoms. When the ring is opened, caprolactam is polymerizable to a nylon resin known as type-6 nylon or polycaprolactam. It is also used as a cross-linking agent for polyurethanes, and a plasticizer. In the late 1960's it was found that caprolactam could be rotationally cast by heating the solid monomer to its melting point (about 162° F) and introducing the molten monomer into a mold along with a catalyst, then heating and rotating the mold in the usual manner. The liquid gradually thickens and gels against the mold as does plastisol, and conversion to Nylon 6 is accomplished within a few minutes.

Caprylic Acid. $CH_3(CH_2)_6COOH$ (octanoic acid, octoic acid, octylic acid, caprilic acid) A plasticizer and an organic intermediate.

Carbamide. See UREA.

Carbamide Phosphoric Acid. $CO(NH_2)_2 \cdot H_3PO_4$ (urea phosphoric acid) A catalyst for acid-setting resins.

Carbamylurea. See BIURET.

Carbathene®. The trade name of Standard Telecommunications Laboratories, Ltd. for a copolymer of ethylene and N-vinyl carbazole. Its structural formula is $C_6H_4 \cdot NCHCH_2 \cdot C_6H_4$.

Carbazole. $(C_6H_4)_2NH$ (dibenzopyrrole, diphenylenimine) A derivative of orthoaminodiphenyl, used in the production of polyvinyl carbazole resins.

Carbolic Acid. See PHENOL.

Carbon 14. (radiocarbon) Radioactive carbon of mass number 14, usually made by irradiating calcium nitrate. It is used as a source of radiation in gauges for measuring the thickness of plastic films.

Carbon Black. A generic term for the family of colloidal carbons. More specifically, carbon black is made by the partial combustion and/or thermal cracking of natural gas, oil, or another hydrocarbon. *Acetylene black* is the type of carbon black derived from the burning of acetylene. *Animal black* is derived from bones of animals. *Channel blacks* are made by impinging gas flames against steel plates or channel irons (from which the name is derived), from which the deposit is scraped at intervals. *Furnace black* is the term sometimes applied to carbon blacks made in a refractory-lined furnace. *Lamp black,* the properties of which are markedly different from other carbon blacks, is made by burning heavy oils or other carbonaceous materials in closed systems equipped with settling chambers for collecting the soot. *Thermal black* is produced by passing natural gas through a heated brick checkerwork where it thermally cracks to form a relatively coarse carbon black. Carbon blacks are widely used as fillers and pigments in PVC, phenolics, polyolefins and several other resins, also imparting resistance to ultraviolet rays. In polyethylene, carbon black acts as a cross-linking agent.

Carbon Fibers. A group of fibrous materials comprising essentially elemental carbon. They may be prepared by (1) growing single crystals in a carbon electric arc under high pressure inert gas, (2) growth from a vapor state by thermal decomposition of a hydrocarbon gas, or (3) pyrolysis of organic fibers, the most widely used method. Cellulosic rayon fibers are most commonly used as starting materials. They are heated in the absence of air and moisture at temperatures ranging from 1300° to 1700°F to form carbon fibers, which may be heated further to about 5000°F to form graphite fibers. Carbon and graphite fibers have been used as reinforcements for ablative plastics. A new type of carbon fiber developed by Union Carbide in 1973 is derived from pitch. Mats and continuous fibers of this new material are cheaper than those made from pyrolyzed rayon or polyacrylonitrile filaments and may broaden the use of carbon fibers, especially in the field of electrically conductive plastics. See also WHISKERS.

Carbon Hexachloride. See HEXACHLOROETHANE.

Carbon Tetrachloride. CCl_4 (tetrachloromethane) A powerful solvent for ethyl cellulose and benzyl resin, and a starting material for synthesis of nylon-7.

Carbonyl. The divalent organic radical CO, found only in combination.

Carboxybenzene. See BENZOIC ACID.

Carboxylic. Term for the COOH group, the radical occurring in organic acids.

Carboxy Nitroso Rubber. (CNR) A fluorocarbon elastomer, synthesized as a terpolymer from tetrafluoroethylene, trifluoronitrosomethane, and nitrosoperfluorobutyric acid. CNR has unique resistance to strong oxidizers and is nonflammable in pure oxygen, hence is finding applications in the aerospace field. The gum can be processed on standard rubber mixing equipment for molding, or dissolved in solvents for application by spraying, dipping or brushing.

Cascade Coating. A process used for coating objects such as electrical resistors and capacitors with epoxy or other thermosetting resins, in which finely powdered resin is poured over the preheated object to be coated. The article is usually rotated while the powder is being applied.

Casein. The protein substance occurring in milk and cheese. It can be obtained by treating skim milk with a dilute acid, but the type used mainly for plastics (rennet casein or paracasein) is made by treating warm skim milk with a rennet extract. See also CASEIN PLASTICS.

Casein Plastics. A family of plastics derived from casein, used widely in the early years of the plastics industry but of less importance today. The first casein plastics were made by precipitating casein from milk with an acid, pressing the curd into shapes, then treating the shaped articles with formaldehyde. These casein plastic articles were slightly flexible but thermosetting in nature. Later, thermoplastic caseins were developed by adding about 2% of an aluminum salt to a rennet casein, forming a mass that could be extruded or pressed into shaped articles. After-treatment with formaldehyde was still necessary for hardening. Alternatively, parts can be machined from sheets, rods and tubes which have been pre-hardened with formaldehyde. Casein plastics have poor water resistance and dimensional stability, which limits their applications.

Cashew Resin. A thermosetting resin produced from the phenolic fraction of cashew nut shell oil.

Casing. A term coined by Bell Telephone Laboratories as an abbreviation for the process of Crosslinking by Activated Species of INert Gases, developed to impart printability and adhesive receptivity to polymers such as PTFE and polyethylene. In this process, the articles are exposed to a flow of activated inert gases in a glow discharge tube, which forms a shell of highly crosslinked molecules on the article surfaces. This shell greatly increases the cohesive strength so that printing inks and adhesives will bond firmly to the articles.

Cast. (v) See CASTING.

Cast Embossing. A term used for the process of casting films against an embossed temporary carrier. Vinyl plastisols, organosols, solutions or latices are used as film formers, which may be backed up with layers of foam or fabric. The temporary carrier is often paper, embossed with the desired pattern and treated so as to be releasable from the fused film or laminate.

Cast Film. Film produced by pouring or spreading a solution, hot-melt or dispersion of plastic material onto a temporary carrier, hardening the material by suitable means, and stripping the hardened film from the surface. Cellulosic, polystyrene and vinyl films are often produced in this manner.

Casting. (v) The process of forming solid or hollow articles from fluid plastic mixtures or resins by pouring or injecting the fluid into a mold or against a substrate with little or no pressure, followed by solidification and removal of the formed object. See also CAST EMBOSSING, CENTRIFUGAL-, FILM-, SLUSH-, SOLID-, ROTATIONAL-, and SOLVENT CASTING; and EMBEDDING, ENCAPSULATION and POTTING.

Casting. (n) The finished product of a casting operation.

Casting Syrups. (casting resins) Liquid monomers or incompletely-polymerized polymers, usually containing catalysts or curing agents, capable of becoming hard after they are cast in molds. The materials most generally used are the acrylics, styrenes, polyesters, epoxies, silicones and nylons. Also called *potting syrups* when used for encapsulating articles such as electrical components.

Castor Oil. (ricinus oil) A pale-yellowish oil derived from the seeds of the castor bean, Ricinus communis, and consisting essentially of ricinolein. It is an important starting material for plasticisers, certain nylons, and alkyd resins; and an ingredient in certain types of urethane foams.

Catalyst. A substance which causes or accelerates a chemical reaction when added to the reactants in minor amount, without being permanently affected by the reaction. A negative catalyst (inhibitor, retarder) decreases the rate of reaction. See also ACCELERATOR, CURING AGENTS, INITIATOR.

Cathode. (1) The positive pole of a battery. (2) In a cell through which current is being forced, the cathode is the negative electrode. (3) In a vacuum tube, the cathode is the electrode from which electrons are emitted.

Cathode Sputtering Process. See VACUUM METALLIZING.

Cation. An atom, molecule or radical which has lost an electron and thus has become positively charged.

Cation Exchange Resins. See ION EXCHANGE RESINS.

Cationic. Pertaining to any positively charged atom, radical or molecule; or to any compound or mixture containing positively charged groups.

Cationic Polymerization. See IONIC POLYMERIZATION.

Caul. A sheet of metal, wood or other material used in laminating to apply and equalize pressure.

Cavity. A depression, or a set of matching depressions, in a plastics-forming mold which forms the outer surfaces of the cast or molded articles. The cavity may surround a *core*, the portion of the mold that forms the inner surfaces of a hollow article.

Cavity Retainer Plates. Plates in a mold which hold the cavities and forces. These plates are at the mold parting line and usually contain the guide pins and bushings. Also called *force retainer plates.*

Cavity Side. (British) The side of an injection mold which is adjacent to the nozzle.

Cavity Side Part. (U.S.A.) The stationary part of an injection mold.

CBA. In the cellular plastics industry, abbreviation for *chemical blowing agent.*

CDP. Abbreviation for CRESYL DIPHENYL PHOSPHATE, which see.

Cell. In cellular plastics terminology, the single void produced by a blowing agent, mechanically entrained gas or by the evaporation of a volatile constituent.

Cell Collapse. A defect in foam plastics characterized by slumping and cratered surfaces, and collapse of internal cells resembling a stack of leaflets when viewed in cross-section under a microscope. The condition is caused by excessively rapid permeation of the blowing gas through the cell walls, or by weakening of the cell walls by plasticization.

Celloidin. (celluidine, photoxylin) A form of cellulose nitrate made by precipitation from an ether-alcohol solution of collodion cotton. See CELLULOSE NITRATE.

Cellophane. Regenerated cellulose, chemically similar to rayon, made by mixing cellulose xanthate with a dilute sodium hydroxide solution to form a viscose, then extruding this viscose into an acid bath for regeneration. The term *rayon* is used when the material is in fibrous form. Cellophane is widely used for packaging, most often with coatings of other polymers to overcome its tendency to absorb moisture and to improve heat sealing characteristics.

Cellular Plastic. (expanded plastic, foamed plastic) A plastic with numerous cells disposed throughout its mass. The terms cellular-, expanded-, and foamed plastic are used synonymously. A cellular plastic may be produced by (1) incorporating a blowing agent which decomposes to liberate a gas; (2) mechanically stirring in a fluid or gas; (3) adding a water-soluble salt or a solvent-extractible agent to the mix prior to forming, then leaching out the agent after forming to leave voids; or (4) other techniques described under EPOXY FOAMS, PHENOLIC FOAMS, POLYSTYRENE FOAMS, SYNTACTIC FOAMS, URETHANE FOAMS. Cellular plastics range in densities from those nearly as great as the basic resin (60 or more pounds per cubic foot) to as low as 0.5 pounds per cubic foot. The cells may be open or closed, depending on the formulation and process. See also STRUCTURAL FOAM.

Cellular Striation. In cellular plastics terminology, a layer of cells differing in size or nature from the majority of cells in the same mass.

Celluloid. Cellulose nitrate compounded with a plasticizer, usually camphor. See CELLULOSE NITRATE.

Cellulose. $(C_6H_{10}O_5)_n$ A carbohydrate polymer of high molecular weight comprised of long chains of D-glucose units joined together by beta-1,4-glucosidic bonds. It is derived from stems of plants and trees, especially cotton, and is used in the production of CELLU-LOSIC PLASTICS, which see.

Cellulose Acetate (CA). An acetic acid ester of cellulose, forming a tough, transparent thermoplastic material when compounded with plasticizers. It is obtained by the action, under rigidly controlled conditions, of acetic acid and acetic anhydride on purified cellulose usually obtained from cotton linters. All three available hydroxyl groups in each glucose unit of the cellulose can be acetylated, but in the material normally used for plastics it is usual to acetylate fully, then to lower the acetyl value by partial hydrolysis, leaving 2.4 hydroxyl groups substituted per C_6 unit. Cellulose acetate compounds are used when toughness, permanence, flame resistance and transparency are required at moderate cost. However, they absorb moisture up to about 2.5%, making them unsuitable for long-term outdoor exposure.

Cellulose Acetate Butyrate (CAB). A mixed ester produced by treating fibrous cellulose with butyric acid, butyric anhydride, acetic acid and acetic anhydride in the presence of sulphuric acid. It is generally supplied in the form of pellets prepared by mixing the molten ester with a plasticizer. CAB is one of the strongest of the cellulosic plastics, and has good transparency, colorability, weatherability, electrical properties and chemical resistance. It can be processed by extrusion, injection molding, blow molding, rotational molding and thermoforming. Applications include pipe, tool handles, housings for instruments and lighting, packaging film and marine hardware.

Cellulose Acetate Propionate. (CAP, cellulose propionate) A thermoplastic formed by treating fibrous cellulose with propionic acid and acetic acid and anhydrides in the presence of sulfuric acid. CAP is easily extruded and injection molded, forming tough, flexible products with shock resistance close to that of ethyl cellulose.

Celluloseacetobutyrate. See CELLULOSE ACETATE BUTYRATE.

Cellulose Esters. Derivatives of cellulose in which the free hydroxyl groups attached to the cellulose chain have been replaced wholly or in part by acidic groups, e.g. nitrate, acetate, propionate, butyrate or stearate groups. Esterification is effected by the use of a mixture of an acid with its anhydride in the presence of a catalyst such as sulfuric acid. Mixed esters of cellulose, e.g. cellulose acetate butyrate, are prepared by the use of mixed acids and mixed anhydrides.

Cellulose Ethers. The cellulose derivates based on the etherification products of cellulose. These include ethyl cellulose, methyl cellulose and sodium carboxymethyl cellulose.

Cellulose Nitrate. (CN, nitrocellulose) Cellulose nitrate, dating back to the work of the French chemist Braconnet in 1833, is the oldest of the synthetic plastics. It is made by treating fibrous cellulosic materials with a mixture of nitric and sulfuric acids, and was first used as a solvent solution. In 1869, John Wesley Hyatt and his brother patented the use of cellulose nitrate as a solid mass, forming the basis of the plastics molding industry. Camphor was first used as a plasticizer for CN, and is still in use today although many camphor substitutes have been developed. Alcohol is normally used as a volatile solvent to assist in plasticization, after which it is removed. Products of CN are extremely tough, but flammable and subject to discoloration in sunlight.

Cellulose Propionate. See CELLULOSE ACETATE PROPIONATE.

Cellulose Triacetate. A member of the cellulosic plastics family made by reacting purified cellulose with acetic anhydride in the presence of a catalyst in such a manner that virtually all of the hydroxyl groups are substituted by acetyl groups. Due to its high softening point this material cannot be molded or extruded. Its major use is for casting films or spinning fibers from solutions, such as in a mixture of methylene chloride and methanol.

Cellulosic Plastics. A family of thermoplastics made by substituting various chemical groups for the hydroxy groups in the cellulose molecules of cotton linters or wood pulp. See the following: CELLULOSE ACETATE, CELLULOSE ACETATE BUTYRATE, CELLULOSE ACETATE PROPIONATE, CELLULOSE ESTERS, CELLULOSE NITRATE, CELLULOSE TRIACETATE, ETHYL CELLULOSE, HYDROXYETHYL CELLULOSE, REGENER-ATED CELLULOSE.

Cementing. The process of joining plastics to themselves or to dissimilar substances by means of solvents (see SOLVENT WELDING), dopes or chemical cements. *Dope adhesives* comprise a solvent solution of a plastic similar to the plastic to be joined. *Chemical cements,* the only type suitable for thermosetting plastics, are based on monomers or semi-polymers which polymerize in the joint to form a strong bond. See also ADHESIVES.

Center Gated Mold. In injection molding, a mold in which each cavity is fed through an orifice at the center of the cavity. This type of gating is employed for items such as cups and bowls.

Centi-. (c) The SI-approved prefix for a multiplication factor to replace 10^{-2} .

Centipoise. One hundredth of a Poise, the old cgs unit of viscosity. Water at $20°C$ has a viscosity of 1.002 centipoise. However, in the new SI system the standard unit of viscosity is the pascal, the conversion factor being one centipoise = 1.000 000*E-03 pascals.

Centistoke. One one-hundredth of a stoke. A stoke is equal to the viscosity in poises times the density of the fluid in grams per cc.

Centrifugal Casting. The process of forming pipes or other hollow cylindrical objects by introducing a measured amount of fluid resin or resin dispersion into a rotatable container or mold, rotating the mold about one axis at a speed high enough to force the fluid against all parts of the mold by centrifugal force, maintaining such rotation while solidifying the plastic by applicable means such as heating, then cooling if necessary and removing the formed part. The fluid resin may be a dispersion such as plastisol, or an A-Stage thermoset with or without reinforcing strands. Should not be confused with ROTATIONAL CAST-ING, which involves rotation at slow speeds about one or more axes and distribution under the force of gravity. See also CENTRIFUGAL MOLDING.

Centrifugal Impact Mixer. A device used for mixing free-flowing dry blends, comprising a conical hopper in which are rotated at high speeds a rotor disc and a peripheral impactor. The material is fed to the center of the rotor which throws it against the impactor blades, which in turn throw the material against fixed impactors at the extremities of the cone. From there, the material flows downward to a discharge orifice.

Centrifugal Molding. A process similar to CENTRIFUGAL CASTING, which see, except that the materials employed are dry, sinterable powders such as polyethylene. Such powders are fused by the application of heat to the rapidly rotating mold.

Ceramic Fibers. This term is being used in the plastics industry for reinforcing fibers made of refractory oxides such as Al_2O_3, BeO, MgO, MgO·Al_2O_3, ThO_2, and ZrO_2. Although glass is also a ceramic material, glass fibers are not generally included. Ceramic fibers are produced by chemical vapor deposition, melt drawing, spinning and extrusion. Their main advantage is high strength and modulus.

Ceraplasts. A coined term for reinforced thermoplastics, particularly polyethylenes, containing ceramic or mineral particles which have been dispersed in the polymer melt to their ultimate size and encapsulated in a band of resin in which there is a gradient in modulus between that of the filler and that of the polymer. Bonding of the encapsulating band to the filler and to the matrix polymer is accomplished by the addition of a small amount of reactive monomer or resin precursor. These composites possess mechanical characteristics better than those of compounds in which the same fillers have been incorporated in the conventional manner.

Cermets. Refractory compositions made by bonding grains of ceramics, metal carbides, nitrides, etc. with metal. Codeposition of cermets with nickel in the electroless nickel process provides excellent wear resistance, chemical resistance to molds, dies, extruder screws and other tooling components used in the plastics industry.

C-Glass. A type of GLASS FIBER REINFORCEMENT, which see, made and used specifically for high chemical resistance.

CGPM. Abbreviation for the French name Conférence Générale des Poids et Mesures, the International group that developed the system of weights and measures intended for worldwide use. The name International System of Units and the international abbreviation SI were adopted by the 11th CGPM in 1960.

Chain Length. The number of monomeric or structural units in a linear polymer. See also DEGREE OF POLYMERIZATION.

Chain Transfer Agent. An agent used in polymerization, which has the ability to stop the growth of a molecular chain by yielding an atom to the active radical at the end of the growing chain, but in turn being left as a radical which can initiate the growth of a new chain. Examples of chain transfer agents are chloroform and carbon tetrachloride. Such agents are useful for holding down molecular weights in polymerization reactions.

Chalk. A soft white mineral consisting essentially of CALCIUM CARBONATE, which see, occurring naturally as the remains of sea shells and minute marine organisms.

Chalking. A specific type of BLOOM, which see, characterized by a dry, chalk-like appearance of the surface of a plastic article.

Change Can Mixer. (Pony mixer) A Planetary type mixer comprising several paddle blades mounted on a vertical shaft rotating in one direction while the can or container rotates in the opposite direction. The paddle shaft usually is mounted on a hinged structure so that it can be swung out of the can, permitting the can to be removed and replaced easily. This type of mixer is employed for relatively small batches (3 to 125 gallons) of fluid dispersions and dry materials.

Channel Black. A type of CARBON BLACK, which see, made by impingement of a natural gas flame against a metal plate, from which the deposit is scraped at intervals.

Channel Depth Ratio. In an extruder screw, the ratio of the depth of the first flight at the hopper end to the depth of the last flight in the metering section. If the pitch of the screw is constant, the channel depth ratio is equal to the channel volume ratio.

Channel Volume Ratio. In an extruder screw, the ratio of the volume of the first flight at the hopper end to the volume of the last flight in the metering section. The term *compression ratio* is frequently used in industry in place of channel volume ratio.

Charge. (n) The amount of material used to load a mold at one time or during one cycle. The amount may be expressed in either weight or volumetric units.

Charles' Law (Gay-Lussac's Law). The volumes assumed by a given mass of a gas at different temperatures, the pressure remaining constant, are directly proportional to the corresponding absolute temperatures.

Charlton White. See LITHOPONE.

Charpy Impact Test. A destructive test of impact resistance, consisting of placing the specimen in a horizontal position between two supports, then applying blows of known and increasing magnitude until the specimen breaks. The result is expressed in terms of foot pounds or kg/cm of energy.

Chase. (shoe) An enclosure of any shape, used to (1) shrink-fit parts of a mold cavity in place, (2) prevent spreading or distortion in hobbing, or (3) enclose an assembly of two or more parts of a split cavity block.

Cheese. A supply of glass fiber wound into a cylindrical mass.

Chelate. A compound comprising metallic ions bound by a CHELATING AGENT, which see.

Chelating Agent. A term derived from the Greek word "chele," meaning claw. Thus, a chelating agent is a substance whose molecules are capable of seizing and holding metallic ions in a clawlike grip. The "claw" is usually a ring structure of nitrogen, oxygen or sulfur ("ligand atoms"), each of which donates two electrons to form a coordinate bond with the ion. See also SEQUESTERING AGENTS.

Chemically Foamed Plastic. A cellular plastic in which the cells are formed by a blowing agent or by the reaction of constituents. See also CELLULAR PLASTIC.

Chemical Resistance. (reagent resistance) The ability of a plastic to withstand exposure to acids, alkalis, solvents and other chemicals. ASTM tests for chemical resistance of plastics in Part 35 of the 1976 Annual Book of ASTM Standards are as follows: C 581-74, Chemical Resistance of Thermosetting Resins Used in Glass Fiber Reinforced Structures; D 543-67, Resistance of Plastics to Chemical Reagents; D 1239-55, Resistance of Plastic Films to Extraction by Chemicals; D 1712, Resistance of Plastics to Sulfide Staining; and D 2299-68, Determining Relative Stain Resistance of Plastics.

Chill Roll. A cored roll, usually temperature controlled with circulating water, which cools an extruded or cast film prior to winding. The surface of the roll may be polished or textured as desired to impart a finish to the film.

Chill Roll Extrusion. A term applied to film extrusion in which the extruded film is drawn over cooled rollers, imparting improved gloss.

China Clay. See KAOLIN.

Chinese White. See ZINC OXIDE.

Chirality. The property of non-identity of a molecule with its mirror image. A molecule in a given configuration or conformation is termed chiral when it is not identical with its mirror image. All asymmetric molecules are chiral; however not all chiral molecules are asymmetric since some molecules having axes of rotation are chiral. Chiral and prochiral atoms are sites or potential sites, respectively, of stereoisomerism. (IUPAC).

Chlorendic Acid. $C_9 H_4 Cl_6 O_4$ (hexachloroendomethylenetetrahydrophthalic acid). A white crystalline powder with uses in fire-retardant polyester resins, plasticizers, fungicides and insectides.

Chlorendic Anhydride. $C_9 H_2 Cl_6 O_3$ (hexachloroendomethylenetetrahydrophthalic anhydride) A di-functional acid anhydride used in the form of a white crystalline powder as a hardening agent and flame retardant in epoxy, alkyd and polyester resins.

Chlorinated Diphenyls. A series of plasticizers ranging from liquids to hard solids, used with polyvinylidene chloride and polystyrene. They are also used in conjunction with DOP as coplasticizers for PVC; and in conjunction with PVA, ethyl cellulose and other thermoplastics, as adhesives.

Chlorinated Hydrocarbons. The term covers a wide variety of liquids and solids resulting from the addition of chlorine to hydrocarbons such as methane, ethylene and benzene. They are employed as solvents, plasticizers and monomers for plastic manufacture.

Chlorinated Paraffins. (chlorocosanes) A family of yellow to light amber liquids produced by chlorinating a paraffin oil, with uses as a secondary plasticizer for vinyls, polystyrene, PMMA and coumarone-indene resins. Chlorinated paraffins also impart flame resistance to polyolefins, polystyrene, PVC, natural rubber, and unsaturated polyester resins.

Chlorinated Polyalkenamer. A specialty elastomer announced by Goodyear in 1976, described as a copolymer of a chlorinated cycloolefin and a nonchlorinated cycloolefin, produced by the "ring opening" polymerization technique. Previously, it is stated, ring opening polymerization has been confined to hydrocarbon materials.

Chlorinated Polyether. A corrosion-resistant thermoplastic obtained by polymerization of the monomer chlorinated oxetane, the oxetane being derived from pentaerythritol, to a high molecular weight (250,000-350,000). The polymer is linear, crystalline, and extremely resistant to degradation at processing temperatures. It may be injection molded, extruded, or applied by fluidized bed techniques. Due to its extremely good resistance to heat and chemicals, chlorinated polyether is used in valves, pumps, meters, etc. for chemical plants.

Chlorinated Polyethylenes. Polyethylenes modified by simple chemical substitution of chlorine on the linear backbone chain. They range from rubbery amorphous elastomers at 25 to 40% of chlorine by weight, to hard, semi-crystalline materials at 68 to 75% chlorine. They are sometimes grouped with chlorinated natural and butyl rubbers under the term *chlorinated rubbers.* Certain types of chlorinated polyethylenes are used as modifiers in PVC compounds to obtain lower brittle points, improved toughness and flexibility, greater latitude in compounding, and superior processing characteristics.

Chlorinated Polyvinyl Chloride. (CPVC) A PVC resin modified by after-chlorination. A series of such resins was introduced by the B.F. Goodrich Chemical Co. in the 1960's under the name "polyvinyl dichloride," abbreviated as PVDC or PVD. This name and its abbreviations were subsequently discontinued, and the series is now known as "Hi-Temp Geon." Compared to conventional rigid PVC, CPVC withstands service temperatures 40° to 60° higher, is stronger, and has better chemical resistance. CPVC is slightly hygroscopic, thus requiring pre-drying before processing. Otherwise, it can be processed by all methods used for rigid PVC with few modifications.

Chlorobenzene. C_6H_5Cl (Chlorobenzol, chlorbenzene, chlorbenzol, phenyl chloride) A solvent, and an intermediate in the production of phenol.

Chlorobutanol. $Cl_3CC(CH_3)_2OH$ (chlorbutanol, trichloro-tert-butyl alcohol, 1,1,1-trichloro-2-methyl-2-propanol, acetone chloroform) A plasticizer for cellulose esters and ethers.

Chlorodiphenyl Resins. Resins made from chlorinated diphenyl, rosin or rosin ester, and the higher fatty acids. They are used as plasticizing and modifying resins in plastics, and in lacquers and varnishes.

Chloroethane. See ETHYL CHLORIDE.

Chloroethene. (chloroethylene) See VINYL CHLORIDE.

Chlorofluorocarbon Resins. Resins made by the polymerization of monomers composed of chlorine, fluorine and carbon only. The principle member is POLYCHLOROTRIFLUOROETHYLENE (PCTFE), which see.

Chlorofluorohydrocarbon Resins. Resins made by the polymerization of monomers composed of chlorine, fluorine, hydrogen and carbon only.

Chlorohydrin. $CH_2OHCHOHCH_2Cl$ (alpha-chlorohydrin; 3-chloropropane-1,2-diol; glyceryl alpha-chlorohydrin) A solvent, especially for cellulosics.

Chlorohydrin Rubbers. See EPICHLOROHYDRIN RUBBERS.

Chloronaphthalene Oils. Nearly colorless oils derived by chlorinating naphthalene, used as plasticizers and flame retardants.

alpha-Chloro-meta-Nitroacetophenone. $NO_2 C_6 H_4 COOH_2 Cl$ A bacteriostat and fungistat for plastics.

Chloroprene Polymers. See NEOPRENE.

Chloropropylene Oxide. See EPICHLOROHYDRIN.

Chlorostyrenated Polyesters. A family of unsaturated polyester resins made by reacting the fluid polyester with monochlorostyrene in place of styrene. (See POLYESTERS, UNSATURATED). Monochlorostyrene is less volatile and more reactive than styrene, providing faster cure rates, increased flexural strength and modulus in glass fiber laminates.

Chlorothene. Dow Chemical Co.'s tradename for a stabilized form of 1,1,1-TRICHLORO-ETHANE, which see.

Chlorotrifluoroethylene. $C_2 ClF_3$ A colorless gas used as the monomer for POLYCHLORO-TRIFLUOROETHYLENE, which see. It is obtained by either dehalogenation or dehydrohalogenation of saturated chlorofluorocarbons or chlorohydrocarbons, e.g. by reacting 1,1,2-trichlorotrifluoroethane with zinc.

Choker Bar. A metal bar incorporated in an extruder die for controlling thickness and reducing stagnation of the melt.

Chopped Strand. A type of glass fiber reinforcement consisting of strands of individual glass fibers which have been chopped into short lengths, such as 1/8″, 1/4″, 3/8″, or 1/2″. The individual fibers are bonded together within the strands so that they remain in bundles after chopping. See also GLASS FIBER REINFORCEMENTS, ROVING.

Chromatography. The process of selective retardation of one or more components of a fluid solution as the fluid uniformly percolates through a column of finely divided substance (or through capillary passageways), the retardation resulting from the distribution of the components of the mixture between one or more thin phases and the bulk fluid, as this fluid moves countercurrent to the thin, stationary phases. The process is used for analysis and separation of mixtures of two or more substances, and for determining many characteristics in research work. The name "chromatography" is derived from the work of the Russian botanist N. Tswett, who first used the process to separate chloroplast pigments. Among the many variations of the process are *gas chromatography* (the specimen in gaseous form is passed through a porous bed, or through a capillary tube line with a liquid or solid phase; *paper chromatography* (a drop of the specimen is placed near one end of a porous paper); *ion exchange chromatography; thin layer chromatography* (the sample is placed on an absorbent cake spread on a smooth glass plate); and GEL PERMEATION CHROMATOGRAPHY, which see.

Chrome Greens. (Brunswick greens) A family of pigments ranging from light yellow green through dark green, based on physical mixtures of chrome yellows and iron blue (a complex ammonium iron hexacyanoferrate). The amount of iron blue determines the shade, about 2% being used for the light yellow green, and up to 64% being used for the dark green.

Chrome Orange Pigments. Pigments based on basic lead chromate, $PbO \cdot PbCrO_4$, which are of deep orange color.

Chrome Oxide Green. A stable pigment based on anhydrous chromium oxide, Cr_2O_3. The form based on hydrated chromium oxide is called *Guignet's green*.

Chrome Yellow Pigments. (primrose chrome, permanent yellow) A family of pigments based on normal lead chromate, $PbCrO_4$, which is characterized by a medium yellow color. Other shades ranging from light greenish-yellow to reddish medium yellow are made by coprecipitating lead chromate with insoluble compounds such as lead sulfate or lead phosphate.

Chromic Chloride. $CrCl_3$ or $CrCl_3 \cdot_6 H_2O$ (Chromium chloride, chromium sesquichloride) A catalyst for polymerizing olefins.

Chromogen. A molecule containing a chromophore.

Chromophore. A group such as $-NO$, $-NO_2$ or $-N=N-$ which when present in a molecule enables the molecule to be transformed into a dye upon the introduction of an acid group.

Chrysotile. $3MgO \cdot 2SiO_2 \cdot 2H_2O$ A hydrated magnesium ortho-silicate, the chief constituent of the serpentine type of asbestos. Chrysotile-bearing asbestos is the most widely used type, accounting for over 90% of the world production. Its fine and silky fibers, and mats and felts made therefrom, are used as fillers and reinforcements for plastics.

Chunk. A term sometimes used for an open face mold.

CIL Flow Test. A test developed by Canadian Industries Limited for measuring rheological properties of thermoplastics. The unit of measurement is the amount of molten resin which is forced through a specified size orifice per unit time when a specified, variable force is applied.

Cinnamene. (cinnamol) See STYRENE.

Circuit. In filament winding, one complete traverse of the fiber feed mechanism of the winding machine; or one complete traverse of a winding band from one arbitrary point along the winding path to another point on a plane through the starting point and perpendicular to the axis.

Cis-. A chemical prefix (Latin: "on this side") denoting an isomer in which certain atoms or groups are on the same side of a plane. See also TRANS-.

Citrate Plasticizers. A family of plasticizers derived from citric acid, noted for their low order of toxicity. Included are triethyl citrate, tri(2-ethylhexyl) citrate, tricyclohexyl citrate, tri-n-butyl citrate, acetyl triethyl citrate, acetyl tri-n-butyl citrate, acetyl tri-n-octyl n-decyl citrate, and acetyl tri(2-ethylhexyl) citrate.

Clamping Area. The largest rated molding area an injection or transfer press can hold closed under full molding pressure.

Clamping Force. In injection molding and transfer molding, the pressure which is applied to the mold to keep it closed, in opposition to the fluid pressure of the compressed molding material, within the mold cavity and the runner system.

Clamping Plate. A plate fitted to a mold and used to fasten the mold to a molding machine.

Clamping Pressure. In injection and transfer molding, the pressure applied to the mold to keep it closed during the molding cycle.

Clamshell Molding. A term applied to a modernized version of the most ancient form of blow molding — preheating two sheets of plastic, placing them between the halves of a split mold, closing the mold, drawing the sheets against their respective mold surfaces by means of vacuum, then completing the forming by means of internal air pressure. The modern process, mechanized and conveyorized, is superior to conventional blow molding in the case of very large parts and those in which uniformity of wall thickness is important.

Clarifier. An additive that increases the transparency of a plastic material.

Clash-Berg Point. The temperature at which the apparent modulus of elasticity of a specimen is 135,000 psi, the end point of flexibility as defined by Clash and Berg in their studies of low temperature flexibility. In a similar test described in ASTM D 1043, the modulus of 60,000 psi is used.

Clays. Naturally occurring sediments rich in hydrated silicates of aluminum, predominating in particles of colloidal or near-colloidal size. There are many types of clays and clay-like minerals. Those of particular interest to the plastics industry are varieties refined by nature and man to a state of good color and particle size distribution, such as kaolin or china clay. They are used as fillers in epoxy and polyester resins, PVC compounds, and urethane foams. *Calcined clays* are those that have been heated to a high temperature to remove the chemically-bound water, sometimes also surface treated to improve their chemical inertness and moisture resistance. They are used primarily in vinyl insulation.

Clear Point. With regard to vinyl plastisols, clear point is the temperature at which an unpigmented plastisol suddenly becomes transparent as it is being heated, denoting that the resin particles have completely dissolved in the hot plasticizer. This test is useful for determining relative fusion temperatures of plastisol compounds.

Clicker Die. A cutting die for stamping out blanks from plastic sheeting.

Clicker Press. A stamping press, utilizing clicker dies for cutting out blanks from plastic sheeting.

Clicking. See DIE CUTTING.

Closed Cell Foamed Plastic. A CELLULAR PLASTIC, which see, in which non-interconnecting cells are present to an extent which renders the mass relatively impervious to air and fluids.

Cloud Point. In condensation polymerization, the temperature at which the first turbidity appears, caused by water separation when a reaction mixture is cooled.

CN. Abbreviation for CELLULOSE NITRATE, which see.

Coacervation. The separation of a polymer solution into two or more liquid phases, one of which is a polymer-rich liquid. The term was introduced to distinguish this phenomenon from the precipitation of a polymer solute in solid form. The process is used in MICRO-ENCAPSULATION, which see, by emulsifying or dispersing the material to be encapsulated with a solution of the polymer. By changing the temperature or concentration of the mixture, or by adding another polymer or solvent, a phase separation is induced and the polymeric portion forms a thin coating on the external surfaces of the particles. After further treatment to solidify the polymeric wall, the solidified capsules can be isolated in powder form by filtration.

Coagulant. A substance which (1) initiates the formation of relatively large particles in a finely-divided suspension, or (2) assists in the formation of a gel; thus accelerating settling of the particles or their deposition on a substrate.

Coagulation. A physical or chemical change inducing transition from a fluid to a semi-solid or gel-like state.

Coal-Tar Resins. See COUMARONE-INDENE RESINS.

Coated Fabrics. Fabrics which have been impregnated and/or coated with a plastic material in the form of a solution, dispersion, hot-melt or powder. The term is sometimes used when a preformed film is applied to the fabric by means of calendering, although such products are more properly termed laminates.

Coathanger Die. A sheet or film extrusion die shaped internally in the form of a coathanger. This type of die is said to yield better distribution of material across the full width of the extruded web, and thus produce sheet without weld lines. Side-fed dies for blow molding and spiral-type blown film dies are also considered to be coathanger dies.

Coating Methods. See:

AIR KNIFE COATING	FLOW COATING
CALENDER COATING	FLUIDIZED BED COATING
CASCADE COATING	FRICTION CALENDERING
CURTAIN COATING	GRAVURE COATING
DECORATING	INTUMESCENT COATING
DIP COATING	KISS-ROLL COATING
ELECTROPHORETIC DEPOSITION	PAINTING ON PLASTICS
ELECTROPLATING	PLASMA SPRAY COATING
ELECTROSTATIC FLUIDIZED BED COATING	PRINTING ON PLASTICS
ELECTROSTATIC SPRAY COATING	REVERSE-ROLL COATING
EXTRUSION COATING	ROLLER COATING
FLAME SPRAY COATING	SILVER SPRAY PROCESS
FLOCKING	SINTER COATING
	SOLUTION COATING

SPRAY COATING   TRANSFER COATING
SPREAD COATING   URETHANE COATINGS
STRIPPABLE COATINGS   VACUUM METALLIZING

Cocatalysts. (promoters) Chemicals which themselves are feeble catalysts, but which greatly increase the activity of a given catalyst.

Coefficient Of Elasticity. See MODULUS OF ELASTICITY.

Coefficient Of Thermal Conductivity. See THERMAL CONDUCTIVITY.

Coefficient Of Thermal Expansion. The fractional change in length (or sometimes in volume, when specified) of a material for a unit change in temperature. Values for plastics range from 0.01 to 0.2 mils per inch per degree C.

Coextrusion. The process of extruding two or more materials through a single die with two or more orifices arranged so that the extrudates merge and weld together into a laminar structure before chilling. Each material is fed to the die from a separate extruder, but the orifices may be arranged so that each extruder supplies two or more plies of the same material. Coextrusion can be employed in film blowing, free film extrusion, and extrusion coating processes. The advantage of coextrusion is that each ply of the laminate imparts a desired characteristic property, such as stiffness, heat-sealability, impermeability or resistance to some environment, all of which properties would be impossible to attain with any single material.

Co-Extrusion Blow Molding. A process using two extruders and a co-extrusion head designed to combine the flow from the two extruders to form a dual-wall parison. The advantages of the process are similar to those of co-extrusion of film, each ply of the composite designed for a property needed for a particular function at minimum cost.

Cogswell Rheometer. See EXTENSIOMETER.

Cohesion. The state in which the particles of a single substance are held together by primary or secondary valence forces. (ISO)

Coining. A term borrowed from the metal stamping industry for a process of forming integral hinges. In the case of a polypropylene article, the integral hinge is produced by molding a thin section between the two parts of the article to be hinged. Such a thin section cannot be molded easily in articles of nylon or acetal resins, because of the difficulty of filling the half of the mold cavity opposite the gated half through the thin section. In the coining process, the area to be formed into a hinge is molded in a thickness suitable for the molding process. Subsequently, the article is placed in a press between bars which compress the plastic to the desired thickness. The material must be deformed beyond the tensile yield point but short of the ultimate tensile strength, so that the material remains essentially stable with little recovery from the deformation. This cold pressing produces a high degree of molecular orientation which imparts high strength and flexibility to the integral hinge area.

Coinjection. A process similar in results to the coextrusion process, but performed by modifications of the injection molding process. By means of various nozzle and valving arrangements, two or three materials can be injected either simultaneously or sequentially to form an article with an outer core of one material with certain desired properties filled with another material to attain other desired properties or reduce costs.

Cold Bend Test. A test for measuring the flexibility of a plastic material at low temperatures. A specimen is bent to a predetermined radius while maintained at the stipulated temperature.

Cold Drawing. (cold stretching) A stretching process employed to improve the tensile properties of thermoplastic filaments and films.

Cold Flow. See CREEP.

Cold Forming. A group of processes by which sheets or billets of thermoplastic materials are formed into three dimensional shapes at room temperature by processes used in the metal working industry such as forging, brake press bending, deep drawing, stamping, heading and coining. The materials used, generally in relatively thick sections, include ABS, polycarbonates, polyolefins, and rigid PVC. When either the material or the forming dies are preheated, the preferred term is SOLID PHASE FORMING, which see.

Cold Heading. A process for forming plastic rods into rivets by uniformly loading the shaft end or projection in compression while holding and containing the shaft trunk. All thermoplastics can be cold headed, but acetal and nylon are particularly suitable.

Cold Molding. A process similar to compression molding, except that no heat is applied during the molding cycle. The formed part is subsequently hardened by curing or heating and cooling. A-stage phenolic resins and bituminous plastics are sometimes molded by this process.

Cold Parison Blow Molding. See BLOW MOLDING.

Cold Pressing. A bonding operation in which an assembly is subjected to pressure without the application of heat.

Cold Runner Injection Molding. (runnerless injection molding) Whereas in the injection molding of thermoplastics the runners are sometimes kept hot to reduce scrap (see HOT RUNNER MOLD), it is advantageous in the case of thermosetting materials to keep the runners cooler than the cavities to prevent the material from curing within the runner system. In the cold runner injection molding process for thermosets the mold is divided into two sections: a heated cavity section, and an insulated manifold section containing the injection nozzle and runners. Material is fed from runners to cavities through very short gates or sub-sprues. The insulated manifold section is maintained at a temperature high enough to soften the uncured material, generally in the area of 200°F, but lower than the curing temperature maintained in the cavity section.

Cold Slug. The first material to enter an injection mold, so called because in passing through the sprue orifice it is cooled below the effective molding temperature.

Cold Slug Well. The space provided directly opposite the sprue opening in an injection mold to trap the cold slug.

Collapse. (1) Inadvertent densification of cellular material during manufacture resulting from breakdown of cell structure. (ASTM D 883-75a). (2) Contraction of the walls of a container, e.g. upon cooling, leading to a permanent indentation.

Colligative Property. A property which is common to all members of a group of facts or substances which vary from each other in some other respects. For example, a property numerically the same for a group of substances independent of their chemical natures.

Collimated Roving. Roving with strands that are more parallel than those in standard roving, usually made by parallel winding.

Collodion. A solution of cellulose nitrate in alcohol and ether.

Collodion Cotton. See CELLULOSE NITRATE.

Colloid. A substance capable of forming a suspension or emulsion with a liquid which will not settle out to a noticeable degree, and will not diffuse readily through vegetable or animal membranes. Colloidal particles are usually of high molecular weight, ranging in diameter from about 10^{-7} to 5×10^{-5} cm.

Colloidal Clay. See BENTONITE.

Colloid Mill. A device for preparing emulsions and reducing particle size, consisting of a high speed rotor and a fixed or counter-rotating element in close proximity to the rotor. The fluid is conveyed continuously from a hopper to the space between the shearing elements, then discharged into a receiver. See also HOMOGENIZER.

Colloxylin. See CELLULOSE NITRATE.

Colorants. Dyes or pigments which impart color to plastics. The dyes are synthetic or natural compounds of submicroscopic or molecular size, soluble in most common solvents, yielding perfectly transparent colors. Their poor heat resistance and tendency to migrate limit their use as additives to a few families that are superior in heat resistance. However, dyes are sometimes used to post-color finished parts such as buttons and fibers. The pigments are organic and inorganic substances with larger particle sizes, rarely less than 1 micron, and usually insoluble in the common solvents. Organic pigments produce translucent and nearly transparent colors, resist migration better than the dyes, and are somewhat more heat resistant. Inorganic pigments are, with few exceptions, opaque and superior in light-fastness, heat resistance and resistance to migration. Colorants are added to plastics by dry coloring (simply tumbling the colorant with the base or compounded resin); by extrusion coloring (extruding a dry colored mixture and chopping it into pellets to be reprocessed); by masterbatching (see COLOR CONCENTRATE); or by stirring colorants or dispersions thereof into liquid plastisols or resin systems. See also: BON PIGMENTS, FLUOROSCENT PIGMENTS, FLUSHED PIGMENTS, GLITTER, INORGANIC PIGMENTS, LIQUID COLORANTS, LUMINESCENT PIGMENTS, METALLIC FLAKE PIGMENTS, ORGANIC PIGMENTS, PEARLESCENT PIGMENTS, PERYLENE PIGMENTS, PHOSPHORESCENT PIGMENTS, PHTHALOCYANINE PIGMENTS, QUINACRIDONE PIGMENTS, RHODAMINES, ULTRAMARINE BLUE PIGMENTS.

Color Concentrate. A plastic compound which contains a high percentage of pigment, to be blended in appropriate amounts with the basic resin or compound so that the correct end concentration is achieved. The concentrate provides a clean and convenient method of obtaining accurate color shades in extrusion compounds, etc. The term *masterbatch* is sometimes used for color concentrate, as well as for concentrates of other additives. More recently, a family of "Multi-functional Concentrates" (MFC) has been developed. Along with, or sometimes without colorants, the MFC pellets may contain UV stabilizers, flame retardants, lubricants, anti-static agents, anti-blocking agents, blowing agents, fillers and other such additives, all in one tailor-made pelletized concentrate.

Colorfastness. See LIGHT RESISTANCE.

Colorimeter. An instrument for matching colors with results approximately the same as those of visual inspection, but more consistently. The sample is illuminated by light from three primary color filters, and scanned by an electronic detecting system. A colorimeter is sometimes used in conjunction with a SPECTROPHOTOMETER, which see, for close control of color in production operation.

Colorimetry. (color identification testing) A method of analysis based on the fact that certain plastics undergo characteristic color changes when exposed to certain chemicals.

Color Migration. The movement of dyes or pigments through or out of a material.

Color Stability. See LIGHT RESISTANCE.

Combination Mold. See FAMILY MOLD.

Combining Weight. See EQUIVALENT WEIGHT.

Combustible Liquid. A liquid that gives off flammable vapors at or below $150°$F and above $80°$F. Most solvents used in plastics processing are combustible according to this definition.

Comminute. (v) To pulverize, or to reduce particles to small sizes as by grinding.

Comonomer. A monomer which is mixed with a different monomer for a polymerization reaction, the result of which is a COPOLYMER, which see.

Compacts. See POWDER COMPACTS.

Compatibility. The ability of two or more substances to mix together without objectionable separation. In plastics technology the term is most often used in connection with plasticizers, but it may apply also to resins and any component of a compound. See also LOOP TEST.

Complexing Agents. See CHELATING AGENTS, SEQUESTERING AGENTS.

Compliance. In tensile testing, the reciprocal of Young's Modulus.

Composite. (n) An article or substance containing or made up of two or more different substances. In the plastics industry the term applies broadly to structures of reinforcing elements (dispersed phase) incorporated in compatible resinous binders (continuous phase). Such composites are subdivided into classes on the basis of the reinforcing constituents: LAMINATES, which see; particulate (the dispersed phase consists of small particles); Fibrous (the dispersed phase consists of fibers); Flake (flat flakes forming the dispersed phase); and Skeletal (composed of a continuous skeletal matrix filled by a resin).

Composite Laminates. A term sometimes applied to a laminated plastic bonded to a non-plastic material such as copper, vulcanized fiber, rubber, asbestos, lead, aluminum and the like. An example of a composite laminate is the copper-clad laminated plastic used for printed circuits.

Composite Mold. A mold in which several different shapes are produced in one cycle.

Composite Molding. A term that has been applied to the process of molding two or more materials in the same cavity in the same shot, by a combination of transfer and compression molding. For example, in the production of a ring gear a loose nylon-filled material is loaded into an open mold around the tooth area, the mold is closed, then nylon molding powder is plasticated and injected by transfer molding means.

Compound. (n) A mixture of resin and the ingredients necessary to modify the resin to a form suitable for processing into finished articles.

Compounding. The step of mixing basic resins with additives such as plasticizers, stabilizers, fillers and pigments in a form suitable for processing into finished articles. In some areas of the industry the term includes fusion of the polymer, for example in the production of molding powders by extrusion and pelletizing. In the plastisol industry, the compounding step ends with the preparation of the dispersion.

Compreg. (compregnated wood) A contraction of "compressed impregnated wood," usually referring to an assembly of veneer layers impregnated with a liquid resin and bonded under high pressure.

Compregnate. (v) To impregnate and simultaneously or subsequently compress, as in the production of compregs.

Compressed-Air Ejection. The removal of a molding from its mold by means of a jet of compressed air.

Compressibility. The change in volume per unit volume produced by a change in pressure. The reciprocal of bulk modulus.

Compression Mold. A mold used in the process of COMPRESSION MOLDING, which see.

Compression Molding. A method of molding in which the molding material, generally pre-heated, is placed in an open heated mold cavity, the mold is closed with a top force or plug member, pressure is applied to force the material into contact with all mold areas, and heat and pressure are maintained until the molding material has cured. The process most often employs thermosetting resins in a partially cured stage, either in the form of granules or putty-like masses, or sometimes in the form of shaped masses ("preforms") roughly conforming to the shape of the mold. Compression molding is also used for some thermoplastic products, most notably phonograph records.

Compression Molding Pressure. The unit pressure applied to the molding material in a mold. The area is calculated from the projected area taken at right angles to the direction of applied force and includes all areas under pressure during the complete closing of the mold. The unit pressure is calculated by dividing the total force applied by this projected area and is expressed in pounds per square inch. Note: The land area in a flash or semipositive mold shall be included in the projected area when material flows over this section and is under pressure when the mold is completely closed. (ASTM D 883-65T).

Compression Ratio. In an extruder screw, the ratio of the volume of material held in the first flight at the hopper end to the volume held in the last flight in the metering section. This ratio is an indication of the compaction performed on the material, and of the amount of work done on the material by the screw. Sometimes called *channel volume ratio* or, in the case of a screw of constant pitch, *channel depth ratio.*

Compression Set. A permanent deformation resulting from compressive stress.

Compression Zone. The portion of an extruder barrel in which melting is completed.

Compressive Modulus. The ratio of compressive stress to compressive strain below the proportional limit. Theoretically equal to Young's Modulus determined from tensile tests.

Compressive Strength. The maximum load sustained by a test specimen in a compressive test divided by the original area of the specimen.

Compressive Stress. The compressive load per unit area of original cross section carried by the specimen during a compression teat.

Concavity Factor. The entire stress-strain curve of rubber and elastomers which have no elastic limit is concave toward the stress axis or away from the strain axis. The concavity factor is the ratio between the energy of the extension curve to that of the straight line curve to the same point.

Condensation. (1) The process of reducing a gas or vapor to a liquid or solid form. (2) A chemical reaction in which two or more molecules combine with the separation or water or some other simple substance. If a polymer is formed the process is called *polycondensation.*

Condensation Agent. A chemical compound which acts as a catalyst and also furnishes a complement of material necessary for a polycondensation reaction to proceed.

Condensation Polymerization. A polymerization process in which water or some other simple substance separates from two or more of the polymer molecules upon their combination. Examples of resins made by this process (condensation resins) are alkyds, phenol-aldehydes and urea-formaldehydes, polyesters, polyamides, polyacetals, and polyphenylene oxides.

Conditioning. Subjecting a material to standard environmental and/or stress history prior to testing. Typical conditions are 40% relative humidity at a temperature of 77°F.

Conductance. The electrical term for the reciprocal of resistance, measured by the ratio of the current flowing through a conductor to the difference of potential between its ends. The old unit of conductance was mho, the reciprocal of the value of a conductor's resistance in ohms. The new SI term is the siemens, expressed as amperes divided by volts. Tests for measuring the D.C. resistance or conductance of insulating materials are given in ASTM D 257-75a.

Conductimetric Analysis. A method of analysis based on the electrical conductivity of a solution of the specimen.

Conductivity, Electrical. The reciprocal of volume resistivity; the conductance of a unit cube of a material.

Configuration. See CONFORMATION.

Configurational Base Unit. A constitutional repeating unit whose configuration is defined at least at one site of stereoisomerism in the main chain of a polymer molecule. (IUPAC) Note: In a regular polymer, a configurational base unit corresponds to the constitutional repeating unit. In the regular polymer polypropylene, the constitutional repeating unit is $-CH(CH_3)CH_2-$ and the configurational base units are

$$
\begin{array}{ccc}
\text{H} & & \text{CH}_3 \\
| & & | \\
-\text{C}-\text{CH}_2- & \text{and} & -\text{C}-\text{CH}_2- \\
| & & | \\
\text{CH}_3 & & \text{H}
\end{array}
$$

These two configurational base units are enantiomeric to each other (mirror images).

Configurational Repeating Unit. The smallest set of one, two or more successive configurational base units that prescribes configurational repetition at one or more sites of stereoisomerism in the main chain of a polymer molecule. (IUPAC).

Configurational Unit. A constitutional unit having one or more sites of defined stereoisomerism. (IUPAC).

Conformal Coatings. See ENCAPSULATION.

Conformation. In polymer terminology, the overall spatial arrangement of the atoms and groups in a polymer molecule, or the general shape of a molecule.

Conical Dry Blender. A device consisting of two hollow cones joined at their bases by a short cylindrical section, mounted on a horizontal shaft passing through the sides of the cylindrical section. Material is charged and discharged at openings in the apexes of the cones. Mixing is accomplished by cascading, rolling and tumbling actions as the cones rotate.

Conjugated. (adj) In chemistry, referring to the regular alternation of single and double bonds between atoms of a molecule. For example, in the following diagram of the benzene molecule each single bond represents a pair of electrons; each double bond, two pairs.

Conjugated Double Bonds. A chemical term denoting double bonds separated from each other by a single bond. An example is 1,3-butadiene, $CH_2=CH-CH=CH_2$.

Consistency. The resistance of a material to flow or permanent deformation when shearing stresses are applied to it; the term is generally used with materials whose deformations are not proportional to applied stress. Note: Viscosity is generally considered to be similar internal friction that results in flow in proportion to the stress applied. See VISCOSITY and VISCOSITY COEFFICIENT.

Consistometer. An instrument for measuring the consistency of semi-fluid substances.

Constantan. A copper-nickel alloy, wires of which are used in conjunction with wires of a different metal, e.g. iron, in thermocouples for measuring temperatures in extruders, injection molding machines, etc.

Constitutional Repeating Unit. The smallest constitutional unit whose repetition describes a regular polymer. (IUPAC).

Constitutional Unit. A species of atom or group of atoms present in a chain of a polymer or oligomer molecule. (IUPAC).

Contact Adhesive. A liquid adhesive which dries to a film that is tack-free to other materials but not to itself. A typical contact adhesive is a neoprene elastomer mixed with either an organic solvent vehicle or an aqueous dispersion medium. The adhesive is applied to both surfaces to be joined and dried at least partially. When pressed together at light to moderate pressure a bond of high initial strength results. Some definitions of contact adhesive stipulate that the surfaces to be joined shall be no further apart than about 0.1 mm for satisfactory bonding.

Contact Laminating. See CONTACT PRESSURE MOLDING.

Contact Pressure Molding. (Contact Molding) This term encompasses processes for forming shapes of reinforced plastics in which little or no pressure is applied during the forming and curing steps. It is usually employed in connection with the processes of *spray-up* and *hand lay-up molding* when such processes do not include the application of pressure during curing.

Contact Pressure Resins. (Contact resins, impression resins, low-pressure resins) Liquid resins which thicken or resinify on heating and, when used for bonding laminates, require little or no pressure. Typical components are an unsaturated monomer such as an allyl ester, or a mixture of styrene or other vinyl monomer with an unsaturated polyester or alkyd.

Contact Resins. See CONTACT PRESSURE RESINS.

Continuous Phase. In a SUSPENSION, which see, the continuous phase refers to the liquid medium in which the solid particles are dispersed. The solid particles are called the *disperse phase*.

Continuous Polymerization. A type of polymerization in which the monomer is continuously fed to a reactor and the polymer is continuously removed.

Continuous Roving. See ROVING.

Contraction Allowance. See SHRINKAGE ALLOWANCE.

Convergent Die. A die for producing hollow articles in which the internal channels of the die leading to the die orifice are converging.

Conversion. A term used primarily in the packaging industry. "Converters" buy plastic film or sheeting in the form of roll stock, and convert it to useful forms by slitting, die cutting, heat sealing into bags, etc., for resale to packaging firms.

Cooling Channels. Passageways provided in molds or platens for circulating water or other cooling media, in order to control the surface temperature of the cavities. Operating cycles in injection molding and blow molding are made shorter by means of such cooling.

Cooling Fixture. A block of wood or metal shaped to hold a part after its removal from a mold, in a manner which avoids distortion of the part while it is cooling. Also called *shrink fixture*.

Coordination Catalysts. Catalysts comprising a mixture of (a) an organometallic compound, e.g. triethylaluminum and (b) a transition-metal compound, e.g. titanium tetrachloride. They are often called Ziegler or Ziegler-Natta catalysts, and are used for the polymerization of olefins and dienes.

Coordination Compound. (Werner complex) A complex compound whose molecular structure contains a central atom bonded to other atoms by coordinate covalent bonds based on a shared pair of electrons, both of which are of a single atom or ion. Such compounds have roles in polymerization catalysts. A CHELATE (which see) is a special type of coordination compound.

Copolycondensation. The copolymerization of two or more monomers by the condensation polymerization process.

Copolymer. This term usually, but not always, denotes a polymer of two chemically distinct monomers. It is sometimes used for terpolymers, quadripolymers, etc. containing more than two monomeric units. Three common types of copolymers are BLOCK COPOLYMERS, GRAFT POLYMER and RANDOM COPOLYMERS, which see. Note: A new IUPAC definition for a polymer derived from two species of monomer is *bipolymer*.

Copolymerization. The simultaneous polymerization of two or more monomers.

Copper-Clad Laminates. Laminated plastics surfaced with copper foil, used for preparing printed circuits.

Core. (1) The central member of a laminate to which the faces of the sandwich are attached. (2) A channel in a mold for circulation of heat-transfer media. (3) Part of a complex mold that forms undercut parts usually withdrawn to one side before the main sections of the mold are opened. Also called *core pin.* (4) The central conductor in coaxial cables.

Core and Separator. The center section of an extrusion die.

Core Pin. A pin inserted in a mold to produce a hole, which may be plain or threaded.

Core Pin Plate. The plate which supports core pins in a mold..

Core-Retainer Plate. See CHASE.

Corfam®. Du Pont's tradename for a tough, leatherlike, nonwoven sheet of urethane polymer fibers deposited in a random manner and held together with a binder of polyester resin to obtain uniform strength in all directions. Each cubic inch contains millions of tiny pores that permit the material to "breathe." If desired, these pores may be saturated with elastomers, resins and waxes. The material was developed as a leather substitute in shoes, and later was put to use in industry for packings and seals in chemical pumps. However, DuPont announced early in 1971 that production of Corfam was being discontinued due to lack of sales volume, attributed to high costs as compared to those of other man-made breathable materials. See also POROMERIC.

Cork. The outer bark of *Quercus Suber,* a species of oak growing in Mediterranean countries. Cork is used as a filler in thermoplastic and thermosetting compounds for special applications such as flooring, ablative plastics, insulating compositions, and shoe inner soles.

Corona Discharge. The flow of electrical energy from a conductor to the surrounding air or gas. The phenomenon occurs when the voltage is high enough (e.g. 5000 or more volts) to cause partial ionization of the surrounding gas. The discharge is characterized by a pale violet glow, a hissing noise, and the odor of ozone formed when the surrounding gas contains oxygen. Corona discharge occurs in high-voltage cables, thus making ozone resistance an important factor in compounding plastics for insulation of electrical wires and cables. See also CORONA DISCHARGE TREATMENT.

Corona Discharge Treatment. A method of rendering inert plastics such as polyolefins more receptive to inks, adhesives or decorative coatings by subjecting their surfaces to a corona discharge. A typical method of treating films is to pass the film over a grounded metal cylinder above which is located a sharp-edged high-voltage electrode spaced so as to leave a small air gap between the film and the electrode. The corona discharge oxidizes the film by means of the formation of polar groups on reactive sites, making the surface receptive to coatings. See also FLAME TREATING.

Corona Resistance. In an insulated electrical conductor, the resistance of the insulation to breakdown caused by ionized air in voids existing in the insulation induced by the current in the conductor.

Corrosion Resistance. A broad term applying to the ability of plastics to resist many environments. See ACID RESISTANCE, ALKALI RESISTANCE, ARTIFICIAL WEATHERING, CHEMICAL RESISTANCE, DETERIORATION, PERMANENCE, SOLVENT RESISTANCE, STAIN RESISTANCE, SULFIDE STAINING, LIGHT RESISTANCE, VOLATILE LOSS, WEATHERING.

Cotton Linters. See LINTERS.

Coulomb. (1) The quantity of electricity which must pass through a circuit to deposit 0.0011180 grams of silver from a solution of silver nitrate. It is equal to one Ampere per second. (2) The quantity of electricity of the positive plate of a condenser of one-farad capacity when the electromotive force is one volt. Note: The coulomb (C) has been accepted in the new SI system of units as the international unit of the quantity of electricity or electric charge, to be computed as amperes per second. (A·s).

Coumarin. $C_9H_6O_2$ (cumarin) The sweet-smelling constituent of white clover, also produced synthetically. It is sometimes copolymerized with styrene to increase the heat distortion temperature.

Coumarone. C_6H_4OCHCH (cumarone, benzofuran) A colorless liquid derived from the coal-tar naphtha fraction boiling between 150 and 200°C, used for producing COUMARONE-INDENE RESINS, which see.

Coumarone-Indene Resins. Thermoplastic resins produced by polymerizing a coal tar naphtha containing coumarone and indene. The naphtha is first washed with sulfuric acid to remove some impurities, then is polymerized in the presence of sulfuric acid or stannic chloride as a catalyst. Remaining impurities determine the quality of the resins, which range from clear, viscous liquids to dark, brittle solids. Coumarone-indene resins have no commercial applications when used alone. They are used primarily as processing aids, extenders and plasticizers with other resins and rubbers.

Coupling Agent. A chemical substance capable of reacting with both the reinforcement and the resin matrix of a composite material to form or promote a stronger bond at the interface. The agent may be applied to the reinforcement, added to the resin, or both. See also SILANE COUPLING AGENTS, TITANATE COUPLERS, ADHESION PROMOTER.

CP. (1) In plastics, abbreviation for CELLULOSE PROPIONATE. See CELLULOSE ACETATE PROPIONATE. (2) In chemistry, abbreviation for *chemically pure.*

CPE. Abbreviation for CHLORINATED POLYETHYLENE, which see.

Cps. Abbreviation for CENTIPOISE, which see.

CPVC. (1) Abbreviation for CHLORINATED POLYVINYL CHLORIDE, which see. (2) In the paint industry, abbreviation for *critical pigment volume concentration,* a source of confusion often encountered in the paint and color literature.

CR. Abbreviation for chloroprene rubber. (British Standards Institution). See NEOPRENE.

Crater. A small, shallow surface imperfection (ASTM D 883-75a).

Crazing. An undesirable defect in plastics articles characterized by distinct surface cracks or minute frost-like internal cracks, resulting from stresses within the article which exceed the tensile strength of the plastic. Such stresses may result from molding shrinkages, or machining, flexing, impact shocks, temperature changes, or the action of solvents. See also STRESS-CRACK.

Creel. The spool and its supporting structure on which continuous strands or rovings of reinforcing material are wound for use in the filament winding process.

Creep. (n) Due to its viscoelastic nature, a plastic subjected to a load for a period of time tends to deform more than it would from the same load released immediately after application, and the degree of this deformation is dependent on the load duration. Creep is the permanent deformation resulting from prolonged application of a stress below the elastic limit. Creep at room temperature is sometimes called *cold flow*. See also ANDRADE CREEP.

Creep Rupture. The rupture of a plastic under a continuously applied stress at a point below the normal tensile strength. This phenomenon is caused by the viscoelastic behavior of plastics. Creep rupture tests are generally conducted over a series of loads ranging from those causing rupture within a few minutes to those requiring very long failure times.

Crepe Rubber. Natural rubber of a pale- to dark amber color prepared by coagulating natural rubber latex with acid, then milling this coagulum into sheets. The other basic form of solid natural rubber (e.g. ribbed sheets) is prepared by drying the latex on rolls in the presence of smoke.

Cresol Resin. A phenolic-type resin obtained by condensing a cresol with an aldehyde.

Cresols. An important family of coal tar derivatives, comprising monomethyl derivatives of phenol and hydroxy derivatives of toluene. Three main isomeric forms exist: ortho-cresol (2-methylphenol), meta-cresol (3-methylphenol), and para-cresol (4-methylphenol). They are used in the production of phenol-formaldehyde resins and tricresyl phosphate, a plasticizer for PVC.

Cresyl Diphenyl Phosphate. (CDP) $(CH_3 C_6 H_4)(C_6 H_5)_2 PO_4$ A plasticizer for cellulosics, vinyl chloride polymers and copolymers, with a high degree of flame resistance and good low temperature properties. It is also acceptable for use in food packaging films. It is most often used in low percentages of the total plasticizer as a flame retardant.

Cresylic Acid. A term sometimes applied to mixtures of meta-, para- and ortho-cresol, which are mildly acidic, but also including wider fractions of phenolic compounds derived from coal tar or petroleum which contain xylenols and other higher-boiling phenols in addition to the cresols. It is used in the production of phenolic resins and tricresyl phosphate.

Crimp. In fibers, the waviness which determines the capacity of the fibers to cohere under light pressure.

Critical Strain. The strain at yield point.

Critical Surface Tension. That value of surface tension of a liquid below which the liquid will spread on a solid.

Crocidolite. (blue asbestos) See ASBESTOS.

Crocking. See BLEEDING.

Crosshead. A device which receives a molten stream of plastic emerging from an extruder, diverts the direction of flow to an angle usually 90° from the axis of the extruder screw, and forms the extrudate to a shape such as a parison for blow molding or a jacket around an electrical wire. An essential element of the crosshead is the mandrel, a tubular core with grooves of various shapes and held in place by a perforated plate, web or a crosspiece. Material emerging from the space between the mandrel and the crosshead housing is given its final shape by means of a die mounted on the end of the crosshead.

Cross-Head Die. See CROSSHEAD.

Cross Laminated. (adj) Pertaining to a laminate in which the reinforcing fibers in some layers are positioned at right angles with respect to the fibers in other layers.

Cross-Linking. Applied to polymer molecules, the setting up of chemical links between the molecular chains. When extensive, as in most thermosetting resins, crosslinking makes one infusible super-molecule of all the chains. Cross-linking can also occur between polymer molecules and other substances. For example, polyethylene can be cross-linked with carbon black filler particles, which have sites to which polyethylene chains can link in the presence of a catalyst. The mixture of resin, filler and catalyst can be molded as a thermoplastic, then transformed to a thermoset by cross-linking in the curing cycle. Cross-linking can be achieved by irradiation with high-energy electron beams, or by means of chemical cross-linking agents such as organic peroxides.

Crosslinking Agent. A substance that promotes or regulates intermolecular covalent bonding between polymer chains. (ISO).

Cross-Linking Index. The average number of cross-linked units per primary polymer molecule in the system as a whole.

Crotonaldehyde. $CH_3 CH=CHCHO$ A colorless liquid synthesized by the aldol condensation of acetaldehyde, accompanied or followed by dehydration. It can be polymerized by triethylamine to a resin with film-forming properties, or copolymerized with many compounds. Other uses include solvent for PVC, short-stopper in the polymerization of vinyl chloride, and plasticizer synthesis.

Crotonic Acid. $CH_3 CH:CHCOOH$ (trans-2-butenoic acid, trans-beta-methylacrylic acid) A white crystalline solid prepared by the oxidation of crotonaldehyde. It forms copolymers with vinyl acetate, used as hot-melt adhesives. Esters of crotonic acid are used as plasticizers for acrylic and cellulosic plastics.

Crotonyl Peroxide. A catalyst for the polymerization of vinyl and vinylidene halides.

Crown. In the calendering industry, a slight increase in diameter at the center of a calender roll to compensate for the downward deflection of the roll under pressure. The term is also used for crowned transmission belt pulleys to keep the belt running in the center of the pulley.

Cryogenic. Pertaining to very low temperatures. The term is usually applied to temperatures below about $-150°C$ $(-238°F)$. Evaluations of plastics at cryogenic temperatures are conducted for potential space applications.

Cryogenic Grinding. (Freeze Grinding) Thermoplastics are difficult to grind to small particle sizes at ambient temperatures because they soften, adhere in lumpy masses and clog screens. When chilled by dry ice, liquid carbon dioxide or liquid nitrogen, the thermoplastics can be finely ground to powders suitable for electrostatic spraying and other powder processes.

Cryptometer. An instrument for measuring the opacity of surface coatings.

Crystal. A homogeneous solid having an orderly and repetitive three-dimensional arrangement of its atoms. It is usually composed of many CRYSTALLITES, which see.

Crystal Lattice. The spatial arrangement of atoms or radicals in a crystal.

Crystallinity. A state of molecular structure in some resins attributed to the existence of solid crystals with a definite geometric form. Such structures are characterized by uniformity and compactness.

Crystallite. A perfect portion of an ordinary crystal; that is, a portion with its atoms and molecules arranged in a perfect crystal lattice. Ordinary crystals are composed of a large number of crystallites, which may or may not be arranged in perfect alignment with one another.

CS. Abbreviation for CASEIN PLASTICS, which see.

C.-Stage. (Resite) The final, fully cured state of a thermosetting resin. See also A-STAGE and B-STAGE.

CTA. Abbreviation for CELLULOSE TRIACETATE, which see.

CTFE Resins. See POLYCHLOROTRIFLUOROETHYLENE.

Cull. (n) (1) A rejected material or product. (2) In transfer molding, the material remaining in the transfer pot after the mold has been filled. A certain amount of cull is usually necessary to enable the operator to know that the cavity is filled.

Cultured Stones. A term applied to decorative embedments of natural stones such as marble, granite, terrazzo and slate in thermosetting resins. They are made by casting the resin, usually a polyester, in molds containing the stones. The embedments are used for counter tops, window sills, wall facings, flooring, giftware and numerous other applications.

Cumarone. See COUMARONE.

Cumene. $C_6H_5CH(CH_3)_2$ (isopropylbenzene, isopropylbenzol, cumol) A volatile liquid in the alkyl aromatic family of hydrocarbons. It is used as a solvent and intermediate for the production of phenol, acetone and alpha-methylstyrene; and as a catalyst for acrylic and polyester resins.

Cumene Hydroperoxide. $C_6H_5C(CH_3)_2OOH$ A colorless liquid derived from an oxidized solution or emulsion of cumene, used as a polymerization catalyst.

Cumylphenol Derivatives. Higher performance-lower cost polymer intermediates have been developed based on cumylphenol. The free phenol – either in the pure state, as a solid – or diluted with appropriate aromatic hydrocarbon has been found to be an effective replacement for nonyl phenol as an accelerator of amine epoxy cures. The pure phenol is roughly two and one-half times as effective on a weight basis as nonyl phenol. Mixtures of cumylphenol and of cumylphenyl acetate with appropriate aldehydes (e.g. furfural and benzaldehyde) may be used with both acid and amine cures to provide non solvent low viscosity phenolics. Comparable substitution for furfuryl alcohol in furan systems likewise results in enhanced physicals and electrical properties, and economics. Cumylphenyl acetate and cumylphenyl glycidyl either have been shown to be reactive monomers in epoxy systems which, in addition to producing the usual reduction in viscosity, also provide increased tensile strength in filled systems without the concomitant significant loss in heat distortion temperatures in both amine and anhydride cured formulations. In addition, the acetate has been found to be an effective plasticizer for urethane systems. Cumylphenyl benzoate has been determined to be an effective extrusion aid in a variety of thermoplastics including PVC.

Cup Flow Test. A British Standard Test (B.S. 771) for measuring the flow properties of phenolic resins. A standard mold is charged with the specimen material, and the mold is closed under specified pressure. The time in seconds for the mold to close completely is the cup flow index.

Cuprammonium Rayon. A regenerated cellulosic fiber made by dissolving cellulosic linters in a solution of copper oxide in ammonia, then forming the fiber by extrusion or spinning.

Cure. (v) To change the properties of a plastic or resin by chemical reaction, which for example may be condensation, polymerization, or addition; usually accomplished by the action of either heat or catalyst or both, and with or without pressure. The term cure is used almost exclusively in connection with thermosetting plastics, vulcanizable elastomers and rubbers. It is not ordinarily used for the hardening of thermoplastics by physical methods such as heating, cooling or evaporation of solvents. However, in the early years of the art some authors used the term cure in connection with the fusion of plastisols before the term fusion came into general use.

Curing Agents. (hardeners) Substances or mixtures of substances added to a plastic or rubber composition to promote or control the curing reaction. An agent which does not enter into the reaction is known as a *catalytic hardener* or *catalyst*. A *reactive curing agent* or *hardener* is generally used in much greater amounts than a catalyst, and actually enters into the reaction. See also ACCELERATOR.

Curing Temperature. The temperature at which a thermosetting or elastomeric material is subjected in order to attain its final stage of cure.

Curing Time. (molding time) In the molding of thermosetting plastics, the interval of time between the instant of cessation of relative movement between the moving parts of the mold and the instant that pressure is released.

Curtain Coating. A coating process in which the substrate to be coated is conveyed rapidly through a free-falling liquid "curtain" of a low-viscosity resin, solution, suspension or emulsion. The coating thickness is governed by the rate of flow of the fluid and the speed of travel of the substrate.

Custom Molder. A firm specializing in the molding of items or components to the specifications of another firm which handles the sale and distribution of the item, or incorporates the custom molded component in one of its own products.

Cut. In the fiber industry, including glass and asbestos, the number of 100 yard lengths of fiber per pound. See also DENIER, GREX NUMBER, TEX.

Cut-Layers. As applied to laminated plastics, a condition of the surface of machined or ground rods and tubes and of sanded sheets in which cut edges of the surface layer or lower laminations are revealed. (ASTM D 883-75a).

Cut-Off. (flash groove, pinch-off) In compression molding, the line where the two halves of a mold come together.

Cyanoguanidine. See DICYANDIAMIDE.

Cyanuric Acid. (tricyanic acid, tricarbimide) An acid evolved from the blowing agent azodicarbonamide when it decomposes. The acid is corrosive, and is the chief cause of plate-out on components of extruders in the structural foam process when azodicarbonamide is used as the blowing agent.

Cycle. The series of sequential operations entering into a process or part of a process. In a molding operation, cycle time is the time elapsing between a particular point in one cycle and the same point in the next cycle.

Cyclized Rubber. A thermoplastic resin produced by reacting natural rubber with stannic chloride or chlorostannic acid. It is claimed (U.S. Patent 3,205,093) that a solution of cyclized rubber in toluene is one of the few known lacquers that will adhere to polyolefins without pre-treatment of after-treatment with radiation. Other uses include films and hot-melt coatings.

Cyclohexane. C_6H_{12} (hexamethylene, hexanaphthene, hexahydrobenzene) A colorless liquid derived from the catalytic hydrogenation of benzene, used as a solvent for cellulosics and as an intermediate in the production of nylon.

Cyclohexanol. $CH_2(CH_2)_4CHOH$ (hexahydrophenol) A colorless, viscous liquid prepared by the oxidation of cyclohexane or by the hydrogenation of phenol. It is used as an intermediate in the production of nylon, a solvent for cellulosic resins, and as an intermediate for the manufacture of phthalate ester plasticizers.

Cyclohexanol Acetate. $CH_3COOC_6H_{11}$ A nonflammable solvent for cellulosics and many other resins.

Cyclohexanone. $CH_2(CH_2)_4CO$ (pimelic ketone, keto hexamethylene) A colorless liquid produced by the oxidation of cyclohexane or cyclohexanol, or by the hydrogenation of cyclohexanol. Its most important use is for the manufacture of adipic acid for nylon 6/6, and caprolactams for nylon 6. It is also an excellent high-boiling, slow evaporating solvent for many resins including cellulosics, acrylics and vinyls. It is one of the most powerful solvents for PVC, and is often used in lacquers to improve their adhesion to PVC.

Cyclohexene Oxide. $C_6H_{10}O$ A highly reactive colorless liquid which resembles ethylene oxide in most of its reactions. It is useful as an intermediate in the production of many organic chemicals used in plastics. Its epoxide structure is especially useful in applications where an HCl scavenger is required.

Cyclohexyl Methacrylate. $H_2C:C(CH_3)COOC_6H_{11}$ A colorless monomer, polymerizable to resins for optical lenses, dental parts and potting of electrical components.

Cyclohexyl Stearate. $C_6H_{11}OOCC_{17}H_{35}$ A plasticizer for polystyrene, ethyl cellulose and cellulose nitrate.

Cyclopentane. C_5H_{10} (pentamethylene) A solvent for cellulose ethers.

DAC. Abbreviation for diallyl chlorendate, which see.

DAF. Abbreviation for diallyl fumarate, a polymerizable monomer.

DAIP. Abbreviation for diallyl isophthalate. See ALLYL RESINS.

DAM. Abbreviation for diallyl maleate. See ALLYL RESINS.

Damping. Hysteresis, or variations in properties resulting from dynamic loading conditions. Damping is related to the fundamental viscoelastic mechanisms of polymers and is characteristic of the plastic as fabricated, the frequency of loading, and the stress. It provides a mechanism for dissipating energy without excessive temperature rise, preventing premature brittle fracture and improving fatigue performance.

Dancer Roll. A roller mounted on an axis which is moveable with respect to axes of other rollers in an apparatus, used to control or measure tension of a continuous length of material as it passes through a series of rollers. Dancer rolls are used as tension-sensing devices in the extrusion coating of wire, and as tension-maintaining devices in film winding.

DAP. Abbreviation for DIALLYL PHTHALATE, which see.

Dart Impact Test. See FREE FALLING DART TEST.

Dash-Pot. A device used for damping down vibrations and cushioning shocks in hydraulic systems. A typical dash-pot consists of a vessel filled with fluid or air and a piston to be attached to the moving machine part to be damped.

Daylight Fluorescent Pigments. See FLUORESCENT PIGMENTS.

Daylight Opening. The clearance between two platens of a molding press when in the open position.

DBEP. Abbreviation for DIBUTOXYETHYL PHTHALATE, which see.

DBP. Abbreviation for DIBUTYL PHTHALATE, which see.

DBPC. Abbreviation for DI-tert-BUTYL-para-CRESOL, which see.

DBS. Abbreviation for DIBUTYL SEBACATE, which see.

DCHP. Abbreviation for DICYCLOHEXYL PHTHALATE, which see.

DCP. Abbreviation for DICAPRYL PHTHALATE, which see.

DDM. Abbreviation for 4,4'-DIAMINODIPHENYL METHANE, which see.

DDP. Abbreviation for DIDECYL PHTHALATE, which see.

DE. Abbreviation for diatomaceous earth. See DIATOMITE.

Dead Fold. A fold which does not spontaneously unfold.

Deaerate. (Deair) To remove air from a substance. Deaeration is an important step in the production of vinyl plastisols, accomplished by subjecting the fluid to a high vacuum with or without agitation, to remove air which would cause objectionable bubbles or blisters in finished products.

Decabromodiphenyl Ether. A fine powder containing about 83% bromine, used as a flame retardant in polystyrene, ABS, polyolefins, polybutylene terephthalate and the like. It is reported to have minimal adverse effect on the environment.

Decabromodiphenyl Oxide. A flame retardant marketed by Dow. Chemical under the name FR-300BA, used in high-impact polystyrene, structural foam furniture, thermosetting polyesters and adhesives. It has a bromine content of 83% and a wide latitude in processability.

Decahydronaphthalene. $C_{10}H_{18}$ A colorless liquid with an aromatic odor, derived by treating of naphthalene in a fused state with hydrogen in the presence of a catalyst. It is a solvent for many resins.

Decalcomania. (Decals) Printed designs on a temporary carrier such as paper, which are used for decorating many materials including plastics. The imprint is adhered to the plastic surface by means of a pressure-sensitive adhesive, solvent welding or heat and pressure.

Decanedioic Acid. See SEBACIC ACID.

deci-. (d) The SI-approved prefix for a multiplication factor to replace 10^{-1}.

Decibel. (dB) A unit for measuring sound or power ratios in telephone engineering. As originally developed, the gain or loss of power expressed in decibels was 10 times the logarithm of the power ratio. The numerical value of one decibel is approximately equal to the smallest change in volume of sound that the normal ear can detect. The scale of decibels is logarithmic, every increase of 10 dB representing an increase of about 300% in sound pressure. Thus, a 100 dB noise is 3 times as intense as a 90 dB noise. Meters for measuring sound levels have several scales for weighting the readings to compensate for frequency ranges, the "A" scale corresponding to the ranges receptive by the human ear. The abbreviation dBA is used to denote that the A scale readings apply. OSHA regulations prescribe maximum sound levels permitted for employee exposure. In plastics operations the major source of excessive noise is often the scrap grinders.

Decorative Boards. A special term for laminates used in the furniture industry, which are defined by The Decorative Board Section of NEMA as "a product resulting from the impregnation or coating of a decorative web of paper, cloth, or other carrying media with a thermosetting type of resin and consolidation of one or more of these webs with a cellulosic substrate under heat and pressure of less than 500 pounds per square inch." This includes all boards that were formerly called low-pressure melamine and polyester laminates, but does not include vinyls.

Deckle. (deckle rod, cut-off plate) In extrusion of film or extrusion coating, a small rod or plate attached to each end of the die to adjust the length of the die opening.

Decorating. The processes used for decorating plastic articles are defined under the following headings:

AIRLESS SPRAYING
ASHING
DECALCOMANIA
ELECTROPHORETIC DEPOSITION
ELECTROPLATING
ELECTROLESS PLATING
ELECTROSTATIC PRINTING
ELECTROSTATIC SPRAY COATING
EMBEDMENT DECORATING
EMBOSSING
FILL-AND-WIPE
FLEXOGRAPHIC PRINTING
FLOCKING
FLOW COATING
GRAVURE PRINTING
GRAVURE COATING
HOT STAMPING
IN-MOLD DECORATION
LETTERPRESS PRINTING
METALLIZING
OFFSET PRINTING
PAINTING OF PLASTICS
PRINTING ON PLASTICS
ROLLER COATING
RUBBER PLATE PRINTING
SCREEN PROCESS PRINTING
SECOND-SURFACE DECORATING
SPRAY-WIPE PAINTING
SPRAY COATING
THERMOGRAPHIC TRANSFER PROCESS
VACUUM METALLIZING
VALLEY PRINTING

Decyl Butyl Phthalate. See BUTYL ISODECYL PHTHALATE.

Decyl-Octyl Methacrylate. $H_2C:C(CH_3)COO(CH_2)_nCH_3$ (n=7−9) A polymerizable monomer for acrylic plastics.

n-Decyl, n-Octyl Phthalate. (NDOP) See n-OCTYL, n-DECYL PHTHALATE.

Decyl Tridecyl Phthalate. $C_6H_{21}OCOC_6H_4COOC_{13}H_{27}$ A plasticizer for vinyls, cellulosics and polystyrene.

Deep Drawing. The process of forming a thermoplastic sheet in a mold involving a high draw ratio.

Deflashing. (defining) The process of removal of flash or rind left on molded plastic articles by spaces between mold cavity edges. Methods include TUMBLING, BLAST FINISHING (both of which see), use of dry or wet abrasive belts, and hand methods using knives, scrapers, broaching tools and files. Soft thermoplastic parts are sometimes deflashed by the cryogenic method, in which the parts are tumbled while chilled by a coolant such as liquid nitrogen. See also ABRASIVE FINISHING, AIRLESS BLAST DEFLASHING.

Deflection Temperature Under Load. See HEAT DISTORTION POINT.

Deflocculation Agent. A substance that breaks down agglomerates into primary particles or prevents the latter from combining into agglomerates. (ISO).

Defoamer. An agent when added in small quantities to a fluid containing gas bubbles causes the small bubbles to coalesce into larger bubbles that rise to the surface and break.

Degassing. (breathing) In injection molding, the momentary opening and closing of a mold during the early stages of the cycle to permit the escape of air or gas from the heated compound.

Degating. The removal of material left on a plastic part formed by the opening through which material was injected into the mold cavity. The operation is sometimes performed automatically by a mold element. Otherwise, the gate can be removed by manual breaking or using a shearing die, followed if necessary by sanding or burnishing.

Degradation. A deleterious change in the chemical structure, physical properties or appearance of a plastic caused by exposure to heat (*thermal degradation*), light (*photodegradation*), oxygen (*oxidative degradation*) or weathering. The ability of plastics to withstand such degradation is called STABILITY. See also ARTIFICIAL WEATHERING, AUTO-CATALYTIC DEGRADATION, BIODEGRADATION, CORROSION RESISTANCE, DETERIORATION, DEW-CYCLE WEATHERING TEST, DISCOLORATION, PINK STAINING, XENON ARC LIGHT AGEING.

Degree of Cure. The extent to which curing or hardening of a thermosetting resin has progressed. See also A-STAGE, B-STAGE and C-STAGE.

Degree of Polymerization. (Chain length) The average number of monomer units per polymer molecule, a measure of molecular weight. In most plastics, the molecular weight must reach several thousand to attain worthwhile physical properties.

Dehydration. The removal of water from a substance either through ordinary drying or heating, or by absorption, adsorption, chemical reaction, condensation of water vapor, or by centrifugal force or hydraulic pressure.

Dehydroacetic Acid. $CH_3C:CHC(O)CH(COCH_3)C(O)O$ (DHA) A colorless crystalline material with uses as a plasticizer, fungicide and bactericide.

Dehydrogenation. The removal of hydrogen from a compound by chemical means.

Dehydrohalogenation. The process of splitting hydrogen chloride from polymers such as PVC, caused by excessive heat and/or light.

Deka. (da) The SI-approved prefix for a multiplication factor to replace 10^1.

Delamination. The separation of ore or more layers in a laminate caused by the failure of the adhesive bond.

Deliquescent. (adj) Capable of becoming liquid by absorbing moisture from the air.

Delustrants. Chemical agents used to produce dull surfaces on synthetic fibers either before or after spinning to obtain a more natural silk-like appearance.

Denier. The weight in grams of 9000 meters of fiber in the form of continuous filament. This is the most widely used unit of weight in the textile industry to indicate the fineness of natural or synthetic fibers. See also CUT, GREX NUMBER, TEX.

Density. (absolute density) Mass per unit volume of a substance, expressed in units such as grams per cubic centimeter, pounds per cubic foot or pounds per gallon. It is an important criterion in specifying some plastics such as polyethylene which vary widely in density. Temperatures at which measurements are taken are usually recorded, the usual standard being 23°C. See also APPARENT DENSITY, BULK FACTOR, GRADIENT TUBE DENSITY DETERMINATION, RELATIVE DENSITY, SPECIFIC GRAVITY.

Deoxy-. (desoxy-) A prefix denoting replacement of a hydroxyl group with hydrogen.

DEP. Abbreviation for DIETHYL PHTHALATE, which see.

Depolymerization. The reversion of a polymer to its monomer, or to a polymer of lower molecular weight. Such reversion occurs in some plastics when exposed to very high temperatures.

Desiccant. A substance capable of absorbing water vapor from air or other gaseous material, used to maintain low humidity in a storage or test vessel.

Destaticization. The treating of a plastics to minimize their tendency to accumulate static electricity charges. See also ANTI-STATIC AGENTS, STATIC ELIMINATORS, SOOT CHAMBER TEST.

Deterioration. A permanent change in the physical properties of a plastic evidenced by impairment of these properties. (ASTM D 883-65T).

Dew-Cycle Weathering Test. An accelerated weathering test in which specimens mounted on panels are alternatively exposed to unfiltered light such as from a carbon arc, and dew caused by spraying cold water on the backs of the panels. This test has been considered by some as a rapid and reliable means of predicting the weatherability of plastic coatings, but its validity for acrylic coatings has been disputed.

D-Glass. Glass with a high boron content, used for fibers in laminates which require a precisely controlled dielectric constant.

DHA. Abbreviation for DEHYDROACETIC ACID, which see.

DHP. Abbreviation for dihexyl phthalate. See DI(2-ETHYLBUTYL) PHTHALATE.

DI-. A prefix meaning two or twice. The terms BI- and BIS- are nearly equivalent, assigned with slight differences in meaning or according to custom.

Diacetin. $CH_3 COOCH_2 CHOHCH_2 OOCCH_3$ (glyceryl diacetate) A water-soluble plasticizer, and a solvent for cellulose nitrate and cellulose acetate.

Diacetone Alcohol. $CH_3 COCH_2 C(CH_3)_2 OH$ (diacetone, 4-hydroxy-4-methyl-2-pentanone, 4-hydroxy-2-keto-4-methylpentane). A pleasant smelling colorless liquid, miscible with water and most organic liquids, used as a solvent for cellulosic, vinyl and epoxy resins.

Diacetyl Peroxide. See ACETYL PEROXIDE.

Diafoam. A term sometimes used for a SYNTACTIC FOAM, which see, which also contains gas bubbles in addition to hollow glass spheres.

Diallyl Chlorendate. (DAC) A reactive monomer used as a flame resisting agent in DAP, epoxy and alkyd resins. It can be used in the monomeric foam (a high-viscosity fluid); or in the polymeric form, alone or in conjunction with other flame retardants.

Diallyl Isophthalate. $C_6 H_4 (COOCH_2 CHCH_2)_2$ A polymerizable monomer, used in laminating and molding.

Diallyl Maleate. $C_4 H_2 O_4 (CH_2 CHCH_2)_2$ A monomer which polymerizes readily when exposed to light or temperatures above $50°C$.

Diallyl Phthalate. $C_6 H_4 (COOCH_2 CH:CH_2)_2$ (DAP) In the monomeric form, DAP is a colorless liquid ester with a viscosity about like that of kerosene, widely used as a cross-linking monomer for unsaturated polyester resins, and as a polymerizable plasticizer for many resins. It polymerizes easily, either gradually or rapidly, increasing in viscosity until it finally becomes a clear, infusible solid. The name DAP is used for both the monomeric and polymeric forms. In the partially polymerized form, DAP is used in the production of thermosetting molding powders, casting resins and laminates.

Di-Alphinyl Phthalate. A plasticizer derived by the esterification of primary aliphatic alcohols in the range C_7 to C_9. It is used in place of DOP in some applications.

Dialysis. The process of separating molecules of one size from those of another size, in which solute molecules are transferred from one liquid to another through a membrane in response to differences in the chemical potentials between the liquids. The membranes used may be parchments or microporous films of polymers, e.g. cellulosic materials.

4,4'-Diaminodiphenyl Methane. (DDM) $NH_2 C_6 H_4 CH_2 C_6 H_4 NH_2$ A silvery crystalline material derived by heating formaldehyde anilide with aniline hydrochloride and aniline. It is used as a curing agent for epoxy resins and as an intermediate in making diisocyanates for urethane elastomers and foams by reaction with phosgene. Possible occupational hazards in the use of DDM were indicated in 1976, including toxic hepatitis and liver damage.

Diamyl Phthalate. $C_6 H_4 (COOC_5 H_{11})_2$ A plasticizer derived by esterification of phthalic anhydride with amyl alcohol, compatible with most vinyls, PMMA and cellulosics.

Diaphragm Gate. (web gate) A gate which forms a solid web across the opening of the part, used in injection molding and transfer molding of annular and tubular articles.

Diatomite. (diatomaceous earth, DE, kieselguhr, infusorial earth, siliceous earth, tripolite) The naturally occurring deposit of skeletons of small aquatic plants called *diatoms,* consisting of from 83 to 89% silica. Its many uses include fillers for plastics.

DIBA. Abbreviation for DI-ISOBUTYL ADIPATE, which see.

Dibasic. Pertaining to acids or salts which have two displaceable hydrogen atoms per molecule. Such substances having one displaceable H atom are called *monobasic,* and those with three are called *tribasic.*

Dibasic Lead Phosphite. $2PbO \cdot PbHPO_3 \cdot \frac{1}{2}H_2 O$ A white, crystalline powder used as a heat and light stabilizer for vinyl resins and other chlorine-containing resins. It has good electrical properties, and acts as a U.V. light screening agent and antioxidant.

Dibasic Lead Phthalates. Heat and light stabilizers for vinyl insulation, opaque film and sheeting, and foam.

Dibasic Lead Stearate. $2PbO \cdot PB(C_{17} H_{35} COO)_2$ A good heat stabilizer with lubricating properties. See also LEAD STEARATE.

Dibenzyl Ether. $C_6 H_5 CH_2 OCH_2 C_6 H_5$ A plasticizer for cellulose nitrate.

Dibenzyl Sebacate. $(C_6 H_5 CH_2 OOC)_2 (CH_2)_8$ A non-toxic plasticizer, often used in vinyl compounds for lining container closures.

Diborane. $B_2 H_6$ (Boron hydride, boro-ethane) A colorless gas used as a catalyst in the polymerization of ethylene.

DIBP. Abbreviation for DIISOBUTYL PHTHALATE, which see.

Dibromobutenediol. A low molecular weight, chemically reactive, brominated primary glycol. It is used as a building block for condensation polymers that can be incorporated into a wide variety of polymers including esters, urethanes and ethers. It is also used as a flame retardant monomer for polyurethanes and thermoplastics, and as a substitute for MOCA in urethane foams.

Dibromoneopentyl Glycol. A high-melting point solid, available in powder or flake form, used as a flame retardant for polyester resins. A more convenient liquid material is made by using dibromoneopentyl glycol to form a polyester alkyd which is dissolved in styrene, the resulting liquid being more easily used in polyester reactors. The material is also adaptable to urethane foams, polymeric plasticizers and coating resins.

2,3-Dibromopropanol. A brominated alcohol used as a component in making fire-retardant urethane foams.

Dibutoxyethoxy Ethyl Adipate. $[C_4H_9OC_2H_4OC_2H_4COO(CH_2)_2]_2$ A plasticizer for cellulose nitrate, ethyl cellulose, polyvinyl butyral and polyvinyl acetate.

Dibutoxyethyl Adipate. $(C_2H_4COOC_2H_4OC_4H_9)_2$ A primary plasticizer for PVC and many other resins, imparting low temperature flexibility and resistance to ultraviolet light. It is widely used as a plasticizer for polyvinyl butyral in the inner layer of safety glass.

Dibutoxyethyl Phthalate. $C_6H_4(COOC_2H_4OC_4H_9)_2$ (DBEP) A primary plasticizer for vinyls, methacrylates, nitrocellulose and ethyl cellulose, imparting low temperature flexibility and resistance to ultraviolet light. Incorporation of up to 20% of DBEP in vinyl calendering compounds eliminates defects such as streaks and blisters. In plastisols, DBEP imparts low initial viscosity and low fusion temperatures.

Dibutoxyethyl Sebacate. $(CH_2)_8(COOC_2H_4OC_4H_9)_2$ A primary plasticizer for PVC and PVAc, with low temperature resistance.

Dibutyl Adipate. $C_4H_9OCO(CH_2)_4COOC_4H_9$ A plasticizer for vinyl and cellulosic resins.

Dibutyl Butyl Phosphonate. $C_4H_9P(O)(OC_4H_9)_2$ A plasticizer and anti-static agent.

Di-tert-Butyl-para-Cresol. $[C(CH_3)_3]_2CH_3H_2OH$ (DBPC, butylated hydroxytoluene, BHT) A white, crystalline solid used as an antioxidant in polyethylene, vinyl monomers and many other substances.

Di-tert-Butyl Peroxide. $(CH_3)_3COOC(CH_3)_3$ A member of the alkyl peroxide family, used as an initiator in vinyl chloride polymerization, polyester reactions, and as a cross linking agent.

Dibutyl Fumarate. $C_2H_2(COOC_4H_9)_2$ A derivative of furamic acid, used as a plasticizer for polyvinyl acetate, polyvinyl chloride and vinyl chloride-vinyl acetate copolymers.

Di-n-Butyl Isosebacate. $C_8H_{16}(COOC_4H_9)_2$ A plasticizer for vinyls and other thermoplastics.

Di-tert-Butyl Peroxide. $(CH_3)_3 COOC(CH_3)_3$ A catalyst for polymerization of vinyl monomers.

Dibutyl Phthalate. $C_6 H_4 (COOC_4 H_9)_2$ (DBP) One of the most widely used plasticizers for cellulose nitrate and other cellulose ester and ether lacquers and coatings. It is a primary plasticizer for many other resins, but its high volatility limits its use in vinyls.

Dibutyl Sebacate. $C_4 H_9 OCO(CH_2)_8 OCOC_4 H_9$ (DBS) A plasticizer, one of the most effective in the sebacate family. It has good low temperature properties, low volatility, and is compatible with vinyl chloride polymers and copolymers, polyvinyl butyral and ethyl cellulose. It is non-toxic, suitable for use in food wrappings.

Dibutyl Succinate. $C_{12} H_{22} O_4$ A plasticizer for cellulosic resins.

Dibutyl Tartrate. $C_4 H_9 OOCCHOHCHOHCOOC_4 H_9$ A lubricant, plasticizer and solvent for cellulosic resins.

Dibutyl Tin Bis(Isooctylmercapto Acetate). A stabilizer for rigid PVC, used primarily in the period from 1953 to 1970. Thereafter, improved butyl tins and methyl tin derivatives, and synergistic mixtures thereof with other stabilizers, replaced this stabilizer for most applications.

Dibutyltin Di-2-Ethylhexoate. $(C_4 H_9)_2 Sn(O_2 CC_7 H_{15})_2$ A white, waxy solid derived from the reaction of dibutyltin oxide with 2-ethylhexoic acid. Used as a catalyst for silicone curing and in polyether foams.

Dibutyltin Dilaurate. $(C_4 H_9)_2 Sn(OOCC_{11} H_{23})_2$ A lubricating-type stabilizer for vinyl resins, catalyst for urethane foams, and catalyst for condensation polymerization. It is used in vinyl compounds when good clarity is needed, and also imparts excellent light stability. However, its properties are reduced at high processing temperatures.

Dibutyltin Maleate. $[(C_4 H_9)_2 Sn(OOCCH)_2]x$. A white amorphous powder used as a condensation polymerization catalyst and a stabilizer for PVC. The molecular weight of the material varies, and grades of lower molecular weight tend to be volatile and produce gases. The higher weight polymers are very effective in rigid PVC.

Dibutyltin Sulphide. $[(C_4 H_9)_2 SnS]_3$ An antioxidant, and stabilizer for vinyl resins.

1,1-Dibutylurea. $NH_2 CON(C_4 H_9)_2$ (N,N-dibutylurea) A polymerizable substance. When copolymerized with simple urea and formaldehyde, permanently thermoplastic resins are obtained.

Dicapryl Adipate. $C_8 H_{17} OOC(CH_2)_4 COOC_8 H_{17}$ A plasticizer for cellulosic and vinyl resins, yielding good low temperature flexibility. Also compatible with PMMA and polystyrene.

Dicapryl Phthalate. $(C_8 H_{17} COO)_2 C_6 H_4$ (DCP) di-(2-octyl) phthalate. A plasticizer for cellulosic and vinyl resins. It is similar to DOP and DIOP, but has low initial viscosity and is preferred in plastisols.

Dicapryl Sebacate. $C_8 H_{17} OOC(CH_2)_8 COOC_8 H_{17}$ A plasticizer for vinyl resins and acrylonitrile rubbers, imparting good low temperature flexibility.

Dicarboxylic Acids. A family of organic acids containing two carboxylic ($-COOH$) groups. Those of greatest importance in the plastics industry are the adipic, azelaic, glutaric, pimelic, sebacic, and succinic acids, esters of which are widely used as plasticizers imparting low-temperature flexibility. They are also used in the production of alkyd and polyester resins, polyurethanes and nylons.

Dicetyl Ether. (dihexadecyl ether) A lubricant used on molds for processing plastics.

2,4-Dichlorobenzoyl Peroxide. A cross-linking agent for silicone elastomers. It is sold as a 40% active paste dispersed in silicone fluid under the trademark Cadox TS-40 by Noury Chemical Corp.

4,4,4-Dichloro-1,2-Butylene Oxide. (TCBO) $Cl_3 C\text{-}CH_2\text{-}CHCl\text{-}CH_2\text{-}OH$ A highly refractive, mobile liquid used for modifying polyols to achieve fire retardance in urethane foams.

Dichlorodifluoromethane. $CCl_2 F_2$ (Freon 12) In addition to its familiar uses as an aerosol propellant and refrigerant, "Freon 12" is a blowing agent for foam plastics, e.g. polystyrene. It has the advantage of being non-hazardous in contrast to the hydrocarbon blowing agents.

1,1-Dichloroethylene. See VINYLIDENE CHLORIDE.

alpha-Dichlorohydrin. $CH_2 ClCHOHCH_2 Cl$ A solvent for cellulosic resins.

Dichloromethane. See METHYLENE CHLORIDE.

2,6-Dichlorostyrene. $C_6 H_3 (CH:CH_2)Cl_2$ A monomer and co-monomer used mainly in plastics research.

Dichlorotetrafluoroethane. $CClF_2 CClF_2$ (Freon 114) A fluorocarbon blowing agent used in applications requiring a low boiling point (3.6°C).

Dichroism. (1) A property possessed by many doubly refracting crystals of exhibiting different colors when viewed from different directions. (2) The exhibition of different colors by certain solutions in different degrees of dilution or concentration.

Dicing. The process of cutting thermoplastic strands or sheets into pellets for further processing.

Dicumyl Peroxide. $[C_6 H_5 C(CH_3)_2 O]_2$ A vulcanizing agent for elastomers, also used as a crosslinking agent for polyethylene.

Dicyandiamide. $NH_2 C(NH)(NHCN)$ The widely used but incorrect name for the dimer of cyanamide, or cyanoguanidine. It is used mainly in the production of melamine, but also as a stabilizer for vinyl resins, and a curing agent for epoxy resins.

Dicyclohexyl Azelate. $C_6H_{11}OCOC_7H_{14}COOC_6H_{11}$ A plasticizer for PVC.

Dicyclohexyl Phthalate. $C_6H_4(COOC_6H_{11})_2$ (DCHP) A plasticizer for PVC and many other resins. It imparts good electrical properties, low volatility, low water and oil absorption, and resistance to extraction by hexane and gasoline. In vinyls DCHP is usually used in combination with other plasticizers. In cellulose nitrate, polystyrene, PMMA and ethyl cellulose it serves as a primary plasticizer.

DIDA. Abbreviation for DI-ISODECYL ADIPATE, which see.

DIDG. See DIISODECYL GLUTARATE.

Didecyl Adipate. $C_4H_8(COOC_{10}H_{21})_2$ A plasticizer for PVC and cellulosics. Its most noteworthy properties are low temperature flexibility, low volatility and good electrical characteristics.

Didecyl Ether. $(C_{10}H_{21})_2O$ A lubricant for plastics molding and processing.

Didecyl Phthalate. $C_6H_4'(COOC_{10}H_{21})_2$ (DDP) A primary plasticizer for vinyl resins, also compatible with polystyrene and cellulosics. It has the lowest specific gravity of the most common phthalate plasticizers, low volatility and resistance to extraction by soapy water.

DIDP. Abbreviation for DI-ISODECYL PHTHALATE, which see.

Die. (1) A steel block containing an orifice through which plastic is extruded, shaping the extrudate to the desired profile. (See also EXTRUDER DIE). (2) The recessed block into which plastic material is injected or pressed, shaping the material to the desired form. The term *cavity* is more often used.

Die Adaptor. (extrusion) See ADAPTOR.

Die Blades. In extrusion, deformable members attached to a die body which determine the slot opening and which are adjusted to produce uniform thickness across the film or sheet produced.

Die Block. That part of an extrusion die which holds the forming bushing and core.

Die Body. In the U.S., same as DIE ADAPTOR. In Great Britain, the outer body or barrel of an extrusion die.

Die Cone. The tapered element in an extrusion die which guides the material to the webs of the spider. Also called TORPEDO, which see, and spreader.

Die Cutting. (blanking, clicking, dinking) The process of cutting shapes from sheets of plastic by pressing a shaped knife edge into one or several layers of sheeting. The dies are often called steel rule dies, and pressure is applied by hydraulic or mechanical presses.

Die Land. In an extruder die, the die land is the portion of the die wherein the dimensions of the opening are constant from a point within the die to the discharge end, which imparts the final shape to the extrudate.

Dielectric. A material with electrical conductivity less than one millionth of a reciprocal ohm per centimeter, thus so weakly conductive that different parts of its surface can have a different electrical charge. In radio-frequency heating operations, the term dielectric is used for the material being heated. The term is also used for the non-conductive material separating the conductive elements of a condenser.

Dielectric Absorption. An accumulation of electrical charges within the body of an imperfect dielectric material when it is placed in an electrical field.

Dielectric Breakdown Voltage. The voltage at which electrical breakdown of a specimen of electrical insulating material between two electrodes occurs under prescribed conditions. The ASTM test for dielectric breakdown voltage and dielectric strength is D 149-75 in Part 35 of the 1976 Book of ASTM Standards.

Dielectric Constant. Between any two electrically charged bodies there is a force (attraction or repulsion) which varies according to the strength of the charges, the distance between the bodies, and a characteristic of the medium separating the bodies (the dielectric) known as the dielectric constant. This force, f, is found by the equation

$$f = \frac{q_1 \cdot q_2}{\epsilon\, r^2}$$

in which q_1 and q_2 are the respective charges, r is the distance, and ϵ is the dielectric constant. For a vacuum ϵ is 1.0000; for air, $\epsilon = 1.00059$. In practice the dielectric constant of a material is found by measuring the capacitance of a parallel plate condenser using the material as the dielectric, then measuring the capacitance of the same condenser with a vacuum as the dielectric, and expressing the result as a ratio between the two capacitances. When the dielectric is a polymeric material whose molecules may readjust their positions in an alternating field, the resulting loss of energy is called the DISSIPATION FACTOR, which see. The ASTM method for measuring dielectric constant (permittivity) of solid electrical insulating materials is given in D 150-74 in Part 35 of the 1976 ASTM Book of Standards.

Dielectric Heating. (electronic heating, RF heating, radio frequency heating, high frequency heating) The process of heating poor conductors of electricity (dielectrics) by means of high frequency electrical currents. Materials that are relatively good conductors of electricity, e.g. the resistance wires in electric stoves, are easily heated by passing through them an alternating current of low frequency, e.g. 60 cycles. As the electrical resistance increases, the voltage required to produce a given current or degree of heating at a constant frequency increases. In the case of poor conductors such as wood and most plastics, the voltage required to produce appreciable heating with low frequency current becomes so high as to be impractical, e.g. in the range of millions of volts. However, upon increasing the frequency of the alternating current the required voltage drops inversely. At frequencies in the "radio" range such as those above 18 million cycles (megacycles) per second, sufficient heat for welding and sealing many plastics can be generated at low, safe voltages. The process of dielectric heating is conducted by placing the material to be heated between two shaped electrodes which are connected to a high frequency alternating current supply. These electrodes act as the plates of a condenser, and the material serves as the dielectric separating them. As the polarity of the electrical charge reverses, molecules of the material tend to change their orientation in an effort to keep their positive poles toward the electrode which is momentarily negative, thus creating molecular and friction stresses which cause heating. Dielectric heating is most effective for materials such as PVC which have a high "loss factor" by virtue of containing polar molecules. Plastics with low loss factors such as polystyrene and polyethylene are much more difficult to heat dielectrically. See also DIELECTRIC HEAT SEALING, MICRO-WAVE HEATING.

Dielectric Heat Sealing. A sealing process widely used for vinyl films and other thermoplastics with sufficient dielectric loss, in which the film is heated by DIELECTRIC HEATING, (which see) and pressed against another film by an applicator serving as one element of a condenser, the other element comprising a platen. The applicator may be a pinpoint electrode as used in "electronic sewing machines," a wheel, moving belt, or contoured blade. Frequencies employed range up to 200 mc per second, although those of 30 mc and below are most often used to avoid technical problems. See also HEAT SEALING.

Dielectric Loss. A loss of energy eventually showing through the rise in heat of a dielectric placed in an alternating electrical field.

Dielectric Loss Angle. (dielectric phase difference) The difference between ninety degrees and the dielectric phase angle.

Dielectric Phase Angle. The angular difference in phase between the sinusoidal voltage applied to the dielectric and the resulting current. The angle is often symbolized by the Greek θ (theta), the cosine of which is the POWER FACTOR, which see.

Dielectric Strength. A measure of the voltage required to puncture a material, expressed in volts per mil of thickness. The voltage figure used is the average root-mean-square voltage gradient between two electrodes at which electrical breakdown occurs under prescribed conditions of test.

Die Lines. In blow molding, vertical marks on the parison caused by damaged die elements or by contamination of the compound.

Diene Polymers. The family of polymers based on unsaturated hydrocarbons or diolefins having two double bonds. When the double bonds are separated by only one single bond, e.g. as in 1,3-butadiene (CH_2=CH–CH=CH_2), the diene is called a conjugated diene. In an unconjugated diene the double bonds are separated by at least two single bonds. The family includes polymers and copolymers of ethylene, propylene, isoprene, butadiene, and cyclopentadiene.

Die Plates. (1) In injection molds, the members which are attached respectively to the fixed and moving heads of the press. (2) In extrusion, especially with regard to pelletizing extruders, the die plate is that part of the die assembly which is bolted to the outlet of the die body and shapes the melt into a continuous strand or filament of desired cross section.

Die Spider. In extrusion, a term used to denote the membranes supporting a mandrel within the head and die assembly.

Die Swell. In extrusion, the increase in diameter of the extrudate over that of the die opening through which it is extruded.

Die Swell Ratio. In extrusion, particularly for blow molding, the ratio of the outer parison diameter (or parison thickness) to the outer diameter of the extrusion die (or die gap). The ratio is affected by the polymer type, head construction, land length, extrusion speed, and temperature.

2,2-Diethoxyacetophenone. (DEAP) A photo-initiator used for curing acrylate coatings, either in an inert atmosphere or in air with UV light. It absorbs impinging light.

Diethoxyethyl Adipate. $C_2H_5OC_2H_4OCO(CH_2)_4COOC_2H_4OC_2H_5$ A plasticizer for cellulosic resins.

Diethoxyethyl Phthalate. $C_6H_4(COOC_2H_4OC_2H_5)_2$ A plasticizer for cellulosic and vinyl resins.

Diethyl Adipate. $C_2H_5OCO(CH_2)_4OCOC_2H_5$ A plasticizer for cellulose acetate, cellulose acetate butyrate and cellulose nitrate.

Diethylaluminum Chloride. $(C_2H_5)_2AlCl$ A colorless liquid which bursts into flame instantly upon contact with air, and reacts violently in contact with water. It is used as a catalyst in the polymerization of olefins.

3-Diethylaminopropylamine. $(C_2H_5)_2NCH_2CH_2CH_2NH_2$ A curing agent for epoxy resins.

Di(2-Ethylbutyl) Azelate. $C_6H_{13}OOC(CH_2)_7COOC_6H_{13}$ A plasticizer for PVC, its copolymers, and cellulose esters. It is very compatible and efficient in vinyls, finding its largest use in high-clarity film and sheeting.

Di(2-Ethylbutyl) Phthalate. $C_6H_4(COOC_6H_{13})_2$ (dihexyl phthalate) A fast-fluxing, highly solvating plasticizer used in vinyl plastisols and extrusion compounds, and for cellulose esters.

Diethyl Carbonate. $(C_2H_5)_2CO_3$ A solvent for cellulosic and many other resins.

Diethylene Glycol Bis(Allyl Carbonate). A thermosetting resin belonging to the family of ALLYL RESINS, which see. It is probably the oldest of the so-called "engineering plastics," having been introduced in the 1940's by Columbia Southern Chemicals Corp. under the commercial name CR-39. Since it is difficult and tricky to handle, the resin has found little commercial use outside of optical applications, in which it is widely used due to excellent optical properties, surface abrasion resistance, and favorable weight (half that of glass). Examples of products made of CR-39 are ophthalmic lenses, bifocals, optical filters, instrument windows, welders' masks, and large windows in atomic energy plants.

Diethylene Glycol Dibenzoate. $C_6H_5CO(OCH_2CH_2)_2OCOC_6H_5$ A plasticizer for cellulosic resins, PMMA, PS, PVC, PVAc and other vinyls. It imparts good stain resistance.

Diethylene Glycol Dipelargonate. $(CH_3(CH_2)_7COOCH_2CH_2)_2O$ A simple ester of pelargonic acid used primarily as a secondary plasticizer for vinyl resins, but also as a plasticizer for cellulosics. Within the limits of its compatibility it provides an economical way of obtaining low temperature flexibility.

Diethylene Glycol Dipropionate. $(CH_3 CH_2 COOH_2 CH_2)_2 O$ A plasticizer for cellulosic plastics.

Diethylene Glycol Distearate. $(C_{17} H_{35} COOC_2 H_4)_2 O$ A plasticizer for cellulose nitrate and ethyl cellulose.

Diethylene Glycol Monoacetate. $HO(CH_2)_2 O(CH_2)_2 OOCCH_3$ A solvent for cellulose nitrate and cellulose acetate.

Diethylene Glycol Monobutyl Ether. $C_4 H_9 OCH_2 CH_2 OCH_2 CH_2 OH$ A solvent with a high boiling point, used in coatings when very slow drying rates are desired. It is also useful as a dispersant in vinyl organosols, and as an intermediate for the production of plasticizers.

Diethylene Glycol Monoethyl Ether. $CH_2 OHCH_2 OCH_2 CH_2 OC_2 H_5$ A solvent for cellulosics.

Diethylene Glycol Monolauiate. See DIGLYCOL LAURATE.

Diethylene Glycol Monoricinoleate. See DIGLYCOL RICINOLEATE.

Di-(2-Ethylhexyl) Adipate. $(C_2 H_4 COOCH_2 CH[C_2 H_5] C_4 H_9)_2$ (dioctyl adipate, DOA) A primary plasticizer for vinyls, cellulose nitrate, polystyrene, and ethyl cellulose. In vinyls, DOA is often used in combination with phthalate and other plasticizers, imparting good low temperature flexibility, resilience, and resistance to extraction by water. It is FDA approved for use in vinyl food packaging films.

Di(2-Ethylhexyl) Azelate. $(CH_2)_7 [COOCH_2 CH(C_2 H_5)C_4 H_9]_2$ (DOZ, dioctyl azelate) A plasticizer for vinyl chloride polymers and copolymers. It is one of the most compatible of the low-temperature monomeric plasticizers, has low volatility, and imparts low extractibility by water and soapy water. This plasticizer is now approved for food-contact use.

Di(2-Ethylhexyl) 2-Ethylhexyl Phosphonate. $C_8 H_{17} PO(OC_8 H_{17})_2$ A plasticizer and stabilizer.

Di(2-Ethylhexyl) Hexahydrophthalate. $C_6 H_{10} [COOCH_2 CH(C_2 H_5)C_4 H_9]_2$ (Dioctyl hexahydrophthalate) A light-colored liquid used as a plasticizer for vinyls.

Di(2-Ethylhexyl) Isophthalate. $C_6 H_4 [COOCH_2 CH(C_2 H_5)C_4 H_9]_2$ (dioctyl isophthalate, DOIP) A primary plasticizer for PVC, most notable for its resistance to marring by nitrocellulose lacquers and low volatility in addition to good general purpose properties. It is also compatible with polyvinyl butyral, vinyl chloride-acetate copolymers, CAB resins with high butyral content, cellulose nitrate, ethyl cellulose, polystyrene and chlorinated rubber.

Di(2-Ethylhexyl) Phthalate. $C_6 H_4 [COOCH_2 CH(C_2 H_5)C_4 H_9]_2$ (DOP, dioctyl phthalate) The most widely used plasticizer for PVC, also compatible with ethyl cellulose, cellulose nitrate, and polystyrene. It is generally recognized as imparting the best all-around good properties to vinyls, and is often used as the standard against which other plasticizers are evaluated. DOP has been approved by the FDA for use in packaging films for non-greasy foodstuffs of high water content.

Di(2-Ethylhexyl) Sebacate. $(CH_2)_8(COOC_8H_{17})_2$ (Dioctyl sebacate) A plasticizer for vinyl chloride polymers and copolymers, cellulosic plastics, polystyrene and polyethylene. It imparts good low temperature properties.

Di(2-Ethylhexyl) Succinate. $C_8H_{17}OCOCH_2CH_2COOC_8H_{17}$ (dioctyl succinate) A plasticizer.

Di-2-Ethylhexyl-4-Thioazelate. $S(CH_2CH_2CH_2COOC_8H_{17})_2$ A plasticizer for cellulose nitrate, ethyl cellulose, PMMA, polystyrene and vinyl resins.

Diethyl Phthalate. $C_6H_4(COOC_2H_5)_2$ (ethyl phthalate) A plasticizer and solvent for nearly all thermoplastics and coumarone resins. It has been approved by FDA for use in food packaging.

Diethyl Sebacate. $C_2H_5OCO(CH_2)_8COOC_2H_5$ A plasticizer with good low temperature properties, compatible with PVC and most other thermoplastics.

Diethyl Succinate. $C_8H_{14}O_6$ A plasticizer for cellulose acetate, cellulose acetate butyrate and cellulose nitrate.

Diethyl Tartrate. $C_4H_4O_6(C_2H_5)_2$ A solvent and plasticizer for cellulose resins.

1,1-Diethylurea. $NH_2CON(C_2H_5)_2$ A white solid polymerizable with simple urea and formaldehyde to form permanently thermoplastic resins.

Differential Scanning Calorimeter. An instrument which measures the rate of heat evolution or absorption of a specimen while it is undergoing a programmed temperature change. A recorder prints out the data as a trace of increase in heat per increase in temperature, versus temperature. The instrument has been utilized to study the curing characteristics and related properties of thermosetting resins.

Differential Thermal Analysis. (DTA) An analytical method in which the specimen polymer and an inert reference material are heated concurrently at a linear rate, each having its own temperature sensing and recording apparatus. The thermal-energy changes, either endothermic or exothermic, which occur in the course of heating, are plotted on a "thermogram" which provides data on the chemical and physical transformations that have occurred, such as melting, sublimation, glass transitions, crystal transitions, and crystallization.

Diffusion. The spontaneous mixing of one substance with another resulting from the movement of molecules of each substance through the empty spaces between the molecules of the other substance. Diffusion can occur in and between gases, liquids and solids. This process accounts for the fact that air and other gases will slowly pass through most nonporous plastic films. See also PERMEABILITY.

Diffusion Couple. An assembly of two materials in such intimate contact that each diffuses into the other.

Diglycol Laurate. $C_{11}H_{23}COOC_2H_4OC_2H_4OH$ (diethylene glycol monolaurate) The lauric acid ester of diethylene glycol, used as a plasticizer for ethyl cellulose, cellulose nitrate, polyvinyl butyral and vinyl chloride-acetate copolymers.

Diglycol Ricinoleate. $C_{17}H_{32}(OH)COOC_2H_4OC_2H_4OH$ (diethylene glycol monoricinoleate) A plasticizer for ethyl cellulose and cellulose nitrate.

Di-n-Hexyl Adipate. $(CH_2)_4(COOC_6H_{13})_2$ An important low-temperature plasticizer for synthetic rubbers and several plastics, including CAB, CN, PVC and PVAc.

Di-n-Hexyl Azelate. (DNHZ) $C_6H_{13}OCOC_7H_{14}COOC_8H_{13}$ A plasticizer for cellulose and vinyl resins. It is approved for food contact use, has low volatility, good compatibility, and is used in food packaging films.

Dihexyl Phthalate. See DI(2-ETHYL BUTYL) PHTHALATE.

Dihexyl Sebacate. $(CH_2)_8(COOC_6H_{13})_2$ A plasticizer for vinyl resins.

Dihydrobiethyl Phthalate. A plasticizer for PVC and cellulose nitrate, with partial compatibility with other thermoplastics.

Di-Isobutyl Adipate. $[C_2H_4COOCH_2CH(CH_3)_2]_2$ (DIBA) A plasticizer for most synthetic resins, including cellulosics, PVC and other vinyls. It has been approved by FDA for use in food packaging films. In vinyls, it is a very active solvent and lowers the processing temperature to a degree permitting the reduction or elimination of stabilizers.

Di-Isobutyl Aluminum Chloride. $[(CH_3)_2CHCH_2]AlCl$ A catalyst for the polymerization of olefins.

Di-Isobutyl Azelate. $[(CH_3)_2CHCH_2OCO](CH_2)_7$ A plasticizer for CAB, CN, ethyl cellulose, PMMA, PVC and PVAc.

Di-Isobutyl Ketone. $(CH_3)_2CHCH_2COCH_2CH(CH_3)_2$ (2,6-dimethyl-4-heptanone) A high boiling ketone with moderate solvent power for cellulose nitrate and vinyl copolymers. Having limited solvency for PVC, it is used as a viscosity modifier in organosols.

Di-Isobutyl Phthalate. $C_6H_4[COOCH_2CH(CH_3)_2]_2$ A plasticizer for vinyls, polystyrene and cellulosic plastics.

Di-Isocyanates. Compounds containing two isocyanate (−NCO) groups, used in the production of polyurethane. Many methods have been reported for synthesizing diisocyanates, but the one most widely used is reacting phosgene with an amine in a solvent solution. Toluene diisocyanate (TDI), the most commonly used, is an 80-20 mixture of 2,4- and 2,6-toluene diisocyanate isomers. Also used are diphenylmethane-4,4'-diisocyanate (MDI); a modified toluene diisocyanate; and polymethylene polyphenyl isocyanate (PAPI). The diisocyanates are key ingredients in the production of urethane foams, fibers, coatings and solid elastomers. See also ISOCYANATES, POLYURETHANES.

Di-Isodecyl Adipate. $(C_2H_4COOC_{10}H_{21})_2$ (DIDA) A plasticizer for PVC in lower concentrations, e.g. up to 30 PHR, at which it imparts low temperature flexibility and resistance to lacquer marring. It is used in combination with other phthalate and phosphate plasticizers. DIDA is completely compatible with vinyl chloride-acetate copolymers, CAB with high butyral content, cellulose nitrate, ethyl cellulose, and chlorinated rubber. In polystyrene, it may be used up to 25 PHR.

Di-Isodecyl-4,5-Epoxy-Tetrahydrophthalate. A plasticizer for PVC which also acts as a stabilizer and as a fungistat. It is also compatible with cellulose nitrate, ethyl cellulose, PMMA, polystyrene and other vinyl polymers and copolymers.

Diisodecyl Glutarate. (DIDG) A new plasticizer introduced by the C. P. Hall Co. in 1977, said to have low temperature flexibility properties nearly identical to those of DOA but with lower volatility and greater resistance to soapy water.

Di-Isodecyl Phthalate. $C_6H_4(COOC_{10}H_{21})_2$ (DIDP) A general-purpose plasticizer for vinyl resins, imparting good water resistance and suitable for processing at high temperatures due to its low volatility. It is also compatible with most synthetic resins, e.g. cellulose nitrate, ethyl cellulose and polystyrene.

Diisononyl Phthalate. (DINP) A plasticizer for PVC, Cellulosics and Polystyrene. It has lower volatility than DOP with equivalent low-temperature performance and poorer efficiency.

Di-Iso-Octyl Adipate. $(C_2H_4COOCH_2C_7H_{15})_2$ (DIOA) A primary plasticizer for vinyls, cellulose nitrate, polystyrene, PMMA and ethyl cellulose. Its performance is similar to that of DOA.

Di-Iso-Octyl Azelate. $C_8H_{17}OCOC_7H_{14}COOC_8H_{17}$ (DIOZ) A diester of azelaic acid, used as a plasticizer for cellulosic resins and polymers and copolymers of vinyl chloride. In vinyls it imparts good low temperature properties and other characteristics similar to those of DOZ.

Di-Iso-Octyl Fumarate. $C_{20}H_{36}O_4$ A derivative of fumaric acid, used as a plasticizer for PVC.

Di-Iso-Octyl Isophthalate. $C_6H_4(COOC_8H_{17})_2$ A plasticizer for polyvinyl chloride, also compatible with cellulose nitrate, ethyl cellulose, polystyrene and other vinyl polymers.

Di-Iso-Octyl Monoisodecyl Trimellitate. A plasticizer for cellulosic plastics and vinyl resins.

Di-Iso-Octyl Phthalate. $C_6H_4(COOCH_2C_7H_{15})_2$ A primary plasticizer for PVC, ethyl cellulose, cellulose nitrate and polystyrene. In vinyls, its performance is similar to that of DOP except that it is slightly less volatile than DOP and produces better viscosity characteristics in plastisols. DIOP is FDA approved for food packing materials and medical applications involving contact with water, but not with fats.

Di-Iso-Octyl Sebacate. $C_8H_{17}OOC(CH_2)_8COOC_8H_{17}$ (DIOS) A plasticizer for vinyl and other resins.

Di-Isopropylene Glycol Salicylate. $C_{13}H_{18}O_5$ The di-isopropylene glycol monoester of salicylic acid, used as an ultraviolet absorber in plastics.

Dilatancy. A rheological flow characteristic evidenced by an increase in viscosity with increasing rates of shear. The opposite of *pseudoplasticity*.

Dilatometer. See PYCNOMETER.

Dilauryl Ether. $(C_{12}H_{25})_2O$ (didodecyl ether) A lubricant for plastics processing.

Dilinoleic Acid. $C_{34}H_{62}(COOH)_2$ A dibasic acid used as a modifier in alkyd, nylon and polyester resins.

Diluent. A substance which dilutes another substance. In an organosol, a diluent is a volatile liquid such as naphtha which has little or no solvating effect on the resin, serving to lower the viscosity of the compound, and which is evaporated during processing. In the paint industry, a diluent is any substance capable of thinning out paints, varnishes, etc. The term is also sometimes used for a liquid added to a thermosetting resin to reduce its viscosity; and for an inert powdered substance added to an elastomer or resin merely to increase its volume.

Dilute Solution Viscosity. The viscosity of a dilute solution of a polymer, measured under prescribed conditions, is an indication of the molecular weight of the polymer and can be used to calculate the degree of polymerization. See INHERENT VISCOSITY, INTRINSIC VISCOSITY, K-VALUE, REDUCED VISCOSITY, RELATIVE VISCOSITY, SPECIFIC VISCOSITY.

Dilution Ratio. As used in the surface coatings industry, dilution ratio is the volume ratio of diluent to solvent in a blend of these two constituents that just fails to completely dissolve 8 grams of nitrocellulose in 100 milliliters of the blend. The procedure is described in ASTM D 1720. It is used to determine the most economical but adequate amount of high-cost active solvent in a nitrocellulose lacquer system.

Dimensional Stability. The ability of a plastic part to retain the precise shape in which it was molded, fabricated or cast.

Dimer. (1) A molecule formed by union of two identical simpler molecules. (2) A substance composed of dimers. For example, C_4H_8 is a dimer of C_2H_4.

Dimer Acid. A coined, generic term for high molecular weight dibasic acids which combine and polymerize with alcohols and polyols to make plasticizers, etc. A trimer acid is similar, having three acid groups.

Dimethoxyethyl Adipate. $CH_3OC_2H_4OCO(CH_2)_4COOC_2H_4OCH_3$ A plasticizer for cellulose acetate, cellulose acetate butyrate and cellulose nitrate.

Di(2-Methoxyethyl) Phthalate. $C_6H_4(COOCH_2CH_2OCH_3)_2$ A plasticizer, especially for cellulose acetate, but also compatible with other cellulosics, polystyrene and vinyls.

Dimethyl Acetamide. $CH_3CON(CH_3)_2$ (DMAC) A colorless liquid used as a solvent for resins and plastics, a catalyst and an intermediate.

2-Dimethylamino Ethanol. (DMAE) $(CH_3)_3 NCH_2 CH_2 OH$ A colorless liquid derived from ethylene oxide and dimethylamine, used as a catalyst for urethane foams. It is of low odor, low toxicity and is non-staining.

3-Dimethylaminopropylamine. $(CH_3)_2 NCH_2 CH_2 CH_2 NH_2$ A colorless liquid used as a curing agent for epoxy resins.

Di(Methylcyclohexyl) Adipate. $CH_3 C_6 H_{10} OCO(CH_2)_4 COOC_6 H_{10} CH_c$ A plasticizer compatible with most thermoplastics.

Di(Methylcyclohexyl) Phthalate.   A plasticizer for cellulosics, polystyrene, PVC and other vinyl resins.

Dimethylformamide. $(CH_3)_2 NCOH$ (DMF) A colorless, high boiling solvent for PVC, nylon, polyurethanes, urea-formaldehyde and other resins. Its strong solvent power makes it useful as a solvent booster in coating, printing and adhesive compositions, and in paint strippers.

Dimethyl Glutarate. (DMG) A liquid chemical intermediate introduced commercially in the 1970's as a source of dicarboxylic acid, and with initial applications as a raw material in the production of plasticizers, polyester resins, synthetic fibers, films, adhesives and solvents.

Dimethyl Glycol Phthalate. $C_6 H_4 (COOCH_2 CH_2 OCH_3)_2$ A solvent and plasticizer for cellulosic resins.

Dimethylisobutylcarbinyl Phthalate. $C_6 H_4 [COOCH(CH_3)CH_2 CH(CH_3)_2]_2$ A plasticizer for most common thermoplastics.

Dimethyl Ketone. See ACETONE.

Dimethylol Urea. $CO(NHCH_2 OH)_2$ A colorless crystalline material resulting from the combination of urea and formaldehyde in the presence of salts or alkaline catalysts, representing the first or "A" stage of urea-formaldehyde resin.

Dimethyl Phthalate. $C_6 H_4 (COOCH_3)_2$ A non-toxic plasticizer for most common thermoplastics, except that its compatibility with PVC is limited.

Dimethyl Polysiloxanes. (dimethyl silicones) A family of fluid silicones of the composition $[(CH_3)_2 SiO]_x$, widely used in aerosol mold releases for plastics that are not to be painted. See also SILICONES.

2,2-Dimethyl-1,3-Propanediol. See NEOPENTYL GLYCOL.

Dimethyl Sebacate. $[(CH_2)_4 COOCH_3]_2$ A solvent and plasticizer for cellulosic and vinyl resins, also compatible with most other thermoplastics.

Dimethyl Terephthalate. (DMT) A white, crystalline solid obtained most generally by the oxidation of p-xylene. The carbomethoxy groups of DMT are typical of those attached to a benzene ring, and their ready participation in alcoholysis reactions is the basis for most uses of the material. DMT is used in the production of polyethylene terephthalate and polyester fibers.

Dimple. See SHRINK MARK.

Dinitrobenzene. $C_6H_4(NO_2)_2$ A yellow crystalline material used as a substitute for camphor in celluloid production.

N,N′-Dinitroso-N,N′-Dimethylterephthalamide. A blowing agent which is a weak explosive in powder form and thus unsafe to handle, but is available in desensitized form by treatment with mineral oil under Du Pont's trademark "Nitrosan." This blowing agent is unique in that its low decomposition temperature permits the expansion of vinyl plastisol prior to gelation (200°F). Subsequent fusion at 350°F produces open cell vinyl foam. Closed cell foam may be produced by heating to fusion in a closed mold, releasing the pressure and subsequently heating in an oven at 212°F.

Dinitrosopentamethylenetetramine. (DNPT) A blowing agent widely used for foam rubber, but of limited use in the plastics industry due to its high decomposition temperature and unpleasant residual odor.

Dinitrosoterephthalamide. (DNTA) A chemical blowing agent for vinyls, liquid polyamide resins and silicone rubbers. It is especially noted for its low decomposition exotherm.

Dinking. See DIE CUTTING.

Dinonyl Adipate. $(CH_2)_4(COOC_9H_{19})_2$ (DNA) An ester of nonyl alcohol, used as a plasticizer for cellulosic, acrylic, styrene and vinyl polymers.

Dinonyl Phthalate. $C_6H_4(COOC_9H_{19})_2$ (DNP) A general-purpose plasticizer for vinyl resins, with low volatility and good electrical properties.

DINP. Abbreviation for DIISONONYL PHTHALATE, which see.

DIOA. Abbreviation for DI-ISO-OCTYL ADIPATE, which see.

Dioctyl-. See Di(2-ETHYL HEXYL)- for several compounds for which this term is more commonly used.

Di-n-Octyl, n-Decyl Adipate. (DNODA) See OCTYL DECYL ADIPATE.

Di(n-Octyl, n-Decyl) Phthalate. (DNODP) A plasticizer for PVC and several other thermoplastics.

Dioctyl Ether. $(C_8H_{17})_2O$ A lubricant used in plastic molding and processing.

Dioctyl Fumarate. $C_2H_2(COOC_8H_{17})_2$ A plasticizer for vinyl resins.

Di(2-Octyl) Phthalate. See DICAPRYL PHTHALATE.

Dioctyl Terephthalate. (DOTP) An ester plasticizer prepared by reacting terephthalic acid with an alcohol containing 10 to 12 carbon atoms. Although the physical properties of DOP and DOTP are similar, DOTP is less volatile, imparts slightly low temperature flexibility, and is more resistant to lacquer marring.

Di-n-Octyltin S, S'(Isooctyl Mercaptoacetate). A stabilizer for PVC (Argus Mark OTM) which has been approved for use in food grade bottles up to 3 PHR when made to certain purity specifications.

Di-n-Octyltin Maleate Polymer. A stabilizer for PVC (Argus Mark OTS) which has been approved for use in food grade bottles up to 3 PHR when made to certain purity specifications.

Dioctyltin Stabilizers. See ORGANOTIN STABILIZERS.

Diol. A term sometimes used for a *dihydric alcohol,* that is an alcohol containing two hydroxyl (OH) radicals.

Diolefin. See DIENE POLYMERS.

DIOP. Abbreviation for DI-ISO-OCTYL PHTHALATE, which see.

DIOS. Abbreviation for DI-ISO-OCTYL SEBACATE, which see.

1,4-Dioxane. $OCH_2CH_2OCH_2CH_2$ (diethylene ether; 1,4-diethylene dioxide; diethylene dioxide; dioxyethylene ether) A solvent for cellulose acetate and other plastics.

DIOZ. Abbreviation for DI-ISO-OCTYL AZELATE, which see.

Dip Coating. A coating process wherein the object to be coated, preheated or cold depending on the nature of the coating material, is dipped into a tank of fluid resin, solution or dispersion, withdrawn and subjected to further heat or drying to solidify the deposit. See FLUIDIZED BED COATING for a similar process employing powdered resins.

Dipentene. $C_{10}H_{16}$ (cinene, limonene) A colorless liquid with a lemon-like odor, used as a solvent for cumar and alkyd resins.

Dip Forming. (dip molding) A process similar to dip coating, except that the fused, cured or dried deposit is stripped from the dipping mandrel. As most frequently used for making vinyl plastisol articles, the process comprises dipping a preheated form shaped to the desired inside dimensions of the finished article in plastisol, which gels in a layer of desired thickness against the form surface; withdrawing the coated form; heating the deposit to fuse the layer; cooling and stripping off the deposit. Some articles may be inverted after stripping so that the shaped inside surface becomes the external surface of the finished article.

Diphenylamine. $(C_6H_5)_2NH$ (DPA) A crystalline solid, used as a stabilizer for several plastics.

Diphenyl Decyl Phosphite. $(C_6H_5O)_2POC_{10}H_{21}$ A nearly colorless liquid used as a stabilizer for vinyl and polyolefin resins.

Diphenyl Mono-o-Xenyl Phosphate. $(C_6H_5O)_2(C_6H_5C_6H_4O)PO$ A plasticizer for cellulosic plastics, polystyrene and, with limited compatibility, vinyls.

Diphenyl Octyl Phosphate. A flame-retardant plasticizer for PVC and cellulosic resins. It has been approved by FDA for use in food packaging.

Diphenyl Phthalate. $C_6H_4(COOC_6H_5)_2$ (DPP) A monomeric ester powder which melts at 156° F. It is used as a solid plasticizer for rigid PVC, cellulosic and other resins.

Diphenyl-Xylenyl Phosphate. $[(CH_3)_2C_6H_3O](C_6H_5O)_2PO$ A plasticizer for CAB, cellulose nitrate, polystyrene, PVC and vinyl chloride-vinyl acetate copolymers.

Dipole. (1) A combination of two electrically or magnetically charged particles of opposite polarity which are separated by a small distance. (2) Any system of charges, such as a circulating electric current, having the properties (a) no forces act on it in a uniform field; (b) a torque proportional to sin θ, where θ is the angle between the dipole axis and a uniform field, does act on it; (c) it produces a potential which is proportional to the inverse square of the distance from it.

Dip Molding. See DIP FORMING.

Dipropylene Glycol. $(CH_3CHOHCH_2)_2O$ A high-boiling glycol with a low order of toxicity, widely used as a solvent and chemical intermediate. As a solvent it is used with cellulose acetate, cellulose nitrate, and is one of the few known solvents for polyethylene. Thus it is used in screening tests to identify polyethylene. As an intermediate, dipropylene glycol reacts with dibasic acids to form alkyd resins, polyester plasticizers and urethane foam intermediates.

Dipropylene Glycol Dibenzoate. $C_6H_5CO_2CH_2CH(CH_3)OCH_2CH(CH_3)OCOC_6H_5$ A plasticizer for PVC, also compatible with most common thermoplastics, imparting good stain resistance.

Dipropylene Glycol Monosalicylate. (Salicyclic acid, dipropylene glycol monoester) A light colored oil used in ultraviolet screening agents and plasticizers.

Dipropyl Ketone. $(CH_3CH_2CH)_2O$ A stable, colorless liquid used as a solvent for many resins.

Dipropyl Phthalate. $C_6H_4(COOC_3H_7)_2$ A plasticizer for cellulose acetate and cellulose acetate butyrate.

Direct Gate. A gate which has the same cross section as that of the runner.

Disc and Cone Agitators. Mixing devices comprising discs or cones rotating at speeds between 1200 and 3600 rpm or higher, which displace fluid contacting their surfaces by centrifugal force. They are used in preparing pastes and dispersions.

Disc Feeders. Horizontal, flat or grooved discs installed at the bottom of hoppers feeding continuous extruders to control the feed rate by varying the speed or rotation of the disc, or by varying the clearance between the disc and the scraper which removes material from it.

Disc Gate. See DIAPHRAGM GATE.

Discoloration. Any change from an initial color possessed by a plastic; a lack of uniformity in color where color should be uniform over the whole area of a plastic object. In the latter sense, where they can be applied, use the more definite terms *mottle, segregation,* or *two-tone.* (ASTM D 883-65T). Discoloration can be caused by overheating, exposure to light, irradiation, or chemical attack.

Dished. Showing a symmetrical distortion of a flat or curved section of a plastic object, so that as normally viewed, it appears concave or more concave. See also WARP. (ASTM D 883-65T).

Dispersant. (n) In an organosol, a liquid component which has a solvating or peptizing action on the resin, so as to aid in dispersing and suspending it.

Disperse Phase. In a SUSPENSION, which see, the disperse phase refers to the particles of solid material dispersed in the liquid medium. The liquid medium is called the *continuous phase.*

Dispersing Agents. Materials added to a suspending medium to promote and maintain the separation of discreet, fine particles of solids or liquids. They are used in the grinding of pigments and for dispersing water-insoluble dyes.

Dispersion. A two-phase or multi-phase system comprising a finely divided material distributed in another material. Types of dispersions are *emulsions* (liquids in liquids), *suspensions* (solids in liquids), *foams* (gases in liquids), and *aerosols* (liquids in gases). In the plastics industry, the term *dispersion* usually denotes a finely divided solid dispersed in a liquid or in another solid. Examples are fillers and pigments in molding compounds, plastisols and organosols.

Dispersion Resins. A special type of PVC resins with very small spherical particles, usually one micron or less in diameter, permitting them to be mixed with plasticizers by simple stirring techniques. They are used in compounding plastisols and organosols.

Displacement Angle. In filament winding, the advancement distance of the winding ribbon on the equator after one complete circuit.

Dissipation Factor. (Electrical) The ratio of the conductance of a capacitor in which the material is the dielectric to its susceptance; or the ratio of its parallel reactance to its parallel resistance. Most plastics have a low dissipation factor, a desirable property because it minimizes the waste of electrical energy as heat.

Dissipation Factor. (Mechanical) The ratio of the loss modulus to the modulus of elasticity.

Distearyl Ether. $(C_{18}H_{37})_2O$ (dioctadecyl ether) A mold lubricant used in plastics processing.

Distearyl Thiodipropionate. (DSTDP) 3,3'-dioctaldecylthiodipropionate, $(C_{18}H_{37}OOCCH_2CH_2)_2S$ An antioxidant widely used in polyolefins, particularly in synergistic combination with a phenolic antioxidant.

Di-Tertiary Butyl Peroxide. $(CH_3)_3COOC(CH_3)_3$ A stable liquid used as a high temperature polymerization catalyst for a variety of olefin and vinyl monomers, e.g. ethylene, styrene and styrenated alkyds.

Ditetrahydrofurfuryl Adipate. $(C_4H_7O\cdot CH_2OCOH_2CH_2)_2$ A plasticizer for cellulose acetate butyrate.

Di-Tridecyl Phthalate. $C_6H_4(COOC_{13}H_{27})_2$ (DTDP) A primary plasticizer for PVC, also compatible with cellulose nitrate, ethyl cellulose, CAB and polystyrene. In vinyls, it imparts high temperature resistance, resistance to extraction by hot soapy water, excellent flexibility retention, and anti-fogging properties.

Ditridecyl Thiodipropionate. $(C_{13}H_{27}OOCCH_2CH_2)_2S$ A stabilizer, plasticizer and softening agent.

Di-Undecyl Phthalate. (DUP) A plasticizer introduced in 1970, based on a C-11, essentially straight chain alcohol. It is characterized by low volatility and good low temperature properties compared to other phthalates. Early applications are expected to be in the wire and cable insulation field.

Divergent Die. A die for hollow articles in which the internal channels leading to the die orifice are diverging.

Divinyl B. See BUTADIENE.

Divinylbenzene. $C_6H_4(CH:CH_2)_2$ (DVB, vinylstyrene) A monomer derived from styrene, used in the production of ion exchange resins, synthetic rubbers and casting resins. It is often used along with styrene as a reactive monomer in the production of polyester resins, to which it imparts higher cross-linking and superior chemical resistance.

Di-ortho-Xenyl Phenyl Phosphate. $(C_6H_5C_6H_4O)_2(C_6H_5O)PO$ A plasticizer for cellulosics, polystyrene and vinyls.

DMEP. Abbreviation for DI(2-METHOXYETHYL) PHTHALATE, which see.

DMF. Abbreviation for DIMETHYLFORMAMIDE, which see.

DMG. Abbreviation for DIMETHYL GLUTARATE, which see.

DMP. Abbreviation for DIMETHYL PHTHALATE, which see.

DMT. Abbreviation for DIMETHYL TEREPHTHALATE, which see.

DNA. Abbreviation for DINONYL ADIPATE, which see.

DNHZ. Abbreviation for DI-n-HEXYL AZELATE, which see.

DNODA. Abbreviation for di-n-octyl, n-decyl adipate. See OCTYL DECYL ADIPATE.

DNODP. Abbreviation for DI(n-OCTYL, n-DECYL) PHTHALATE, which see.

DNP. Abbreviation for DINONYL PHTHALATE, which see.

DNPT. Abbreviation for DINITROSOPENTAMETHYLENETETRAMINE, which see.

DNTA. Abbreviation for DINITROSOTEREPHTHALAMIDE, which see.

DOA. Abbreviation for dioctyl adipate or DI(2-ETHYLHEXYL) ADIPATE, which see.

Doctor. (v) To spread a coating on a substrate in a layer of uniform, controlled thickness.

Doctor Bar. (doctor blade, doctor knife) A flat bar used for regulating the amount of liquid material on the rollers of a coating machine, or to control the thickness of a coating after it has been applied to a substrate.

Doctor Roll. A roll which operates at a different speed or in the opposite direction as compared to the primary roll of a coating machine, thus regulating the uniformity and thickness of material on the roll before it is applied to a substrate.

4-Dodecycloxy-2-Hydroxy-Benzophenone. An ultraviolet inhibitor for polyethylene and polypropylene, also suggested as suitable for PVC, polystyrene, polyesters, and surface coatings such as those based on CAB, CN and acrylic resins.

Doily. In filament winding, the planar reinforcement that is applied to a local area between windings to provide extra strength in an area where a cut-out is to be made; for example, port openings.

DOIP. Abbreviation for dioctyl isophthalate. See DI(2-ETHYL HEXYL) ISOPHTHALATE.

DOP. Abbreviation for dioctyl phthalate or DI(2-ETHYLHEXYL) PHTHALATE, which see.

Dope. A solution of cellulosic plastic, usually cellulose nitrate, used for treating fabrics.

DOS. Abbreviation for dioctyl sebacate. See DI(2-ETHYLHEXYL) SEBACATE.

Dosimeter. An apparatus worn by workers to measure the amount of an environmental agent such as radioactivity, noise or gases of questionable toxicity to which they are exposed during a working shift or other unit of time.

DOTP. Abbreviation for DIOCTYL TEREPHTHALATE, which see.

Double Bond. A type of molecular structure in which a pair of valence bonds joins a pair of carbon or other atoms, or a covalent linkage in which atoms share two pairs of electrons. Double bonds are represented in formulas by the symbols "=" or ":", as in ethylene ($H_2C=CH_2$) or ($H_2C:CH_2$).

Doubler. In filament winding, a local area with extra wound reinforcement, either wound integrally with the part or wound separately and fastened to the part.

Double Ram Press. A press for injection or transfer molding in which two distinct systems of the same kind (hydraulic or mechanical), or of a different kind, create respectively the injection or transfer force and the clamping force.

Double-Screw Extruders. See EXTRUDER, TWIN SCREW.

Double-Shot Molding. A process for production of two-color or two-component parts by means of successive molding methods. The basic process includes the steps of injection molding one part, transferring this part to a second mold as an insert, and molding the second component against the first. Examples of parts made by double-shot molding are typewriter keys, pushbuttons, telephone dials, and other such products in which indicia must remain permanently legible. In a more recent modification of the process, cups and the like with differently colored insides and outsides are made automatically by means of a machine equipped with two injection molders and a swinging platen carrying two cup cores indexed around a central tie-bar, bringing the molds into position for each successive shot. This arrangement permits simultaneous molding of both shots. Other terms sometimes used for the process are *two-shot molding, insert molding, two-color molding* and *over-molding.*

Dough. (dough molding compound) A term sometimes used for a reinforced plastic mixture of dough-like consistency in an uncured or partially cured state. A typical dough molding compound consists of polyester resin, glass fiber, calcium carbonate, lubricants and catalysts. The compounds are formed into products by hand lay-up processes and compression molding.

Dowel. (n) (Dowel pin) A pin used to maintain alignment between two or more parts of a mold.

Dowel Bushing. A hardened steel insert in the portion of a mold which receives the dowel pin.

DOZ. Abbreviation for dioctyl azelate. See DI(2-ETHYLHEXYL) AZELATE.

DPA. Abbreviation for DIPHENYLAMINE, which see.

DPCF. Abbreviation for diphenyl cresyl phosphate, a plasticizer.

DPOF. Abbreviation for DIPHENYL OCTYL PHOSPHATE, which see.

Draft. A slight taper in a mold wall designed to facilitate removal of the molded object from the mold. When the taper is in the opposite direction, tending to impede removal of the article, the term *back draft* is employed.

Drag Flow. In the metering section of an extruder screw, drag flow is the component of total material flow caused by the relative motion between the screw and the cylinder; the volumetric forward displacement of the material in the screw channel. The extruder output is equal to the drag flow less the sum of the PRESSURE FLOW and the LEAKAGE FLOW, both of which see.

Drape. A term used with reference to plastic films and coated fabrics, denoting their ability to form graceful folds when hung as draperies, shower curtains and the like.

Drape Assist Frame. In sheet thermoforming, a frame made from thin wires or thick bars shaped to the peripheries of the depressed areas of the mold and suspended above the sheet to be formed. During forming, the assist frame drops down, drawing the sheet tightly into the mold and thereby preventing webbing between high areas of the mold and permitting closer spacing in multiple molds.

Drape Forming. (drape vacuum forming, drape thermoforming) The method of forming a thermoplastic sheet into three-dimensional articles in which the sheet is clamped in a movable frame, heated, and lowered to drape over the high points of a male mold. Vacuum is then applied to complete the forming operation. See also THERMOFORMING.

Drawdown. In extrusion, the process of pulling the extrudate away from the die at a linear speed higher than at which the melt is emerging, thus reducing the cross section dimensions of the extrudate. The term is also used in the blow molding industry to denote the decrease in parison diameter and wall thickness due to gravity.

Draw Down Ratio. In extrusion or fiber spinning, the ratio of the thickness of the die opening to the final thickness of the product. See also DRAW RATIO.

Drawing. The process of stretching a thermoplastic filament, sheet or rod to reduce its cross-sectional area and/or to improve its physical properties by ORIENTATION, which see.

Draw Ratio. A measure of the degree of stretching during the orientation of a fiber or filament, expressed as the ratio of the cross-sectional area of the undrawn material to that of the drawn material.

Draw Resonance. A phenomenon occurring in extrusion processes in which the extrudate is drawn into a quenching bath at a certain critical speed which creates a cyclic pulsation in the cross sectional area of the extrudate. The pulsation increases with increasing drawing speed until it eventually breaks at the interface of the cooling medium and air. This phenomenon has been noted in the extrusion of polypropylene, polyethylene and polystyrene.

Drop-Weight Test. An impact resistance similar to the Izod test, except that the weights are dropped on the specimen from varying heights.

Drum Coloring. See DRY COLORING.

Drum Tumbler. A device used to mix plastic pellets with color concentrates and/or regrind. The materials are charged into cylindrical drums which are tumbled end-over-end or rotated about an inclined axis for a time sufficient to thoroughly blend the components.

Dry Blend. (n) A dry, free-flowing mixture of resin with plasticizers and other additives prepared by blending the components under high shear at temperatures below the fluxing point. Dry blends are generally more economical than molding powders and pellets made by plasticating and extrusion, but in many cases are not as easily processed.

Dry Coloring. The process of adding colorants to molding compounds and resins in particulate form by tumble blending them with dyes, pigments or color concentrates. This process enables custom molders or extruders to carry a large inventory of uncolored compound, preparing smaller batches of colored compounds to customers' specifications.

Dry Laminate. A laminate containing insufficient resin for complete bonding of the reinforcement.

Dry Layup. The process of construction of a laminate by the layering of preimpregnated, partly cured reinforcements in or on a mold, usually followed by bag molding or autoclave molding.

Dry Spinning. See SPINNING.

Dry-Spot. An imperfection in reinforced plastics, an area of incomplete surface film where the reinforcement has not been wetted with resin. (ASTM D 883-75a).

Dry Strength. The strength of an adhesive joint determined immediately after drying under specified conditions or after a period of conditioning in a standard laboratory atmosphere. See also WET STRENGTH.

Dry Winding. A term used to describe filament winding using pre-impregnated roving, as differentiated from wet winding in which unimpregnated roving is pulled through a resin just prior to winding on a mandrel.

DTA. Abbreviation for DIFFERENTIAL THERMAL ANALYSIS, which see.

DTDP. Abbreviation for DITRIDECYL PHTHALATE, which see.

Ductility. The amount of plastic strain that a material can withstand before fracture. Ductile materials generally have a yield point in the stress-strain curve.

DUP. Abbreviation for DI-UNDECYL PHTHALATE, which see.

Duplicate Cavity-Plate. A removable plate that retains cavities, used where two-plate operation is necessary for loading inserts, etc.

Durene. (Durol; 1,2,4,5-tetramethylbenzene) $C_{10}H_{14}$ A substance occurring in coal tar, usually prepared from xylene and methyl chloride in the presence of $AlCl_3$. It has been patented (U.S. 4,000,120) as an additive to make packaging films of polyolefins and polystyrene photodegradable under direct action of sunlight.

Durometer. An instrument used for measuring the hardness of a material. See also INDENTATION HARDNESS.

Durometer Hardness. See INDENTATION HARDNESS.

Dutch Liquid. (dutch oil) See ETHYLENE DICHLORIDE.

DVB. Abbreviation for DIVINYLBENZENE, which see.

Dwell. (1) A pause in the application of pressure to a mold, made just before the mold is completely closed, to allow the escape of gas from the molding material. (2) In filament winding, the time that the traverse mechanism is stationary while the mandrel continues to rotate to the appropriate point for the traverse to begin a new pass.

Dye. An intensely colored substance which imparts color to a substrate to which it is applied. Retention of the dye in the substrate may be by means of adsorption, solution, mechanical bonding, or by ionic or covalent chemical bonding. The dye substance is devoid of crystal structure. Dyes used for coloring plastics usually dissolve in the plastic melt, as opposed to PIGMENTS (which see), which remain dispersed as undissolved particles.

Dynamic Stress Relaxometer. An instrument which measures the relaxation response of an elastomeric material to a prescribed shear deformation at any desired temperature from room temperature to 200°C. Basic elements of the instrument are a cone-shaped stator cavity and a rotor conical plunger, both electrically heated. The sample is placed in the stator, which rises to a position to form a constant specimen thickness, forcing out the excess material. After heating to the desired temperature, the rotor is rotated for a short time through a small angle, resulting in a shear deformation which increases within the time and then remains constant. The load variation, which results from the sample relaxation, is measured as a torque-time curve.

Dynamic Viscosity. See ABSOLUTE VISCOSITY.

Dyne. In the old cgs system, the force necessary to give acceleration of one centimeter per second to one gram of mass. In the new SI, the dyne is replaced by 1.000 000*E-05 newtons.

Dypnone. $C_6H_5COCHC(CH_3)C_6H_5$ (phenyl alpha-methyl-styryl ketone; 1,3-diphenyl-2-buten-1-one) A plasticizer and U.V. light absorber.

EC. Abbreviation for ETHYL CELLULOSE, which see.

E-CTFE. See POLYETHYLENE-CHLOROTRIFLUOROETHYLENE.

Eddy Current. The current induced in a mass of conducting material by a varying magnetic field. Also called *Foucault current.*

EEA. Abbreviation for ETHYLENE ACRYLATE COPOLYMERS, which see.

E-Glass. (electrical glass) A low alkali borosilicate glass. This type is the most widely used in fibers for reinforcing plastics. Its high resistivity makes E-glass suitable for electrical laminates.

EHMW Polyethylenes. Abbreviation for extra-high molecular weight polyethylenes, those with molecular weights in the 250,000 to 1,500,000 range. See also POLYETHYLENES.

Ejector Pin. (ejector sleeve) A rod, pin or sleeve which pushes a molding off of a force or out of a cavity of a mold. It is attached to an ejector bar or plate which can be actuated by the ejector rod(s) of the press or by auxiliary hydraulic or air cylinders.

Ejector Pin Retainer Plate. A retainer into which the ejector pins are assembled.

Ejector Plate. A plate which backs up the ejector pins and holds the ejector assembly together.

Ejector Return Pins. Projections that push back the ejector assembly as the mold closes. Also called *return pins, surface pins, safety pins,* and *position pushbacks.*

Ejector Rod. A bar that actuates the ejector assembly when a mold is opened.

Elastic Deformation. A change in dimensions of an object under load that is fully recovered when the load is removed. That part of the total strain in a stressed body which disappears upon removal of the stress. See also PLASTIC DEFORMATION.

Elasticity. The ability of a material to quickly recover its original dimensions after removal of a load that has caused deformation. When the deformation is proportional to the applied load, the material is said to exhibit *Hookean elasticity* or *ideal elasticity.*

Elasticizer. A term sometimes used for a compounding additive which contributes elasticity to a resin. For example, chlorinated polyethylenes and chlorinated copolymers of ethylene and propylene are blended with PVC compositions for this purpose.

Elastic Limit. The greatest stress which a material is capable of developing without any permanent strain remaining upon complete release of the stress.

Elastic Melt Extruder. See EXTRUDER, ELASTIC MELT.

Elastic Memory. A characteristic of certain plastics evidenced by their tendency upon reheating to revert to a shape or dimension previously existing during their manufacture. For example, a film which has been stretched or oriented under certain conditions will, upon reheating, return to its unstretched condition due to elastic memory.

Elastic Modulus. See MODULUS OF ELASTICITY.

Elastic Nylon. See NYLON 6/10.

Elastic Recovery. That fraction of a given deformation which behaves elastically. A perfectly elastic material has an elastic recovery of 1; a perfectly plastic material has an elastic recovery of 0. Elastic recovery is an important factor in films used for stretch packaging, because it relates directly to the ability of a film to hold a load together. Retention of the elastic recovery factor over a period of time is also important.

Elastodynamic Extruder. See EXTRUDER, ELASTIC MELT.

Elastomer. A material which at room temperature can be stretched repeatedly to at least twice its original length and, upon immediate release of the stress, will return with force to its approximate original length. This definition is one criterion by which materials called plastics in commerce are separated from elastomers and synthetic rubbers. Another criterion is that, unlike thermoplastics which can be repeatedly softened and hardened by heating and cooling without substantial change in properties, most elastomers are given their final properties by mastication with fillers, processing aids, antioxidants, curing agents, etc., followed by vulcanization (curing) by heating. Polymers usually considered to be elastomers, at least in some of their forms, are listed below.

Abbreviation	Common or Trade Name	Chemical Name
NBR	Buna N, Nitrile Rubber	Butadiene-acrylonitrile copolymer
–	Plaskon CPE	Chlorinated polyethylene
CR	Neoprene	Chloroprene polymers
CSM	Hypalon (HYP)	Chlorosulfonyl polyethylene
–	Hytrel	Copolyester
EOT	Thiokol B	Ethylene ether polysulfide
ET	Thiokol A	Ethylene polysulfide
EPM	EP Elastomer	Ethylene propylene copolymer
EPDM	EP Elastomer	Ethylene propylene terpolymer
FPM	Viton, Fluorel, Kel-F	Fluorinated hydrocarbon
FVSi	Silastic LS	Fluorosilicone
IIR	Butyl, chloro-butyl rubber	Isobutylene-isoprene
Si	Silicone rubber	Organopolysiloxane
ABR	Acrylic rubber	Polyacrylate
BR	CBR, PBd	Polybutadiene
CO	Hydrin (CO, ECO0	Polyepichlorohydrin
NR	Natural rubber	Polyisoprene, natural
IR	Synthetic rubber	Polyisoprene, synthetic
AU	Urethane rubber (UR)	Polyurethane (polyester)
EU	Urethane rubber (UR)	Polyurethane (polyether)
SBR	GR-S, Buna S, Solprene	Styrene-butadiene copolymer
TPO		PE-Butyl graft copolymer
	Kraton "G"	SBS Block Copolymer
TPR®	Thermoplastic Rubber®	
	Telcar	Polyolefin
TNP	Thermoplastic Nordal	
	Profax SB814	Olefinic
	X-414	Polyisoprene

Elastomeric Adhesives. See ADHESIVES.

Elayl. See ETHYLENE.

Electrets. Disks of polymeric material which have been electrically polarized so that one side has a positive charge and the other side a negative charge, much like permanent magnets. Electrets may be formed of poor conductors such as PMMA, polystyrene, nylon, and polypropylene, by heating and cooling them in the presence of a strong electromagnetic field.

Electrically Conductive Plastics. Business machine housings, structural components and static control accessories often require plastics that have some degree of electrical conductivity. Additives that impart such conductivity are metallic powders, carbon black, carbon fibers and mats, and (more recently) metallized glass fibers and spheres.

Electrical Resistance. See ARC RESISTANCE, BREAKDOWN VOLTAGE, CORONA RE-

SISTANCE, DIELECTRIC, DIELECTRIC CONSTANT, DIELECTRIC STRENGTH, IN-
SULATION RESISTANCE, SURFACE RESISTANCE, SURFACE RESISTIVITY, TRACK-
ING, VOLUME RESISTIVITY.

Electrochemical Equivalent. (of an ion) The mass liberated by the passage of a unit quantity
of electricity.

Electrode. A terminal member in an electrical circuit designed to promote an electrical field
between it and another electrode. In the plastics industry electrodes are used in radio-
frequency heat sealing and surface treating of films. One of the electrodes may be a press
platen or a roll.

Electroformed Mold. A mold made by electroplating a model which is subsequently removed from the metal deposit. The deposit is sometimes reinforced with cast or sprayed
metal backings to increase its strength. Such molds are used in slush casting of vinyl plastisols and other forming processes involving low pressures.

Electroforming. A process used for making molds for plastics processes, usually those employing low or moderate pressures, comprising electroplating a pattern which is usually of
wax or flexible material.

Electroless Plating. The deposition of metals on a catalytic surface from solution without an
external source of current. This process is used as a preliminary step in preparing plastic
articles for conventional electroplating. After cleaning and etching, the plastic surface is
immersed in solutions that react to precipitate a catalytic metal in situ, for example first in
an acidic stannous chloride solution, then into a solution of palladium chloride. Palladium is
reduced to its catalytic metallic state by the tin. Another way of producing a catalytic
surface is to immerse the plastic article in a colloidal solution of palladium followed by
immersion in an accelerator solution. The plastic article thus treated can now be plated with
nickel or copper by the electroless method, which forms a conductive surface which then
can be plated with other metals by the conventional electroplating method.

Electrolysis. When a current i flows for a time t and deposits a metal whose electrochemical
equivalent is e, the mass deposited is: $m=eit$. The value of e is usually given for mass in
grams, i for current in amperes, and t in seconds.

Electromagnetic Adhesive. A mixture of an electromagnetic energy absorbing material and a
thermoplastic of the same composition as the sections to be bonded. The adhesive is applied
in the form of a liquid, ribbon, wire or molded gasket to one of the surfaces to be joined.
The surfaces are placed in contact, then the adhesive is rapidly heated by hysteresis and
eddy currents induced by a high frequency induction coil placed close to the joint. This heat
welds the abutting surfaces together.

Electromechanical Vibration Welding. A welding method developed by Branson Sonic
Power Co. It is similar to VIBRATION WELDING (which see) except that it employs a
spring-suspended vibration table which is "tuned" to the resonance of the table's mass. The
table is driven by two electromagnets on opposite sides, and due to the "tuning" the
conversion of potential energy to kinetic energy takes place with very low inertia. In effect,
the table vibrates as if its mass were close to zero, regardless of its actual weight.

Electromotive Force. The force which causes a flow of current. The electromagnetic unit of potential difference is that against which one erg of work is done in the transfer of electromagnetic unit quantity. The volt is that potential difference against which one joule of work is done in the transfer of one coulomb which is also equal to 10^8 electromagnetic units of potential.

Electron. A small particle having a unit negative electrical charge, a small mass, and a small diameter. Its charge is about 4.80294×10^{-10} absolute electrostatic units; its mass is 1/1837 that of the hydrogen nucleus; and its diameter about 10^{-12} cm. Every atom consists of one nucleus and one or more electrons. Cathode rays and Beta rays are electrons.

Electronic Heating. See DIELECTRIC HEATING.

Electron Spin Resonance (ESR) Spectroscopy. Sometimes called electron paramagnetic resonance (EPR) spectroscopy, ESR is a form of spectroscopy very similar to NMR except that the species studied is an unpaired electron rather than a magnetic nucleus. The magnetogyric frequency of an electron differs greatly from that of nuclei, being in the cm range. The sample is placed in a resonant cavity rather than being surrounded by a coil of wires as in NMR instruments. The energy for the process is created by a special electron tube (Klyston) and requires waveguides rather than coaxial cables for transportation. The principle utility of ESO to organic chemists is that the structure of organic free radicals can be studied because of the presence in the radical of magnetic nuclei, usually hydrogen.

Electronic Treating. A method of rendering printability to a plastic film such as polyethylene which is not ordinarily receptive to printing inks, by oxidizing its surface by passing the film between electrodes and subjecting it to a high voltage corona discharge. See also CASING, FLAME TREATING.

Electron-Volt. (EV) The energy acquired by any charged particle carrying unit electronic charge when it falls through a potential difference of one volt. 1 ev is equal to 1.60207×10^{-12} ergs. Multiples of this unit also in common use are the kilo (10^3), million (10^6) and billion (10^9) ev.

Electrophoretic Deposition. A process somewhat similar to electroplating, used to coat electrically-conductive articles with plastics, deposited from aqueous latices or dispersions, by means of a d.c. current. The cathode may be a non-corrodible metal such as stainless steel, generally serving as the container in which the process is performed. The d.c. potential is usually under 100 volts. The deposited coatings are baked to remove residual water. Among available polymer latices suitable for the process are PVC, polyvinylidene chloride, acrylics, nylons, polyesters, PTFE and polyethylene.

Electroplating On Plastics. Articles of almost any of the common plastics can be plated by conventional processes used for metals, after their surfaces have been rendered conductive by precipitation of silver or other conductive substance. (See ELECTROLESS PLATING). A layer of copper is usually applied first, followed by a final plating of gold, silver, chrome or nickel. ABS resins have been most widely used for articles to be electroplated. Others in commercial use for the process include cellulose acetate, some grades of polypropylene, polysulfones, polycarbonate, polyphenylene oxide, nylon, and rigid PVC. See also METALLIZING.

Electrostatic Fluidized Bed Coating. A process combining elements of the fluidized bed coating method and electrostatic spraying. Pointed electrodes are inserted through the porous bottom of a fluidized bed container. When the bed is aerated in the usual manner, a potential of about 100,000 volts is applied between the electrodes and ground. This charge repels the fluidized plastic particles into the space above the bed, from which they are attracted to a grounded article to be coated. The article may be at room temperature when inserted in the powder bed, the coating temporarily adhered by the electrostatic charge. Subsequent heating is then employed to fuse the coating.

Electrostatic Printing. A printing process under development in 1966, employing electrostatic charges to transfer powdered ink from an electrically charged stencil to a plastic film or sheet. The film to be printed is interposed between a grounded metal plate and the stencil. Areas corresponding to those not to be printed are blocked off on the stencil as in conventional screen printing. The powdered ink is brushed on the far side of the screen, where it receives a charge propelling it toward the grounded plate as an image cloud until intercepted by the film. After-heating is usually required to fuse the ink to the substrate.

Electrostatic Spray Coating. A spraying process which employs electrical charges to direct the path of atomized particles to the work surface. Dry plastic powders are charged with static electricity as they emerge from a spray gun, the nozzle of which is attached to the negative terminal of a high voltage d.c. power supply. The charged particles are attracted to the grounded object, which must be at least slightly electrically conductive. The powder coating is subsequently heated to obtain a smooth, homogeneous layer.

Elongation. In tensile testing, elongation is the increase in length of a specimen at the instant before rupture occurs. *Percentage elongation* is expressed as the increase in distance between two gauge marks at rupture divided by the original distance between the marks, the quotient being multiplied by 100. For example, if original gauge marks are 1 inch apart before stretching and three inches apart at the instant of rupture, the percentage elongation is $\frac{3-1}{1}$ x 100 or 200%.

EMA. Abbreviation for copolymers of ethylene and methacrylic acid.

Embedding. The process of encasing an article in a resinous mass performed by placing the article in a mold, pouring a liquid resin into the mold to completely surround the article, curing the resin and removing the encased article from the mold. In the case of electrical components, the lead wires or terminals may protrude from the embedment. The main difference between embedding and potting is that in potting the mold is a container which remains fixed to the resinous mass. See also ENCAPSULATION, IMPREGNATION, POTTING.

Embedment Decorating. A technique for decorating reinforced plastic articles, in which a mat or web of fibrous material printed with the desired design is embedded in the surface of the article and covered with a transparent gel coat. The technique can be adapted to use in hand lay-up, continuous laminating, pultrusion, matched die molding or other reinforced plastics processes.

Embossing. Techniques used to create depressions of a specific pattern in plastics film or sheeting. In the case of cast films, embossing can be accomplished directly by casting on a textured belt or roll. Calendered films are frequently embossed by rollers just after the calendering process. Other films or coated fabrics can be embossed subsequent to manufacture by reheating and passing through embossing rollers, or compressing between plates. Extruded sheets, up to 1/8″ or thicker, are commonly embossed as the sheets emerge from the extruder.

EMI. Abbreviation for 2-ETHYL-4-METHYLIMIDAZOLE, which see.

Emulsifying Agent. A substance used to assist the formation of an EMULSION, which see, from two or more immiscible liquids, and/or to promote the stability of the emulsion. Emulsifiers act as surface-active agents, of which they are a sub-class, in reducing interfacial tensions between the two phases. They also act as protective colloids to promote stability.

Emulsion. The strict definition of an emulsion is a two-phase, substantially permanent mixture of two incompletely miscible liquids, one of which is dispersed as finite globules in the other. However, in plastics and other industries the term is sometimes broadened to include colloidal dispersions of solids such as waxes and resins in liquids. In an emulsion, the liquid which is broken up into globules is known as the *dispersed, discontinuous* or *internal phase.* The surrounding liquid is called the *continuous* or *external phase.* The dispersed phase may be held in suspension by means of mechanical agitation or by the addition of small amounts of additional substances known as emulsifying agents.

Emulsion Polymerization. A polymerization process in which the monomer or mixture of monomers is emulsified in a low viscosity aqueous medium by means of soaps and/or surface active solubilizing and emulsifying agents. The emulsion does not require intensive stirring as in suspension polymerization, and produces polymers of higher molecular weight than those produced by bulk or suspension polymerization. The polymers remain in emulsion, and must be recovered from the latex by freezing or chemical precipitation. The polymerization medium usually contains also a water-soluble initiator, catalyst or chain transfer agent. Examples of polymers produced by emulsion polymerization are PVAc, ABS, PVC, PE, PMMA, polystyrene, and acrylics.

Enamel. A dispersion of pigment in a liquid which forms a solid adherent film on the surface to which it is applied by means of oxidation, polymerization or other chemical reaction. The liquid vehicle of an enamel usually contains a thermosetting resin and a solvent. An initial soft film is formed by evaporation of the solvent, then the film hardens or cures at room temperation or during baking.

Enantimorphs. Molecules that are identical in every way except that one is the mirror image of the other.

Enantiomeric Configurational Unit. Either of two configurational units that are mirror images at the plane containing the main-chain bonds. (ISO).

Enantiotropic. Crystal forms capable of existing in reversible equilibrium with each other.

Encapsulation. The process of applying a conformal coating, that is one conforming approximately in external shape to the shape of the article being encapsulated. The coating, of either thermoplastic or thermosetting resins, may be applied by brushing, dipping, spraying or thermoforming. The process is widely used for the protection and insulation of electrical components. See also EMBEDDING, IMPREGNATION, POTTING,MICRO-ENCAPSULATION.

Encapsulization. The enclosure of adhesive particles with a protective film which prevents adhesive particle coalescence until such time as proper pressure or solvation is applied.

End. (1) A strand of roving consisting of a given number of filaments gathered together. The group of filaments is considered to be an end or strand before twisting, and a yarn after the twist has been applied. (2) An individual warp yarn, thread, fiber or roving.

End Groups. The terminal groups at ends of polymer chains. Although they make up a minute portion of the polymer as a whole, the end groups may vary considerably from the chemical structure of the main chain of the polymer molecule to which they are attached, and may exert an effect on the properties of the polymer that is out of proportion to the number of end groups.

ENDO-. A chemical prefix denoting an inner position, for example in a ring rather than in a side chain, or attached as a bridge within a ring. The opposite of EXO-, which see.

Endothermic. Pertaining to a reaction which is accompanied by the absorption of heat, as opposed to *exothermic.*

Endurance Limit. (fatigue limit) The stress level below which a specimen will withstand cyclic stress indefinitely without exhibiting fatigue failure. Rigid, elastic, low damping materials such as thermosetting plastics and some crystalling thermoplastics do not exhibit an endurance limit.

Engineering Plastics. A broad term covering all plastics, with or without fillers or reinforcements, which have mechanical, chemical and thermal properties suitable for use in construction, machine components and chemical processing equipment. Included are ABS, acetal, acrylic, fluorocarbon, nylon, phenoxy, polybutylene, polyaryl ether, polycarbonate, polyether (chlorinated), polyether sulfone, polyphenylene, oxide, polysulfone, polyimide, rigid PVC, polyphenylene sulfide, thermoplastic urethane elastomers, and many other reinforced plastics.

Engraved Roll Coating. See GRAVURE COATING.

Enthalpy. (Heat content) A thermodynamic quantity of heat, equal to the sum of the internal energy of a system plus the product of the pressure-volume work done on the system. The formula for calculating enthalpy is: $H = E + pv$, wherein H = enthalpy or heat content; E = internal energy of the system; p = pressure; v = volume.

Entrance Angle. In an extrusion die, the total included angle of the converging surfaces of the flow channel leading to the land area of the die. This angle is 180° or less.

Entropy. A measure of the unavailable energy in a thermodynamic system, commonly expressed in terms of its changes on an arbitrary scale, the entropy of water at 32° F being zero. The increase in entropy of a body is equal to the amount of heat absorbed divided by the absolute temperature of the body.

Envenomation. The process by which the surface of a plastic close to or in contact with another surface is deteriorated. Note: Softening, discoloration, mottling, crazing, or other effects may occur. (ASTM D 883-65T).

Environmental Stress Cracking. The formation of external or internal cracks in a plastic caused by tensile stresses less than that of its short-time mechanical strength, when such strength has been reduced by ageing or exposure to some environmental condition.

EP. (1) Abbreviation for EPOXY RESINS or EPOXIDES, both of which see. (2) Abbreviation sometimes used for copolymers of ethylene and propylene.

EPDM. Abbreviation for ethylene-propylene-diene monomer. See ETHYLENE PROPYLENE RUBBERS.

EPI-. (1) A prefix denoting an intramolecular bond or the presence of condensed double closed chain nucleus substituted in the 16 positions. (2) Abbreviation for EPICHLORO-HYDRIN, which see.

Epichlorohydrin. CH_2OCHCH_2Cl (chloropropylene oxide, EPI) A solvent for cellulosic and other resins, and one of the key reactants for epoxy resins as they were first made. It is highly reactive with polyhydric phenols such as bisphenol A.

Epichlorohydrin Rubbers. (CO, ECO) A group of elastomers comprising polymers and copolymers of epichlorohydrin, with good high temperature resistance, low temperature flexibility, and resistance to fuels, oils and ozone. The homopolymer (CO) is a saturated, aliphatic polyether with a chloromethyl side chain. The ECO type is a 1:1 mole copolymer of epichlorohydrin and ethylene oxide.

EPM. Abbreviation for ethylene-propylene copolymer. See ETHYLENE PROPYLENE RUBBERS.

Epoxidation. A chemical reaction in which an oxygen atom is joined to an olefinically unsaturated molecule to form a cyclic, three-membered ether. The products of epoxidation are known as *oxirane compounds* or EPOXIDES, which see.

Epoxide Equivalent. The weight of resin in grams which contains one gram equivalent of epoxy.

Epoxides. Compounds containing the oxirane structure, a three-membered ring containing two carbon atoms and one oxygen atom. The most important members are ethylene oxide and propylene oxide.

Epoxidized Soybean Oil. See EPOXY PLASTICIZERS.

Epoxy-. (epoxy group, oxirane group) A prefix denoting an oxygen atom joined to each of two other atoms which are already united in some way, as in -C-O-C- or O .
$$\overset{O}{\underset{-C-C-}{\bigwedge}}$$

beta-(3,4-Epoxycyclohexyl) Ethyltrimethoxy Silane. A silane coupling agent for reinforced polyester, epoxy, phenolic, melamine and many thermoplastics.

Epoxy Foams. Two basic types of epoxy foams are in use. Foamable powder compositions contain the resin, curing agent, blowing agent, wetting agent and an organic compound such as toluene to control the exothermic heat produced in curing. Liquid epoxy foam systems contain the same ingredients, but the curing agent is withheld until all other ingredients have been mixed, to be added just prior to casting. Liquid systems may also contain amine-terminated polyamide resins to impart resiliency to the foam. Epoxy foams are used in casting, potting and encapsulating of electrical components, insulating coatings for chemical storage tanks, and cores in laminates for aircraft and boats.

Epoxy-Novolak Resins. 2-step resins made by reacting epichlorohydrin with phenol formaldehyde condensates. They are also defined as linear, thermoplastic B-stage phenolic resins that are in a partial stage of cure. Whereas normal bisphenol-based epoxy resins contain up to two epoxy groups per molecule, the epoxy-novolaks may have seven or more such groups, producing a more tightly cross-linked structure in the cured resins. Thus, they are stronger and superior in many properties.

Epoxy Plasticizers. (epoxide plasticizers) A family of plasticizers obtained by the epoxidation of vegetable oils or fatty acids. The two main types are (a) epoxidized unsaturated triglycerides, e.g. soy bean oil and linseed oil; and (b) epoxidized esters of unsaturated fatty acids, e.g. oleic acid, or butyl-, octyl- or decyl esters. Most epoxy plasticizers have a heat-stabilizing effect, and they are often used for stabilization in conjunction with other stabilizers. Epoxidized oils in general have good extraction and migration resistant properties and low votality, but they cannot be used as sole plasticizers in unfilled vinyl compounds and hence are not considered to be primary plasticizers. Certain epoxidized soybean oils have been FDA approved for food contact use.

Epoxy Resins. A family of thermosetting resins containing the group

$$\overset{\displaystyle O}{\underset{\displaystyle -C-C-}{\diagup\diagdown}}$$

Originally made by condensing epichlorohydrin and bisphenol A, epoxy resins are now more generally formed from low molecular weight diglycidyl ethers of bisphenol A and modifications thereof; or, as another type, by the oxidation of olefins with peracetic acid. Depending on molecular weight, the resins range from liquids to solid resins. The liquids, used for casting, potting, coating and adhesives, are cured with amines, polyamides, anhydrides or other catalysts. The solid resins are often modified with other resins and unsaturated fatty acids. Epoxy resins are also widely used in the reinforced plastics field, having good adhesion to glass fibers. In 1974 a new family of epoxy-type resins known as *hydantoin epoxy resins* with the following heterocyclic formula was introduced:

These resins are useful in electrical composites because their thermal expansion can be tailored to match that of copper. Their low viscosities are effective in wetting the various reinforcing materials used with them. Later, a rapid curing epoxy resin, diglycidyl ether of 4-methylol-resorcinol (DGEMR), was developed at the Harry Diamond Laboratories in three high-yield steps from resorcinol. The built-in methylol group appears to effectively catalyze the curing reactions. The resin is curable with all types of conventional epoxy hardeners including aliphatic and aromatic amines, anhydrides, and amidoamines. In general, DGEMR cures approximately thirty times faster than a conventional bisphenol A-type epoxy and two to five times faster than the known commercial rapid-gelling epoxies. Usually, a much lower temperature is necessary for a complete cure. The inherent rapid gelation and the relatively low viscosity of DGEMR allow formulation with flexibilizers and fillers without prolonging gel time. These same properties make the resin ideal for adhesives, coatings, and low temperature applications. See also EPOXY-NOVOLAK RESINS.

Epoxy Stabilizers. (epoxide stabilizers) Most EPOXY PLASTICIZERS, which see, also serve as stabilizers due to the ability of the epoxide group to accept hydrochloric acid, or, according to some authors, to serve as an intermediate in the presence of metallic salts in the conversion of hydrogen chloride to a metallic chloride. Epozy stabilizers are most often used in conjunction with barium-cadmium and other stabilizers, with which they have a synergistic effect.

EPR. Abbreviation for ETHYLENE PROPYLENE RUBBER, which see.

Equivalent Weight. (combining weight) The atomic or formula weight of an element or ion divided by its valence. Elements entering into combination always do so in quantities proportional to their equivalent weights. In oxidation-reduction reactions the equivalent weight of the reacting substances is dependent upon the change in oxidation number of the particular substances.

Erosion Breakdown. In an electrical conductor insulation, erosion breakdown is caused by chemical attack of corrosive chemicals such as ozone and nitric acid which are formed by corona discharge from a high voltage cable. This breakdown can occur even in the most chemically resistant polymers such as PTFE after long term exposure to the conditions.

Erucyl Alcohol. $C_{22}H_{43}OH$ A fatty alcohol used as a lubricant on plastics molds.

Erythrene. See BUTADIENE.

E.S.C. Abbreviation for ENVIRONMENTAL STRESS CRACKING, which see.

ESO. Abbreviation for epoxidized soybean oil. See EPOXY PLASTICIZERS.

Ester. An organic compound corresponding in structure to a salt in inorganic chemistry. Esters are formed by reacting an acid with an alcohol, or by the exchange of a replaceable hydrogen atom of an acid for an organic alkyl radical. Esters of many monofunctional alcohols and organic acids are oily liquids, forming an important family of plasticizers. When the alcohol selected is polyfunctional, that is containing two or more reactive groups, and the acid is polyfunctional or dibasic, an infinite number of repeating units can be formed between the alcohol and the acid. The product of such a reaction is called a POLYESTER, which see.

Ester Exchange. A reaction between an ester and another compound in which there occurs an exchange of alkoxy or acyl groups, resulting in the formation of another ester. When an ester is reacted with an alcohol, the process is called *alcoholysis;* reaction between an ester and an acid is called *acidolysis.* Ester exchange reactions are used in the production of plasticizers, polyvinyl alcohol, acrylics, polyesters, and polycarbonates.

ETFE Fluoropolymer. A modified copolymer of ethylene and tetrafluoroethylene, marketed by DuPont under the trade mark Tefzel. ETFE is readily processed by conventional methods, including extrusion and injection molding.It has good thermal properties, abrasion resistance, impact strength, chemical resistance, and electrical properties.

Ethanal. See ACETALDEHYDE.

Ethanol. See ETHYL ALCOHOL.

Ethanolurea. $NH_2 COHHCH_2 CH_2 OH$ A white substance solidifying at $71-74°C$. It condenses with formaldehyde to form permanently thermoplastic, water soluble resins. Simple urea can be incorporated in the condensation reaction to give modified resins with any desired degree of water-solubility and flexibility, both of which properties increase with increasing amounts of simple urea.

Ethene. See ETHYLENE.

Ethenoid Plastics. (1) Plastics made from monomers containing the polymerizable double bond group C:C, for example, ethylene. Thermosetting ethenoid resins are made from monomers or linear polymers capable of giving cross-linked structures as a result of double bond polymerization. (2) A generic term (British) for acrylic, vinyl and styrene plastics.

Ethers. Organic compounds in which an oxygen atom is interposed between two carbon atoms or organic radicals in the molecular structure. They are often derived from alcohols, by elimination of one molecule of water from two molecules of alcohol.

Ethyl Acetanilide. $C_6 N_5 NC_2 H_5 COCH_3$ (ethyl phenylacetamide) A substitute for camphor in the manufacture of celluloid.

Ethyl Acetate. $CH_3 COOC_2 H_5$ (acetic ether, acetic ester, vinegar naphtha) A colorless liquid made by heating acetic acid and ethyl alcohol in the presence of sulfuric acid, and distilling. It is a powerful solvent for ethyl cellulose, polyvinyl acetate, cellulose nitrate, CAB, acrylics, polystyrene and coumarone-indene resins. Although highly flammable, ethyl acetate is the least toxic of industrial solvents.

Ethylacetic Acid. See BUTYRIC ACID.

Ethyl Acrylate. $CH_2 :CHCOOC_2 H_5$ A polymerizable monomer, used for acrylic resins used in paints.

Ethyl Alcohol. (alcohol, grain alcohol, ethanol, ethyl hydroxide) $C_2 H_5 OH$ An alcohol used, in denatured form, as a solvent for ethyl cellulose, polyvinyl acetate and polyvinyl butyrate.

Ethyl Aldehyde. See ACETALDEHYDE.

Ethyl Aluminum Dichloride. $C_2H_5AlCl_2$ A clear, yellow flammable liquid, used as a catalyst for olefin polymerization.

Ethyl Aluminum Sesquichloride. $(C_2H_5)_3Al_2Cl_3$ A catalyst for olefin polymerization.

Ethyl Benzoate. $C_6H_5CO_2C_2H_5$ (benzoic ether) A colorless liquid derived by heating ethyl alcohol and benzoic acid in the presence of sulphuric acid. It is used as a solvent for cellulosics.

Ethyl ortho-Benzoylbenzoate. $C_6H_5COC_6H_4COOC_2H_5$ A yellowish white solid, used as a plasticizer for cellulose nitrate.

2-Ethylbutyl Acetate. $C_2H_5CH(C_2H_5)CH_2OOCCH_3$ A solvent for cellulose nitrate.

Ethyl n-Butyl Ketone. $C_2H_5CO(CH_2)_3CH_3$ (2-heptanone) A stable, colorless liquid with a medium evaporation rate, used in solvent mixtures for cellulosic and vinyl resins. When used in vinyl organosols it imparts good viscosity stability on ageing.

Ethyl Butyrate. $C_3H_7CO_2C_2H_5$ (ethyl butanoate) A solvent for cellulosics.

Ethyl Carbamate. See URETHANE.

Ethyl Cellulose. (EC) An ethyl ether of cellulose formed by reacting cellulose steeped in alkali with ethyl chloride. Being an ether, it is chemically different from other cellulosics, which are esters, and therefore, is not compatible with them. EC resin can be injection molded, extruded, cast into film or used as a coating material. It has the lowest density of all cellulosic plastics, good toughness and shock resistance, and is dimensionally stable over a wide temperature range.

Ethyl Chloride. C_2H_5Cl (chloroethane) A colorless gas at ordinary pressure, used in the production of ethyl cellulose by reaction with sodium cellulose.

Ethyl Citrate. See TRIETHYL CITRATE.

Ethylene. $H_2C{:}CH_2$ (olefiant gas, bicarburetted hydrogen, elayl, ethene) A colorless, flammable gas derived by cracking of petroleum and natural gas. In addition to serving as the monomer for polyethylene, its many uses in the plastics industry include the synthesis of ethylene oxide, ethyl alcohol, ethylene glycol (used in the production of alkyd and polyester resins), ethyl chloride and other ethyl esters.

Ethylene Acrylate Copolymers. (EEA) Resins similar in appearance to low-density polyethylene, but possessing elastomeric properties similar to those of rubber and flexible vinyls. Copolymers containing about 20% of ethyl acrylate are easily processed by injection molding, blow molding, calendering, extrusion and flame spraying.

N,N′-Ethylene Bis Stearamide. (Acrawax c) A lubricant used in ABS, PVC and styrene resins.

Ethylene-Bis Tris (2-Cyanoethyl) Phosphonium Bromide. (ECPB) A flame retardant for thermoplastics. 20% of ECPB in PMMA produced opacity and reduced the burning rate to zero.

Ethylene Carbonate. $(CH_2O)_2CO$ (glycol carbonate, dioxolone-2) A solvent for many polymers and resins.

Ethylene Carboxylic Acid. See ACRYLIC ACID.

Ethylene Chloride. See ETHYLENE DICHLORIDE.

Ethylene-CTFE Copolymer. Marketed by Allied Chemical Corp. under the trademark Halar, this copolymer of ethylene and chlorotrifluoroethylene is said to have properties previously unavailable in a single fluoroplastic. These include mechanical, thermal, electrical, processing and resistance properties.

Ethylene Dichloride. CH_2ClCH_2Cl (sym-dichloroethane; 1,2-dichloroethane; ethylene chloride; Dutch liquid; Dutch oil) A colorless, oily liquid used in the production of vinyl chloride monomer and as a solvent for phenolic and cellulosic resins.

Ethylene Glycol. CH_2OHCH_2OH (ethylene alcohol, glycol) A clear, syrupy liquid used as a solvent for cellulosics, particularly cellophane, and in the production of alkyd resins and polyethylene terephthalate.

Ethylene Glycol Diacetate. $CH_3COOCH_2CH_2OOCCH_3$ A very slowly evaporating solvent for cellulosic and acrylic resins, and sometimes used as a fugitive plasticizer for vinyls and acrylics.

Ethylene Glycol Dibenzoate. $C_{16}H_{14}O$ A plasticizer for cellulosic resins, with limited compatibility for some vinyl resins.

Ethylene Glycol Dibutyrate. $(CH_2OCOC_3H_7)_2$ (glycol dibutyrate) A plasticizer for cellulosic plastics.

Ethylene Glycol Dipropionate. $(CH_2OCOC_2H_5)_2$ (glycol propionate) A plasticizer for cellulosic plastics.

Ethylene Glycol Monoacetate. $HOCH_2CH_2OOCH_3$ (glycol monoacetate) A solvent for cellulose nitrate and cellulose acetate.

Ethylene Glycol Monobenzyl Ether. $C_6H_5CH_2OC_2H_4OH$ A solvent for cellulose acetate.

Ethylene Glycol Monobutyl Ether. $HOCH_2CH_2OC_4H_9$ (2-butoxyethanol) A colorless liquid used as a solvent for cellulosic, phenolic, alkyd and epoxy resins, especially in varnish and other coating formulations.

Ethylene Glycol Monobutyl Ether Acetate. $C_4H_9OCH_2CH_2OOCCH_3$ A colorless liquid with a fruity odor, used as a high-boiling solvent for cellulose nitrate, epoxy resins, and as a film coalescing aid for polyvinyl acetate latex.

Ethylene Glycol Monobutyl Ether Laurate. $C_{11}H_{23}COO(CH_2)_2OC_4H_9$ A plasticizer for cellulosic plastics, polystyrene and vinyls.

Ethylene Glycol Monobutyl Ether Oleate. $C_{17}H_{33}COOCH_2CH_2OC_4H_9$ A plasticizer for cellulose nitrate, ethyl cellulose and PVC.

Ethylene Glycol Monobutyl Ether Stearate. $C_{17}H_{35}COOC_2H_4OC_4H_9$ A plasticizer for cellulose nitrate, ethyl cellulose, polystyrene and polyvinyl butyral.

Ethylene Glycol Monoethyl Ether. $HOCH_2CH_2OC_2H_5$ A solvent for cellulose nitrate, phenolic, alkyd and epoxy resins. It is colorless, nearly odorless, has a low evaporation rate, and imparts good flow properties to coatings.

Ethylene Glycol Monoethyl Ether Acetate. $CH_3COOCH_2CH_2OC_2H_5$ A solvent for cellulose nitrate, ethyl cellulose, vinyl polymers and copolymers, PMMA, polystyrene, epoxy, coumarone-indene and alkyd resins.

Ethylene Glycol Monoethyl Ether Laurate. $C_{11}H_{93}COO(CH_2)_2OC_2H_5$ A plasticizer for cellulose acetate butyrate, cellulose nitrate, ethyl cellulose, polystyrene and vinyl resins.

Ethylene Glycol Monoethyl Ether Ricinoleate. $C_{17}H_{32}(OH)COO(CH_2)_2OC_5H_5$ A plasticizer.

Ethylene Glycol Monomethyl Ether. $CH_3OCH_2CH_2OH$ (2-methoxyethanol) A solvent for cellulose acetate, cellulose nitrate and cellulose acetate butyrate.

Ethylene Glycol Monomethyl Ether Acetate. $CH_3COOCH_2CH_2OCH_3$ A high-boiling solvent for cellulose nitrate, cellulose acetate and CAB resins.

Ethylene Glycol Monomethyl Ether Myristate. A plasticizer for cellulosic plastics, polyvinyl butyral and PVC.

Ethylene Glycol Monomethyl Ether Oleate. $C_8H_{17}CH:CH(CH_2)_7COOC_2H_4OCH_3$ A plasticizer for cellulosic and vinyl resins.

Ethylene Glycol Monomethyl Ether Stearate. $C_{17}H_{35}COOCH_2CH_2OCH_3$ A plasticizer for cellulosics and polystyrene, with limited compatibility for other thermoplastics.

Ethylene Glycol Monooctyl Ether. $C_4H_9CHC_2H_5CH_2OCH_2CH_2OH$ A solvent for cellulose esters, and plasticizer.

Ethylene Glycol Monophenyl Ether. $C_6H_5OCH_2CH_2OH$ A solvent for cellulosics, vinyls, phenolic and alkyd resins.

Ethylene Glycol Monoricinoleate. $C_{17}H_{32}(OH)COO(CH_2)_2OH$ A plasticizer and an intermediate for urethane polymers.

Ethylene Glycol Ricinoleate. $C_{17}H_{32}(OH)COOCH_2CH_2OH$ A plasticizer for cellulose nitrate, ethyl cellulose and polyvinyl butyral.

Ethylene Oxide. CH_2CH_2O A colorless gas at room temperatures, important as a raw material for the production of ethylene glycol, higher alcohols, acrylonitrile, and ethanolamines.

Ethylene Plastics. See POLYETHYLENE.

Ethylene Propylene Rubbers. (EPR, EPM, EPDM) A group of elastomers obtained by the stereospecific copolymerization of ethylene and propylene (EPM), or of these two monomers and a third monomer such as diene (EPDM). Their properties are similar to those of natural rubber in many respects, and they have been proposed as potential substitutes for natural rubber in tires.

Ethylene Urea Resin. A type of AMINO RESIN, which see.

Ethylene-Vinyl Acetate Copolymers. (EVA) Copolymers of major amounts of ethylene with minor amounts of vinyl acetate, that retain many of the properties of polyethylene but have considerably increased flexibility, elongation and impact resistance. They resemble elastomers in many ways, but can be processed like thermoplastics.

Ethyl Formate. $HCOOC_2H_5$ A solvent for cellulose acetate and cellulose nitrate.

2-Ethyl-1,3-Hexanediol. $C_3H_7CH(OH)CH(C_2H_5)CH_2OH$ A stable, colorless, nearly odorless, high-boiling liquid with weak solvent action. In two-package urethane systems, the material acts as a viscosity reducer at room temperature. When the urethane package is heated to cure it, the diol reacts into the urethane matrix to eliminate solvent emissions.

2-Ethyl Hexyl. An eight-carbon radical of the formula $CH_3(CH_2)_3CH(C_2H_5)CH_2$-, often called *octyl* in the plastics industry. For example, the common plasticizer di-2-ethylhexyl phthalate is often referred to as di-octyl phthalate and its abbreviation, DOP.

2-Ethylhexyl Acetate. $CH_3COOCH_2CHC_2H_5C_4H_9$ (octyl acetate) A high-boiling, retarder solvent with low evaporating rate and limited water solubility, used primarily in coating formulations based on cellulose nitrate. It is also used as a dispersant in vinyl organosols.

2-Ethylhexyl Acrylate. $CH_2:CHCOOCH_2CHfd(C_2H_5)C_4H_9$ One of the monomers for acrylic resins, especially for those used in water-base paints.

2-Ethylhexyl Alcohol. $CH_3(CH_2)_3CHC_2H_5CH_2OH$ (2-ethylhexanol, octyl alcohol) A high-boiling specialty solvent with many uses in the plastics industry. As a solvent, it is used in coatings for stenciling, silk screening and dipping. As an intermediate, the alcohol is an important raw material for the production of the 2-ethylhexyl esters of dibasic acids used as plasticizers, such as DOP, DOA and DOZ.

Ethylhexyl-Decyl Phthalate. A plasticizer for cellulosics, polystyrene, PVC and PVAc.

2-Ethylhexyl Epoxytallate. An epoxy ester used mainly as a combination plasticizer and stabilizer in vinyl compounds. At concentrations as low as 5 PHR it reacts synergistically with many metallic stabilizers to provide stability comparable to similar combinations based on epoxidized soybean oils. As a partial replacement for other plasticizers, it imparts good low temperature flexibility. It is also compatible with vinyl chloride-acetate copolymers, CAB resins with a high butyral content, cellulose nitrate, ethyl cellulose, polystyrene and chlorinated rubbers.

Ethylhexyl-Isodecyl Phthalate. $C_6H_4(COOC_8H_{17})(COOC_{10}H_{21})$ (octyl-isodecyl phthalate) A mixed ester of 2-ethylhexyl and isodecyl alcohols, compatible with PVC, vinyl chloride-acetate copolymers, cellulose acetate butyrates with higher butyrate contents, cellulose nitrate, and, in lower concentrations, polyvinyl butyral. In vinyls, it is somewhat less volatile than DOP and has equivalent low temperature properties.

2-Ethylhexyl-p-Oxybenzoate. $HOC_6H_4COOC_8H_{17}$ A plasticizer for polyamides.

Ethyl alpha-Hydroxyisobutyrate. $(CH_3)_2COHCOOC_2H_5$ A solvent for cellulosic nitrate and cellulose acetate.

Ethylidene Acetobenzoate. $C_6H_5COOCH(COCH_3)CH_3$ (ethylidene benzoacetate) A solvent for cellulose nitrate, cellulose acetate and other synthetic resins.

Ethyl Lactate. $CH_3CHOHCOOC_2H_5$ A solvent for cellulosic and other resins.

Ethyl Levulinate. $CH_3CO(CH_2)_2COOC_2H_5$ A solvent for cellulose acetate.

Ethyl Methacrylate. $H_2C:CCH_3COOC_2H_5$ A readily polymerizable monomer used for certain types of acrylic resins.

2-Ethyl-4-Methylimidazole. (EMI) A curing agent for epoxy resins of the types made from epichlorohydrin and Bisphenol A or —F, and for novolac epoxy resins. It is said to produce ease of compounding, long pot life, low viscosity, and non-staining characteristics; and to yield castings with excellent mechanical and electrical properties.

Ethyl Oleate. $C_{17}H_{33}COOC_2H_5$ A solvent, lubricant and plasticizer.

Ethyl Oxalate. $(COOC_2H_5)_2$ (diethyl oxalate) A solvent for cellulosics and many synthetic resins.

Ethyl Phthalate. See DIETHYL PHTHALATE.

Ethyl Phthalyl Ethyl Glycolate. $C_2H_5OCOC_6H_4COOCH_2COOC_2H_5$ A plasticizer compatible with PVC and most common thermoplastics. It has been approved by FDA for food packaging.

Ethyl Propionate. $C_2H_5COOC_2H_5$ (propionic ester) A solvent for cellulose ethers and esters.

N-Ethyl-p-Toluene Sulfonamide. A solid plasticizer for rigid PVC.

EU. Abbreviation for the polyether type of polyurethane rubber.

Eutectic. Pertaining to a specific mixture of two or more substances which has a lower melting point than that of any of its constituents alone or of any other percentage composition of the constituents.

EVA. Abbreviation for ETHYLENE-VINYL ACETATE COPOLYMERS, which see.

EVE. Abbreviation for *ethyl vinyl ether.* See VINYL ETHYL ETHER.

Exa-. (E) The Si-approved prefix for a multiplication factor to replace 10^{18}.

Exo-. A prefix denoting attachment to a side chain rather than to a ring. See also ENDO-.

Exotherm. (1) The temperature/time curve of a chemical reaction giving off heat, particularly the polymerization of casting resins. (2) The amount of heat given off. The term has not been standardized with respect to sample size, ambient temperature, degree of mixing, etc.

Exothermic. Pertaining to a reaction which is accompanied by the evolution of heat. An example in the plastics industry is the isocyanate-polyol reaction by which polyurethanes are made.

Expanded Plastic. See CELLULAR PLASTIC.

Expanding Agents. See BLOWING AGENTS.

Expansivity. See COEFFICIENT OF THERMAL EXPANSION.

Extender. (n) (1) A substance, generally having some adhesive action, added to an adhesive to reduce the amount of the primary binder required per unit area. (See BINDER, FILLER). (2) In plastics compounding, a substance added to the mixture to reduce its cost. The substance may be a resin, plasticizer or filler. See also BLENDING RESIN.

Extensibility. The ability of a material to extend or elongate upon application of sufficient force. It is expressed as a percentage of the original length.

Extensiometer. A rheometer for measuring flow properties of molten polymers with low shear-rate viscosities. In its most simple form, an apparatus known as the *Cogswell Rheometer,* unidirectional tensile force is provided by a dead weight acting through a cam and pulley, is used.

External Plasticizer. A plasticizer which is added to a resin or compound, as opposed to an internal plasticizer which is incorporated in a resin during the polymerization process.

Extraction. The transfer of a constituent of a plastic mass to a liquid with which the mass is in contact. The process is generally performed by means of a solvent selected to dissolve one or more specific constituents, or it may occur as a result of environmental exposure to a solvent.

Extrudate. The product or material delivered from an extruder, for example film, pipe, profiles, wire coatings, etc.

Extruded-Bead Sealing. A method of welding or sealing continuous lengths of thermoplastic sheeting or thicker sections by extruding a bead of the same material between two sections and immediately pressing the sections together. The heat in the extruded bead is sufficient to cause it to weld to the adjacent surfaces.

Extruder. A machine for producing more or less continuous lengths of plastics sections such as rods, sheets, tubes, profiles and cable coatings. Its essential elements are a tubular barrel, usually electrically heated; a revolving screw, ram or plunger within the barrel; a hopper at one end from which the material to be extruded is fed to the screw, ram or plunger; and a die at the opposite end for shaping the extruded mass. Extruders may be divided into three general types: single screw, twin- or multiple screw, and ram; each type in turn having several variations.

Extruder Barrel. (extruder cylinder) A cylindrical steel tube, either nitrided to resist wear or fitted with a special hard alloy liner, which forms the housing around the extruder screw and contains the plastic material as it is conveyed through the extruder. Barrels are usually heated by means of external electrical electrical heater bands, cast-in aluminum heaters, induction heating, or by steam or hot fluids circulating through attached jackets.

Extruder Breaker Plate. A heavy metal plate perforated with many closely-spaced holes, generally 1/8″ to 3/16″ diameter, flared at each end to prevent stagnation of the melt. Its function, along with the EXTRUDER SCREEN PACK (which see) which it supports, is to prevent contaminants from reaching the die and to help stabilize the flow of material. The breaker plate is placed between the end of the screw and the entrance to the adaptor.

Extruder Die. The orifice-containing element mounted at the end of an extruder, which gives the extrudate its final shape. Elements of the die assembly are (1) the die block, (2) an adaptor which attaches the die to the extruder, (3) a manifold, or reservoir, in the die for distributing the melt to the shaping section, (4) in the case of dies for hollow sections, a mandrel inserted in the flow channel to form a hollow extrudate, (5) a spider to hold the mandrel in position, and (6) the land area, which gives the extrudate its final shape. Extrusion dies are classified in four ways according to the flow direction: *Straight (in-line), off-set, angle,* and *crosshead.* In a straight or in-line die the axis of the die discharge channel is the same as the axis of the extruder. In an off-set die the axis of the die discharge channel is parallel to, but not coaxial with the extruder. In an angle die the axis of the die discharge channel is not parallel to that of the extruder. A cross-head die is one in which the axis of the die is perpendicular to the axis of the extruder. Dies are also classified according to type of feed. In a *center feed die* the plastic melt enters in the center of the die manifold, or channel, and the melt flows to the sides of the die. In the *end feed* or *side feed die* the plastic melt enters at one end of the die manifold or channel, and the melt must flow to the opposite side of the die.

Extruder Drive. The system comprising motor, controls, speed regulator, and coupling which drives an extruder screw.

Extruder, Elastic Melt. (Elastodynamic extruder) A type of extruder in which the material is fed into a fixed gap between a stationary and a rotating plate, is melted by frictional heat and flows in a spiral path towards the center of rotation, from which it is discharged through a die. Only rubbery polymers with certain viscoelastic properties are suitable for the process.

Extruder, Hydrodynamic. A device similar to the elastic melt extruder (see EXTRUDER, ELASTIC MELT) in that the principle involves shearing the material between rotating discs. However, the discs in a hydrodynamic extruder are shaped to provide positive driving force independent of the flow properties of the melt.

Extruder, Planetary Screw. A multiple screw device in which a number of satellite screws, generally six, are arranged around one longer central screw. The portion of the central screw extending beyond the satellite screws serves as the final pumping screw as in a single-screw extruder, while the planetary screws permit the discharge of volatiles toward the hopper end. This arrangement is used primarily for processing powders such as dry-blended PVC.

Extruder, Ram. An extruder in which the material is advanced through the barrel and die by means of a ram or plunger rather than by a screw. The ram extruder was the earliest type to be used in the plastics industry, dating back to 1870 when cellulose nitrate was extruded into rods. See also EXTRUDER, RECIPROCATING RAM.

Extruder, Reciprocating Ram. A modification of the old ram extruder employing two units placed end-to-end. Pulsating flow from the alternately operated rams, aided by a valving system, is combined in a single smooth-flowing stream. This type of extruder has been developed to overcome degradation problems with heat-sensitive materials such as PVC.

Extruder Screen Pack. A woven metal wire screen placed between the end of the screw and the breaker plate to prevent contaminants from being passed on to the die, and also to create back pressure to help stabilize the flow of material through the extruder. The pack may comprise several screens of different mesh sizes, e.g. one of 100 mesh for fine screening, backed up by one or more coarser screens to provide strength. Devices known as screen pack changers are available for removing and replacing screen packs without the need for shutting down or purging the extruder.

Extruder Screw. A solid or hollow shaft with a continuous helical channel cut into its surface, usually extending from the feed throat of the extruder barrel to the die end of the screw. The channel varies in depth and helical pitch throughout the length of the screw to accomplish several functions in different zones, e.g. feeding, compressing, venting, mixing and metering of the material. Extruder screws are made of hardened alloy steel, and are usually chrome plated. The Extrusion PAG of SPE has defined various types of extruder screws as follows:

Constant Lead Screw (uniform pitch screw): A screw with a flight of constant helix angle.

Constant Taper Screw: A screw of constant lead and a uniformly increasing root diameter over the full flighted length.

Cored Screw: A screw with a hole in the center of the root for circulation of a heat transfer medium, or installation of a heater.

Decreasing Lead Screw: A screw in which the lead decreases over the full flighted length (usually of constant depth).

Metering Type Screw: A screw which has a metering section.

Multiple Flighted Screw: A screw having more than one helical flight such as double flighted, double lead, double thread, two starts, and triple flighted, etc.

Single Flighted Screw: A screw having a single helical flight.

Two Stage Screw: A screw constructed with an initial feed section followed by a restriction section, and then an increase in the flight channel volume to release the pressure on the material while carrying it forward, such as a screw used for venting at an intermediate point in the extruder.

Vented Screw: A two stage screw with a screw vent in the second stage.

Water Cooled Screw: A cored screw suitable for the circulation of cooling water.

Extruder, Single Screw. An extruder with one barrel in which a solid or cored screw is rotated.

Extruder Size. The nominal inside diameter of the extruder barrel.

Extruder Surging. See SURGING.

Extruder, Tandem. A pair of extruders used together in tandem arrangement for the production of foamed polystyrene sheet. The first extruder, usually with a two-stage screw operating at high shear rate to provide good dispersion, melts and mixes the resin, blowing agent and nucleating agent. It feeds directly into the second extruder, usually a larger diameter unit operating at a lower temperature than the first, which reduces the temperature of the melt and forces it through the die orifice.

Extruder, Twin Screw. (extruder, double screw) An extruder with a barrel of figure-8 cross section formed by two intersecting holes, each of which contains a screw with tangential or intermeshing flights. When the flights are tangential, the screws may be rotated in the same or opposite directions regardless of screw design, and the extruder functions much the same as a single screw extruder. When the flights are intermeshing, as is most often the case, the relative direction of the screw rotation depends on whether the screws are of the same or opposite "hands." Intermeshing screws rotating in the same direction transfer the material from one screw flight to the other, producing considerable shear stress. Counter-rotating screws create even higher shear stresses by milling the material between them. Both types of intermeshing screw extruders provide a positive conveying action and efficient mixing at temperatures lower than those necessary for single screw extruders. The largest application for twin screw extruders is for compounding and extruding PVC, especially large-diameter pipes. Other applications are compounding light polyolefin powders, processing volatile-containing polymers, compounding incompatible mixtures such as wax and resin, and processing difficult-to-feed materials such as film scrap.

Extruder, Vented. An extruder provided with a vent hole, usually in the metering zone where the material has attained the molten condition, for the escape of gases and air. Such vents require two-stage screws and, usually, valves arranged to control the flow of the melt from the metering zone to the vent zone.

Extrusion. The process of forming continuous shapes by forcing a molten plastic material through a die. Typical shapes extruded are hose, tubing, flat films and sheets, jackets around electrical wires, parisons for blow molding, filaments and fibers, strands for pelletizing, and webs for coating and laminating. See also COEXTRUSION.

Extrusion, Autothermal. A method of extrusion in which the plastic material is heated solely by frictional heat derived from the energy of the screw. The process is very commonly called *adiabatic extrusion,* although this use of the term ADIABATIC (which see) is not strictly correct.

Extrusion Blow Molding. The process of BLOW MOLDING, which see, in which the parison is formed by extrusion.

Extrusion Casting. A term sometimes employed in the literature for the process of extruding unsupported film, especially a composite of two or more integral resin layers formed by extruding separate molten streams into a single die assembly in which the streams are combined under pressure before they emerge from the die. Such extrusion-cast composite films possess desired properties on each of the respective sides, e.g. heat-sealability on one side and stiffness on the other side, or different slip levels.

Extrusion Coating. The process of coating a substrate by extruding a layer of molten resin onto the substrate with sufficient pressure to bond the two together without the use of an adhesive. A common application of the process is the coating of paper and fabrics with polyethylene, by extruding a web directly into the nip of a pair of rolls through which the substrate is passing.

Extrusion Coloring. The method of adding colorants to a plastic compound comprising dry blending the colorant with the solid granular resin, extruding the mixture into continuous strands, and chopping these strands into pellets to be used in subsequent molding operations.

Extrusion Laminating. A laminating process in which a plastic layer is extruded between two layers of substrate. See also EXTRUSION COATING.

Extrusion Rheometer. (extrusion plastometer) A type of viscometer used for determining the melt index of a polymer. It consists of a vertical cylinder with two longitudinal bored holes, one for measuring temperature and one for containing the specimen, the latter having an orifice of stipulated diameter at the bottom and a plunger entering from the top. The cylinder is heated by external bands, and a weight is placed on the plunger to force the specimen polymer through the orifice.

Exudation. The undesirable appearance on the surface of an article of one or more of its constituents, which have migrated or exuded to the surface. In vinyls, such constituents may be residual emulsifier from the resin, stabilizer, lubricant, or plasticizer. Secondary plasticizers in particular have a tendency to exude when used in excessive amounts. Exudation may appear on a product shortly after it is made, but more often is delayed for periods ranging from several weeks to years. Products that do not exude for long periods under ideal storage conditions can be caused to exude by exposure to pressure, heat, high humidity, light, or other environmental agents that can extract plasticizer or other ingredients.

Fabricate. In the broadest sense, this term means to manufacture, devise, or to make an assembly of parts and sections. In the plastics industry it refers to the assembly or modification of preformed plastics articles by processes such as welding, heat-sealing, adhesive joining, machining, and joining by mechanical devices. It is not generally used for basic manufacturing processes such as extrusion, calendering, molding and the like.

Fading. See LIGHT RESISTANCE.

Fadometer. An apparatus for determining the resistance of resins and finished products to fading by subjecting the articles to high density ultraviolet rays of approximately the same wave length as those found in sunlight.

Falling Dart Impact Test. In addition to the ASTM D1709 test described under FREE FALLING DART TEST, several similar tests exist for products such as pipe and bottles as well as sheeting. In the *Bruceton Staircase Method,* a dart is dropped on the sample from an arbitrary height. If it fails to break the sample the drop height is raised until one sample is broken. Then the height of the next drop is lowered. By repeating such steps the impact energy at which 50% of the samples fail is determined and used as the measure of impact strength. In the *Probit Method,* groups of samples are tested at specific energy levels between the limits at which respectively none or all of the specimens in a given set fail. In a "Fail Sensitivity" method developed by Borg Warner Chemicals combines elements of the Bruceton and Probit methods. The Bruceton method is used to select the best out of ten, and at this level the remainder of the samples are tested to obtain impact energy profiles.

False Body. The deceptively high apparent viscosity of a pseudoplastic fluid at a low rate of shear, which disappears upon higher degrees of agitation. See also THIXOTROPY.

False Neck. In the blow molding of containers, a neck construction which is additional to the neck finish of the container and which is only intended to facilitate the blow molding operation. Afterwards the false neck part is removed from the container.

Family Mold. A multiple-cavity mold containing variously-shaped cavities, each of which produces a component of an item which is assembled from these components. For example, a family mold for a model airplane construction kit would contain a cavity for each part, and components of a complete kit would be produced in one shot.

Fan Gate. A shallow gate somewhat wider than the runner from which it extends.

Fan-Tail Die. An extrusion die of divergent form.

Farad. (F) The unit of electrical capacitance. A capacitance of one farad requires one coulomb of electricity to raise its potential one volt.

Fatice. Sometimes called "artificial rubber" or a "rubber substitute," fatice is a substance made by vulcanizing with sulfur a vegetable oil such as soybean, rapeseed or castor oil. It is used as a processing aid and extender in natural rubber compounds and synthetic elastomers.

Fatigue Failure. The failure or rupture of a plastic article under repeated cyclic stresses, at a point below the normal static breaking strength.

Fatigue Life. The number of cycles of deformation required to bring about failure of a test specimen under a given set of alternating stresses.

Fatigue Limit. The stress below which a material can be stressed cyclically for an infinite number of times without failure.

Fatigue Notch Factor. The ratio of the fatigue strength of a specimen with no site of stress concentration (notch) to the fatigue strength of a similar specimen with a notch.

Fatigue Ratio. The ratio of fatigue strength to tensile strength. The mean stress and alternating stress must be stated.

Fatigue Strength. The maximum cyclic stress a material can withstand for a given number of cycles before failure occurs.

Fatty Acids. Monobasic organic acids obtained by the hydrolysis (saponification) of natural fats and oils. The linoleic, linolenic, oleic, palmitic and stearic acids are used in the synthesis of many plasticizers and stabilizers for plastics.

FDA. Abbreviation for Food and Drug Administration, the U.S. agency under the Department of Health, Education and Welfare which is concerned with the safety of products marketed for consumer use.

Feed Bushing. The hardened steel bushing in an injection mold which forms a seal between the mold and the injection nozzle.

Feed Zone. The first zone of an extruder screw which is fed from the hopper, terminating at the beginning of the compression zone.

Felt. (felting) A fibrous material made up of interlocked fibers held together by mechanical or chemical action, heat or moisture.

Femto-. (f) The SI-approved prefix for a multiplication factor to replace 10^{-10}.

FEP. Abbreviation for FLUORINATED ETHYLENE PROPYLENE, which see.

Ferrocene. (dicyclopentadieneyl iron) $(C_5H_5)Fe$ A coordination compound of ferrous iron and cyclopentadiene in which the organic portions have typically aromatic chemical properties. Its uses include smoke reduction in rigid PVC, curing agent for silicone resins, intermediate for high temperature polymers and ultraviolet absorber. The UV-radiation resistance of ferrocene has prompted investigations in the use of its polymeric derivatives in extraterrestrial shielding uses.

Festooning Oven. An oven used to dry, cure or fuse plastic-coated fabrics with uniform heating. The substrate is carried on a series of rotating shafts with long loops or "festoons" between the shafts.

FF. Abbreviation for furan formaldehyde copolymers.

FFF. Abbreviation for phenol-furfural copolymers.

Fiber. (fibre) A single homogeneous strand of material having a length of at least 5 mm, which can be spun into a yarn or roving or made into a fabric by interlacing in a variety of methods. Fibers can be made by chopping filaments (converting). *Staple fibers* may be ½ to a few inches in length and usually 1 to 5 denier. The natural fibers used exclusively by mankind from the beginning were first supplemented by rayon and acetate, both of which are derived from cellulosic materials. The first commercially successful wholly-synthetic fiber was nylon, introduced in 1939. Then followed acrylic fibers in 1950, polyesters in 1951, and various other polymeric fibers in subsequent years. In 1967 the wholly-synthetic "man-made" fibers surpassed the natural fibers in volume produced. See also MAN-MADE FIBER.

Fiberfil Molding. A term used for an injection molding process employing as a molding material pellets containing short bundles of fiber surrounded by resin.

Fiber Glass. See GLASS FIBER REINFORCEMENTS.

Fiber Optics. A term employed for light-transmitting fibers of glass and, more recently, transparent plastics such as PMMA. Each fiber is coated with a material with a refractive index lower than that of the fiber itself, and many fibers may be gathered in a bundle jacketed in polyethylene or other flexible plastic. Such bundles transmit light from one end to the other even though curved. Applications are being made in aircraft and automobiles (e.g. for lighting instrument panels), electronics, displays, medicine and packaging.

Fiber Show. (fiber prominence) In reinforced plastics, a condition in which ends of reinforcement strands, rovings or bundles unwetted by resin appear on or above the surface. It is believed to be caused by a deficiency in the glass, and may not appear until the resin is fully cured. Remedies include measures to improve wet-out, use of resins of optimum viscosity, and holding down exotherm rates which cause stresses within the laminate.

Fiber Spinning. See SPINNING.

Fibre. See FIBER.

Fibrid. A generic name for fibers made of synthetic polymers.

Fibrillation. The phenomenon wherein a filament or fiber shows further evidence of basic fibrous structure or fibrillar crystalline nature, by a longitudinal opening-up of the filament under rapid, excessive tensile or shearing stresses. Separate fibrils can then often be seen in the main filament trunk. The whitening of polyethylene when unduly strained at room temperture is a manifestation of fibrillation.

Fibrous Glass Reinforcements. See GLASS FIBER REINFORCEMENTS.

Filament. A variety of fiber characterized by extreme length, which permits its use in yarn with little or no twist and usually without the spinning operation required for fibers.

Filament Winding. A method of forming reinforced plastic articles comprising winding continuous strands of resin-coated reinforcing material onto a mandrel. Reinforcements commonly used are single strands or rovings of glass, asbestos, jute, sisal, cotton and synthetic fibers. Polyester resins are most widely used, followed by epoxies, acrylics, nylon, and various other resins. To be effective, the reinforcing material must form a strong adhesive bond with the resin. The mandrels may be permanent structures remaining in the finished article, or of flexible or destructible material capable of being removed after curing. The process is performed by drawing the reinforcement from a spool or "creel" through a bath of resin, then winding it on the mandrel under controlled tension and in a predetermined pattern. The mandrel may be stationary, in which event the creel structure rotates about the mandrel, or it may be rotated on a lathe about one or more axes. By varying the relative amounts of resin and reinforcement, and the pattern of winding, the strength of filament wound structures may be controlled to resist stresses in specific directions. After sufficient layers have been wound, the structure is cured at room or elevated temperatures.

Filing. Manual filing is sometimes used to bevel, smooth, deburr and fit the edges of plastic moldings and sheets. The process is limited to parts that cannot be tumbled easily, and to plastics with suitable hardness and heat resistance.

Fill-And-Wipe. A decorating process for articles molded with depressed designs, wherein the general area containing the designs is coated with paint by brushing, spraying or rolling, then surplus paint is wiped from the raised areas surrounding the depressions.

Filler. A relatively inert substance added to plastic compound to reduce its cost and/or to improve physical properties, particularly hardness, stiffness and impact strength. A filler differs from a REINFORCEMENT (which see) in two respects. Filler particles are generally small, and they do not markedly improve the tensile strength of a product, whereas reinforcements are fibrous and do markedly improve the tensile strength. The most commonly used general purpose fillers are clays, silicates, talcs, carbonates, asbestos fines and paper. Some fillers also act as pigments, e.g. carbon black, chalk and titanium dioxide. Graphite, molybdenum disulfide and PTFE are used as fillers to impart lubricity. Magnetic properties can be obtained by incorporating magnetic mineral fillers such as barium sulfate. Other metallic fillers such as lead or its oxides are used to increase specific gravity; powdered aluminum imparts higher thermal and electrical conductivity, as do other powdered metals such as copper, lead and bronze. More detailed descriptions of fillers and their specific applications are given under:

ACETYLENE BLACK	KAOLIN
ALPHA CELLULOSE	KERATIN
ALUMINA TRIHYDRATE	LIGNIN
ALUMINUM SILICATES	LITHOPONE
ASBESTOS	MICA
BARIUM SULPHATE	MICROBALLOONS
BENTONITE	MICROSPHERES
BLANC FIXE	MOLYBDENUM DISULFIDE
CALCIUM CARBONATE	NEPHELINE SYENITE
CALCIUM SILICATE	NOVACULITE
CALCIUM SULPHATE	PEANUT HULL FLOUR
CARBON BLACK	PYROGENIC SILICA
CELLULOSE	QUARTZ
CERAPLASTS	SHELL FLOUR
CLAYS	SILICA
CORK	SOYBEAN MEAL
DIATOMITE	TALC
FUMED SILICA	TERRA ALBA
GLASS FLAKE	VERMICULITE
GLASS SPHERES	WOOD FLOUR

Filler Rod. (welding rod) A rod of material for use in HOT GAS WELDING, which see, made of the same material as the plastic to be welded.

Filler-Specks. Visible specks of a filler used, such as wood flour or asbestos, which stand out in color contrast against a background of plastic binder. It should be stated whether the specks are visible before or only after removal of the surface film. (ASTM D883-65T).

Filling Yarn. (weft, woof) The transverse threads or fibers in a woven fabric; those fibers running perpendicular to the warp.

Film Blowing. The process of forming thermoplastic film wherein an extruded plastic tube is continuously inflated by internal air pressure, cooled, collapsed by rolls and wound up on subsequent rolls. The tube is usually extruded vertically upward, and air is admitted as the hot tube emerges from the die. An AIR RING, which see, is often employed to control cooling. Air is·retained within the blown bubble by means of a pair of pinch rolls which collapse the film. Thickness of the film is controlled by varying the internal air pressure and rate of extrusion. Extremely thin films with some degree of orientation can be produced by this method.

Film Casting. The process of making unsupported film or sheet by casting a fluid resin or plastic compound on a temporary carrier, usually an endless belt or circular drum, followed by solidification and removal of the film from the carrier. The plastic may be a solution, dispersion or an incompletely polymerized fluid resin. The term *film casting* has been used also for the process of extruding a molten polymer through a slot die onto a chill roll. This usage of the term is likely to cause confusion.

Films. Films are distinguished from sheets in the plastics and packaging industries only according to their thicknesses. A web under 10 mils (.010″) thick is usually called a film, whereas one 10 mils and over is usually called a sheet. Films are most commonly made by extrusion, casting and calendering.

Film Slitting. See SLITTING.

Fin. See FLASH.

Fines. In the classification of powdered or granular materials such as molding compounds according to particle sizes, fines are the portion of the material composed of particles which are smaller than a specified size.

Finish. (1) In reinforced plastics, a compound containing a coupling agent and (optionally) a lubricant and/or binder, used to treat glass fibers. (2) In the container industry, the plastic forming the opening of a container shaped to accommodate a specific closure. (3) The surface texture of a finished article.

Finishing. The removal of flash, gates and defects from plastic articles, and also the development of desired surface textures. See also DEFLASHING, DEGATING, TUMBLING, SANDING.

Finish Insert. In blow molding of bottles, a removable part of the mold that forms a specific neck finish of the bottle. Sometimes called *neck insert*.

Fish Eye. A fault in transparent or translucent plastics materials such as films or sheets, appearing as a small globular mass and caused by incomplete blending of the mass with surrounding material.

Fissure. A term used in the cellular plastics industry to denote a crack, separation or split in a formed cellular article.

Flake. A term used to denote the dry, unplasticized base of cellulosic plastics.

Flame-Retardant. (adj.) Having the ability to resist combustion. A flame retardant plastic is considered to be one that will not continue to burn or glow after the source of ignition has been removed.

Flame Retardants. Materials that reduce the tendency of plastics to burn. They are usually incorporated as additives during compounding, but sometimes are applied to surfaces of finished articles. Some plasticizers, particularly the phosphate esters and chlorinated paraffins, also serve as flame retardants. *Inorganic flame retardants* include antimony trioxide, monoammonium phosphate, dicyandiamide, isano oil, zinc borate, boric acid and ammonium sulfamate. Another group called *reactive-type flame retardants* includes bromine-containing polyols, phosphorus-containing polyols, chlorendic acid and anhydride, tetrabromophthalic anhydride, tetrabromo bisphenol A, tetrachloro phthalic anhydride, diallyl chlorendate, and unsaturated phosphonated chlorophenols.

Flame Spray Coating. A coating process utilizing powdered metals or plastics, in which the powdered materials are heated to sintering temperature in a cone of flame enroute from a spray gun orifice to the work being coated.

Flame Treating. A method of rendering inert thermoplastics, particularly polyolefins, receptive to inks, lacquers, paints, adhesives and the like by bathing the surface of the article in a high oxidizing flame. This treatment oxidizes the surface layer of the article, making it receptive to coatings. See also CASING, ELECTRONIC TREATING.

Flammability. With respect to plastics, the word flammability is a very broad term. The behavior of various plastics when burning, and tests designed to evaluate flammability, encompass six categories: (1) ignitability, (2) burning rate, (3) heat evolution, (4) smoke production, (5) products of combustion, and (6) endurance of burning. For the convenience of users, the most commonly used tests for these aspects of flammability are listed under this topic rather than in alphabetical order throughout the dictionary.

ARC Ignition Test. (UL high current) A flammability test for plastics used in electrical applications certified by UL. High current electrodes resting on the sample are repeatedly moved together until they arc then apart until the arc is ruptured. The number of ruptures required to ignite the sample is recorded.

ARC Ignition Test. (UL high voltage) A flammability test for plastics used in electrical applications certified by UL. Two electrodes rest on the surface of the sample 4.0 mm apart. Current is applied to cause a continuous arc. Time required for ignition is measured.

ASTM D 568-74 Test for Flammability of Flexible Plastics. A small-scale screening procedure for comparing the relative flammability of plastics in the form of flexible, thin sheets or films, tested in the vertical position. The specimen, of standard length and width, is suspended vertically and exposed to a gas flame at its lower end. Time and extent of burning are measured and reported if the specimen does not burn 38 cm. An average burning rate is reported if the specimen burns to the 38 cm mark.

ASTM D 635-74. Test for Flammability of Self-Supporting Plastics. (also recognized by FTM 2021) This method covers a small-scale laboratory screening procedure for comparing the relative flammability of self-supporting plastics in the form of bars, molded or cut from sheets, plates, or panels, tested in the horizontal position. A bar of the material to be tested is supported horizontally at one end. The free end is exposed to a specified gas flame for 30 seconds. Time and extent of burning are measured and reported if the specimen does not burn 101 mm. An average burning rate is reported for a material if it burns to the 100 mm mark from the ignited end.

Flammability Tests and Nomenclature continued.

This method is used to establish relative burning characteristics and should not be used as a fire hazard test method.

ASTM D 757-75. Test for Incandescence Resistance of Rigid Plastics in a Horizontal Position. This test provides for laboratory comparisons of the resistance of rigid plastics to an incandescent surface at 950 ± 10 °C. It may supplement tests using a flame source of ignition such as D 635. The specimen is moved into contact with a silicon carbide rod at the specified temperature for three minutes.

ASTM D 1433-74. Test for Flammability of Flexible Thin Plastic Sheeting. This method covers the determination of the relative flammability of flexible plastics in the form of film or thin sheeting. It is not to be used for materials that shrink excessively upon ignition, or that melt to cause the flame to be carried away while dripping. A specimen 76 x 228 mm is placed in a cabinet protecting the specimen and igniting flame from air currents, but vented to provide combustion air. The igniting flame is produced by butane connected to a No. 22 hypodermic needle. A specimen holder track is positioned at a 45 ° angle, and the specimen is supported by spring clips depressed prior to ignition by threads. Time for burning to the threads is automatically recorded.

ASTM D 1692-74. Rate of Burning or Extent of Burning of Cellular Plastics Using a Supported Specimen by a Horizontal Screen. This test covers a small-scale horizontal laboratory screening procedure for measuring the rate of burning or extent of burning of rigid and cellular plastics that do not exhibit pronounced shrinking, curling, or melting away upon heating. The specimen to be tested is placed on a wire cloth support with one end bent to form a right angle, placed above a bunsen burner with a wing tip. Under specified distances, the burner flame is allowed to burn to 60 seconds, after which it is moved further away. The time in seconds for flame to reach a gauge mark 125 mm from the raised end is recorded. If the flame goes out before reaching the gauge mark, the distance of burning is recorded. A similar test is UL Subject 94.

ASTM D 1929-69. Test for Ignition Properties of Plastics. Sometimes called the Setchkin Technique, this test covers a laboratory determination of the self-ignition and flash-ignition of plastics using a hot air furnace. The sample is placed in a vertical refractory tube, which is inside a furnace tube that is vertically heated by electrical curent passing through nichrome wire in an asbestos sleeve wound around the tube. Air is admitted at a controlled rate, and air temperature is measured by thermocouples.

ASTM D 2843-70. Smoke Generation Test. (Rohm & Haas XP2) Used for testing plastics for compliance with building codes. The sample is placed in a chamber and is ignited by a propane flame, and smoke density is measured by a photocell across a horizontal 12-in. light path. Most building codes permit materials if maximum light absorption is less than 50%. Uniform Building Code accepts up to 75%.

ASTM D 2863-74. Test for Flammability of Plastics Using the Oxygen Index Method. This test is a procedure for determining the relative flammability of plastics by measuring the minimum concentration of oxygen in a flowing mixture of oxygen and nitrogen that will just support flaming combustion. The mixture, based on an initial concentration based on experience with similar materials, is caused to flow upwards in a test column. The vertically-positioned specimen, supported by a frame

Flammability Tests and Nomenclature continued.

if necessary, is exposed to an ignition flame at its top. If ignition does not occur at the starting concentration, the oxygen percentage is increased until ignition occurs. The volume percent of oxygen is reported as the oxygen index. This test is regarded by some as useful only for materials that barely burn in air.

ASTM D 3014-74. Test for Flammability of Rigid Cellular Plastics. This test is a screening procedure for comparing relative flammability of rigid cellular plastics, and should not be used as a fire hazard classification. The specimen is mounted in a vertical chimney with a glass front and ignited with a bunsen burner for 10 seconds. The height and duration of flame and the weight percent retained by the specimen are recorded.

ASTM E 84. 25-Foot Tunnel Test. This is probably the most widely used test for surface flame spread of burning plastics. The specimen used is a 25 foot long by 20 inches wide slab mounted face down so as to form the roof of a 25-foot long tunnel 17½ inches wide. A fire source comprising two gas burners is located 12 inches from the fire end and 7½ inches below the surface of the specimen. The flame is adjusted so that a sample of red oak flooring will spread flame 19½ inches from the end of the igniting fire in 5½ minutes. Flame spread rating is based on a formula yielding zero for asbestos-cement board and 100 for red oak flooring. A thermocouple located 1 foot from the vent end of the sample is used to measure heat release, which is recorded at least every 30 seconds. The flame end point is taken as the time for the vent-end thermocouple to reach 980°F, which in the case of red oak takes 5.5 minutes. The area under the time-temperature curve for the 10 minute test is used as a measure of heat release or fuel contribution. A light source mounted on a horizontal section of the vent pipe with the light beam directed downward is used to measure smoke density. Model codes require a rating of 25 for interior finishes in stairwells. The 25 ft. Tunnel Test is also recognized in UL 723 and NFPA 225. It is also called the Steiner Tunnel Test.

ASTM E119. Fire Endurance Test. UL 263, MFPA 251. Used for testing walls, floors, ceilings, roofs, etc. required by various building codes. A full-size wall section is used as a partition in a room-size furnace. In ASTM E119 one side of the partition is exposed to gas fire with temperatures reaching 1000°F at 5 min., 1300°F after 10 min., 1700°F after 1 hr., and 2000°F after 4 hrs. To pass, temperature on the far side of the specimen should not exceed 250°F. Similar tests specify different temperatures and time limits.

ASTM E162. Radiant Panel Test. This test, developed at the National Bureau of Standards, employs a radiant heat source consisting of a 12″ x 18″ vertically mounted porous refractory panel maintained at 1238° ± 7°F. A specimen measuring 6″ x 18″ is supported in front of the panel with the 18″ dimension inclined 30° from the vertical. A pilot burner ignites the top of the specimen, so that the flame front progresses downward along the underside exposed to the radiant panel. The temperature rise recorded by stack thermocouples, above their base level of 356 to 446°F, is used as a measure of heat release. A smoke sampling device which collects smoke particles on glass fiber filter paper is used to measure smoke density.

Bureau of Mines Flame Penetration Test. Used for plastic foams in mines. Time required to burn through a 1 inch thick layer of foam exposed to a continuous flame from a propane torch, temperature of which is 2150°F.

FAA Horizontal Flammability Test. Required by FAA for components of aircraft. Same conditions as in FAA Vertical test except samples are horizontal. Maximum

Flammability Tests and Nomenclature continued.

acceptable burn rate is 2½ inches per minute for acrylic windows, instrument assemblies and seat belts. Small molded parts are acceptable if burn rate is less than 4 inches per minute.

FAA Vertical Flammability Test. Required by FAA for materials including surface finishes and decorative components of aircraft crew and passenger compartments. Bunsen burner flame is applied to a vertical specimen for 60 seconds. Flame time, burn length and burn time of drips are noted. To pass, burn must be less than 6 inches, flame time must not exceed 15 seconds, and drippings must go out before 3 seconds.

Factory Mutual Calorimeter Heat Contribution Test. Used by insurance underwriters to test building components. Sample is burned in a gasoline-fired furnace and time-temperature curve is recorded. A non-combustible sample is substituted and time-temperature curve is reproduced using propane. Heat added to reproduce curve gives value of sample.

Flame Propagation Test, UL Subject 94. A test for self-extinguishing polymers in sheet form for applications certified by UL. Vertically oriented sample is exposed to a bunsen burner flame for 10 sec. If burning ceases in less than 30 sec., a second 10-sec. application of flame is required. Flaming droplets are allowed to fall on cotton. If average burning time is less than 5 sec. and drips don't ignite cotton, material is Self-Extinguishing, Group 0. If time is less than 25 sec. and drips don't ignite cotton, material is rated Self-Extinguishing, Group I, and if cotton is ignited, material is Self-Extinguishing, Group II.

Flame-Retardant. (adj.) This term has been used in the past to indicate the ability of a plastic to resist combustion as shown by a small-scale laboratory test, or one that will not continue to burn or glow after a source of ignition has been removed. However, it is now considered necessary to warn users that this term does mean that flame-retarded materials will not burn in large-scale fire conditions.

Flashover. (flame-over) (1) A term used in literature pertaining to the flammability of plastics and other materials in buildings. Flashover occurs when hot, combustible gases are generated in fire areas of a building, become mixed with sufficient oxygen upon spreading to non-burning areas, and ignite to cause total surface involvement without a progressive flame-spreading stage. (2) In the electrical industry, an electric discharge around the edge or over the surface of electrical insulation. (ASTM D 149-75).

Horizontal Burn Test MVSS302. Used by the Department of Transportation for all materials used in automotive interiors. The horizontal specimen is ignited by a 15-second application of a bunsen burner flame. When flame has burned 1½ inches of the sample, time is measured until material ceases to burn or until burning has progressed 10 inches, and rate must not exceed 4 inches per minute.

Hot Wire Ignition Test. A flammability test used for plastics in electrical applications certified by UL. A bar-shaped sample is wrapped with resistance wire which is heated electrically to red heat. Time for the sample to ignite is measured.

Methenamine Pill Test. A flammability test for carpets and floor coverings, for compliance with DOC standards for carpets exceeding 24 sq. ft. A methenamine pill is ignited on the sample and allowed to burn out. The burned area is measured. This test is described in ASTM D 2859.

Flammability Tests and Nomenclature continued.

NBS Smoke Chamber Test. The sample is placed in a completely closed cabinet, supported vertically in a frame and exposed to an electric heat source. Smoke production is measured by a vertical photometer within the chamber, results being expressed in terms of specific optical density.

Flash. The thin, surplus web of material which is forced into crevices between mating mold surfaces during a molding operation, and which remains attached to the molded article. For methods of removing flash, see DEFLASHING.

Flash Gate. A long, shallow rectangular gate in an injection mold, extending from a runner which runs parallel to an edge of a molded part along the flash or parting line of the mold.

Flash Groove. (spew groove) A groove ground in a force to allow the escape of excess material during a molding operation.

Flash Line. See PARTING LINE.

Flash Mold. A mold in which the mating surfaces are perpendicular to the clamping action of the press so that as the clamping force increases the distance between the mating surfaces decreases, thus permitting excess molding material to escape as flash during closing of the mold.

Flash Point. The lowest temperature at which a combustible liquid will give off a flammable vapor that will burn momentarily.

Flash Ridge. That part of a flash mold along which the excess material escapes until the mold is closed.

Flexibilizer. A term seldom used in the plastics industry for an additive which makes a resin or elastomer more flexible. See PLASTICIZER.

Flexible Molds. Molds made of rubber, elastomers or flexible thermoplastics, used for casting thermosetting plastics or non-plastic materials such as concrete and plaster. They can be stretched to permit removal of cured pieces with undercuts.

Flexographic Printing. A rotary process employing flexible rubber or elastomeric printing plates adhered to a roll, inked by a screen roll which in turn is coated from a feed roll immersed in ink.

Flexural Modulus. The ratio, within the elastic limit, of the applied stress on a test specimen in flexure to the corresponding strain in the outermost fibers of the specimen.

Flexural Strength. (modulus of rupture) The maximum stress in the outer fiber at the moment of crack or break. In the case of plastics, this value is usually higher than the straight tensile strength.

Flight. In an extruder screw, the outer surface of the helical ridge of metal left after machining the screw channels. The diameter of the screw flights is usually the same as the inner diameter of the extruder barrel minus a specified clearance.

Flitter. See GLITTER.

Floating Chase. A mold member, free to move vertically, which fits over a lower plug or cavity, and into which an upper plug telescopes.

Floating Platen. In compression molding, a platen located between the main head and the press table in a multi-daylight press and capable of being moved independently of them.

Floating Punch. A male mold member attached to the head of a press in such a manner that it is free to align itself in the female part of the mold when the mold is closed.

Floats. (n) A term once used for asbestos filler in the form of very fine short fibers with associated dust.

Flock. Short fibers of cotton or synthetic fibers such as polyester, acrylic or nylon. They are used as reinforcements in phenolic, allylic and other thermosetting molding compounds, and also for decorating plastics by the process of FLOCKING, which see.

Flocking. (flock coating) A method of finishing sometimes employed for plastics articles whereby an article is coated with a tacky, slow-drying adhesive, then is dusted with a fibrous material cut into very short lengths to give a finish resembling suede or other fabrics. Fibers for flocking are available in a wide range of materials including acrylic, nylon, polyester, polyolefins and natural fibers. Machinery for flocking films and fabrics includes gravure printing stations for applying the adhesive in desired patterns; and flock heads which distribute a predetermined layer of flock to the web, and retrieve and recirculate surplus flock.

Flooding. An undesirable separation of pigment in a wet paint film, causing a color change or mottled effect. The condition is caused by improper milling of pigment, excess of solvent or low viscosity.

Flow-Coating. A painting process in which the article to be painted is drenched with paint, either by pouring or spraying with a mist in a closed or semi-closed chamber. The parts are sometimes rotated during and after drenching to avoid sags and runs. The process is used for coating metallized parts and other irregularly shaped articles which are difficult to paint by conventional spraying methods.

Flowers of Antimony. See ANTIMONY TRIOXIDE.

Flow Line. See WELD MARK.

Flow Marks. Defects in a molded article characterized by a wavy surface appearance, caused by improper flow of the resin into the mold.

Flow Molding. (1) A variation of the INJECTION MOLDING process, which see, used for thick walled parts. Additional molten compound is forced into the mold during the cooling of the initial charge, to overcome shrinkage. (2) A term applied to the process of heating a material such as cloth-backed PVC by high frequency alternating current while pressing the material into a mold made of silicone rubber. In a 10 to 15 second cycle the PVC is formed to the contours and surface texture of the mold, which can resemble hand-tooled leather or any such effect.

Flow Properties. See MELT INDEX, VISCOSITY, PSEUDOPLASTIC FLUID, RHEOLOGY.

Fluidity. The reciprocal of viscosity. The cgs unit of fluidity is the *rhe*, the reciprocal of the poise.

Fluidization. A gas-solid contacting process in which a stream of gas is passed upwards through a bed of finely divided particles, causing them to lift and behave like a boiling fluid. The process is widely used for performing chemical reactions in many industries. In the plastics industry, the principal application is for FLUIDIZED BED COATING, which see.

Fluidized Bed Coating. The process of applying plastics coatings to objects of other materials, often metals, wherein a powdered resin is placed in a container provided with a porous or perforated bottom through which a gas is discharged upward to keep the resin particles in a state of flotation. The part to be coated is preheated and lowered into the fluidized bed until a deposit of the desired thickness is formed, then withdrawn. Subsequent heating is usually necessary to fuse the resin particles to a smooth, homogeneous layer. In a variation of the process recently developed in France, the particles in the fluidized bed are electrostatically charged so that they adhere to the part without preheating, when the part is at least slightly electrically conductive and grounded. See also ELECTROSTATIC FLUIDIZED BED COATING.

Fluorescent Brightening Agents. See BRIGHTENING AGENTS.

Fluorescent Pigments. Pigments which absorb light at certain frequencies and re-emit light at lower frequencies, thus making articles containing these pigments appear to possess an actual glow of their own. A type known as *daylight fluorescent pigments* responds to radiation in both the ultraviolet and visible ranges, causing the effect of glowing in normal daylight. These pigments comprise fluorescent dyes incorporated in a resin matrix, ground to powder form. Urea and melamine resins have been used as matrices until recently, when a modified sulfonamide resin base came into use.

Fluorinated Ethylene Propylene. (FEP) This member of the fluorocarbon family of plastics is a copolymer of tetrafluoroethylene and hexafluoropropylene, possessing most of the desirable properties of PTFE but having a melt viscosity low enough for processing in conventional thermoplastic molding or extrusion equipment. It is available in pellet form for molding and extrusion, and as dispersions for spraying and dipping.

Fluorocarbon Blowing Agents. A family of inert, noncorrosive liquid compounds containing carbon and fluorine. They are compatible with all resins and leave no residue in the mold. They are widely used in the structural foam process, in which they are incorporated with the polymer by direct injection into the barrel. Fluorinated hydrocarbons are numbered by a system developed by Du Pont for use with its registered trademark Freon. The first number on the right is the number of fluorine atoms in the molecule. The next number on the left gives one more than the number of hydrogen atoms. The next number on the left gives one less than the number of carbon atoms. When this number is 0, as in the case of one-carbon compounds, it is omitted. Other details of the numbering system are used for compounds other than blowing agents. The types of fluorocarbons most commonly used as blowing agents for plastics are Trichloromonofluoromethane (Freon 11), Dichlorodifluoromethane (Freon 12), Trichlorotrifluoroethane (Freon 113), and Dichlorotetrafluoroethane (Freon 114).

Fluorocarbon Resins. Thermoplastic resins chemically similar to the polyolefins, with all of the hydrogen atoms replaced with fluorine atoms. They are made from monomers composed only of fluorine and carbon. The main members of the fluorocarbon resin family are POLYTETRAFLUOROETHYLENE, FLUORINATED ETHYLENE PROPYLENE, and POLYHEXAFLUOROPROPYLENE, which see.

Fluoroethylene. See VINYL FLUORIDE.

Fluorohydrocarbon Resins. Resins made by polymerizing monomers composed of fluorine, hydrogen and carbon only. Included are POLYVINYLIDENE FLUORIDE, POLYVINYL FLUORIDE, POLYTRIFLUOROSTYRENE, (which see) and copolymers of halogenated and fluorinated ethylenes.

Fluoroplastics. The ASTM term for plastics based on polymers with monomers containing one or more atoms of fluorine or copolymers of such monomers, the fluorine-containing monomer(s) being in greatest amount by mass. (ASTM D 883-75a). Included in this family are POLYTETRAFLUOROETHYLENE, FLUORINATED ETHYLENE-PROPYLENE, POLYCHLOROTRIFLUOROETHYLENE, and POLYVINYLIDENE FLUORIDE (all of which see); and a variety of copolymers of both halogenated and fluorinated hydrocarbons. See also FLUOROCARBON RESINS, FLUOROHYDROCARBON RESINS, CHLOROFLUOROCARBON RESINS, CHLOROFLUOROHYDROCARBON RESINS, ETFE FLUOROPOLYMER, ETHYLENE-CTFE COPOLYMER, PERFLUOROALKOXY RESINS.

Flushed Pigment. A pigment which has been transferred from an aqueous medium in which it was first manufactured to an oil medium. The flushing process comprises mixing the aqueous pigment paste with oil in such a manner that the pigment preferentially transfers to the oil phase, pouring off the free water, and removing the remaining water by heating under vacuum.

Fluted Core. An integrally woven reinforcing material consisting of ribs between two skins for unitized sandwich construction.

Flux. (n) (1) In chemistry and metallurgy, a substance, e.g. borax or fluorspar, used to promote the fusion of metals or minerals. (2) In plastics compounding, the term flux is sometimes used for an additive to improve flow properties, e.g. coumarone-indene resins in the milling of vinyl compounds.

Flux. (v) To melt, fuse or make fluid. In the early years of the vinyl plastisol art this term was often used for "fuse" before the latter term came into general use.

Fluxing Temperature. See FUSION TEMPERATURE.

Foam. See CELLULAR PLASTIC.

Foam Backs. A term used for fabrics laminated to or coated with a layer of rubber or plastic foam.

Foam Casting. (foam molding) A process with many variations, depending on the polymers used. In general, a fluid resin or pre-polymer/catalyst system is foamed before or during molding by mechanical frothing, or by gas dissolved in the mixture or released from a low-boiling point liquid. See also REACTION INJECTION MOLDING.

Foamed Plastic. See CELLULAR PLASTIC.

Foam Fabrication. The process of cutting large slabs or "buns" of foamed plastics into sections of desired dimensions. The raw slab is conveyed through a series of saws or knives which in turn remove the top, bottom and side skins, then cut longitudinal sections, and finally cross cut the sections into lengths.

Foaming Agents. See BLOWING AGENTS.

Foam-in-Place. Refers to the deposition of foams at the site of the work, as opposed to bringing the work to the foaming machine.

Foil Decorating. See IN-MOLD DECORATING.

Folding Machines. Machines designed for folding sheet plastics such as cellulose acetate into shapes such as identification card holders, sheets for ring binders, visible indexes and the like. An electrically heated blade softens the plastic and folds it into a 180° U-type bend.

Foot-Candle. The old unit of lighting intensity, formerly defined as 1 lumen of incident light per square foot, or 1.076 milliphots. In the new SI system it is expressed in lux (lx), the conversion factor being 1 foot-candle = 1.076 391 $\bar{E}+01$ lux.

Force. (1) Either half of a compression mold (top force or bottom force), but most often the half which forms the inside of the molded part. (2) The male half of a mold, which enters the cavity and exerts pressure on the resin causing it to flow. (3) That which changes the state of rest or motion in matter, measured by the rate of change of momentum. See DYNE, NEWTON.

Force Plate. The plate that carries the plunger or force plug of a mold and the guide pins or bushings. Since the force plate is usually drilled for steam or water lines, it is also called the *steam plate*.

Force Plug. (plunger, piston) The portion of a mold that enters the cavity block and exerts pressure on the molding compound, designated as the *top force* or *bottom force* by position in the assembly.

Ford Viscosity Cups. A series of three cylindrical cups with conical bottoms, differing only in orifice diameter, holding approximately 100 ml of sample liquid. The interval of time for the sample to flow from the bottom orifice is the measure of viscosity. The Ford test is described in ASTM D 1200.

Forging. See COLD FORMING, SOLID PHASE FORMING.

Formaldehyde. HCHO. (oxymethylene, formic aldehyde, methanal) A colorless gas with a pungent, suffocating odor, obtained most commonly by the oxidation of methanol or low boiling petroleum gases such as methane, ethane, propane and butane. The gas is difficult to handle, so it is sold commercially in the form of aqueous solutions, solvent solutions, as the low polymer "paraformaldehyde," and as the cyclic trimer "s-trioxane" or "alpha-trioxymethylene." High molecular weight polymers of formaldehyde are called polyoxymethylene or ACETAL RESINS, which see. Formaldehyde is also used in the production of other resins such as PHENOLIC (phenol formaldehyde) and AMINO (urea formaldehyde), which see.

Formalin. (formol) An aqueous solution of formaldehyde, usually 37 to 40% in strength.

Form Grinding. A method of forming cylindrical or spherical parts from plastic rod or tubing, employing a grinding wheel shaped to the desired contour, a smaller hardened steel regulating wheel which presses the plastic rod or tube against the grinding wheel, and a work rest blade which supports the work between the two wheels. Water is usually employed as the coolant, from which the scrap powder can be recovered and reused.

"Formica." A registered trademark of The Formica Company for high-pressure laminates of melamine-formaldehyde, phenolic and other thermosetting resins with paper, linen, canvas, glass etc. This trademark is often misused by the public in a generic manner.

Forming. A general term encompassing processes in which the shape of plastics pieces such as sheets, rods or tubes is changed to a desired configuration. The term is not usually applied to operations such as molding, casting or extrusion in which shapes or pieces are made from molding materials and liquids. See FABRICATE, JOINING, THERMOFORMING.

Forming Cake. In filament winding, the collection (package) of glass fiber strands on a mandrel during the forming operation.

Formol. See FORMALDEHYDE.

Fossil Resin. A natural resin obtained from fossilized remnants of plant or animal life. An example is amber, a fossil resin derived from an extinct variety of pine.

Foucault Current. See EDDY CURRENT.

Foundry Resins. Thermosetting resins used as binders for sand in foundry operations. The types most commonly used are water-soluble phenol-formaldehyde resins which become insoluble when cured, and cold setting furfuryl alcohol resins that cure in the presence of an acid catalyst.

Fractionation. A method of determining the molecular weight distribution of polymers based on the fact that polymers of high molecular weight are less soluble than those of low molecular weight. Two basic methods in use are (1) precipitation fractionation, in which phases are separated from a solvent solution of the polymer by the incremental additions of nonsolvents, lowering of temperature, or volatilization of the solvent; and (2) extraction fractionation, in which fractions of increasing molecular weight are preferentially extracted from a layer of polymer which has been deposited on a substrate. In either method, a series of fractions are obtained which must be recovered and characterized with respect to molecular weight.

Fracture. The separation of a body, usually characterized as either brittle or ductile. In brittle fracture, the crack propagates rapidly with little accompanying plastic deformation. In ductile fracture, the crack propagates slowly, usually following a zig-zag path along planes on which a maximum resolved shear stress occurred.

Free Falling Dart Test. A method of measuring the impact resistance of thermoplastic films by dropping a dart with a hemispherical head onto a film specimen held in a clamping frame. As described in ASTM D 1709, the dart is dropped from a fixed height onto each of 10 specimens and the per cent failure is noted. Another increment of weight is added and another 10 specimens are tested. This process is repeated until 50% of the specimens fail, and the weight of the dart at this point is the measure of impact strength. See also FALLING DART IMPACT TEST.

Free Phenol. The uncombined phenol existing in a phenolic resin after curing, the amount of which is indicative of the degree of cure. The presence of such free phenol can be detected by the GIBBS INDOPHENOL TEST, which see.

Free Radical. An atom or group of atoms having at least one unpaired electron. Most free radicals are short-lived intermediates with high reactivity and high energy, difficult to isolate. They play a role in many polymerization processes.

Free Radical Polymerization. A reaction initiated by a free radical (molecule in which one of the carbon valences is not satisfied) derived from a polymerization catalyst. Polymerization proceeds by the chain reaction addition of monomer molecules to the free radical ends of growing chain molecules. Major polymerization methods such as bulk, suspension, emulsion and solution polymerization are types of free radical polymerization. The free radical mechanism is also useful in copolymerization reactions, in which alternating monomeric units are promoted by the presence of free radicals.

Freeze Grinding. See CRYOGENIC GRINDING.

French Mold. A two-piece mold for irregular shapes: tall, topheavy, leaning to one side, or with extremely fine detail.

Freon 11. DuPont's trademark for TRICHLOROFLUOROMETHANE, which see.

Freon 12. DuPont's trademark for DICHLORODIFLUOROMETHANE, which see.

Freon 113. DuPont's trademark for TRICHLOROTRIFLUOROETHANE, which see.

Freon 114. DuPont's trademark for DICHLOROTETRAFLUOROETHANE, which see.

Friction. The resistance to the relative motion (sliding or rolling) of surfaces of solid bodies in contact with each other. The coefficient of friction, $k = \dfrac{F}{W}$ in which F is the force required to move one surface over another, and W is the weight pressing the surfaces together.

Friction Calendering. A process in which an elastomeric compound is forced into the interstices of woven or cord fabrics while passing through the rolls of a calender. The heated compound is fed into the top opening of three adjacent rolls, so that it will cling to the middle roll. The fabric to be treated is fed into the lower opening between the rolls. The distance between the rolls is regulated so as to squeeze the fabric without brushing it, and the rolls are operated at different speeds so that the compound is wiped by friction into the meshes of the fabric.

Friction Welding. (angular welding) A term encompassing SPIN WELDING, which see, and the newer process of applying rapid angular oscillations to heat the plastic parts to be joined. This variation of the spin welding process is used for parts that are not symmetrical about an axis of rotation. The equipment must be programmed to stop when the parts are properly positioned for joining.

Friedel-Crafts Catalysts. Strongly acidic metal halides such as aluminum chloride, aluminum bromide, boron trifluoride, ferric chloride and zinc chloride, used in the polymerization of unsaturated hydrocarbons, e.g. olefins. Friedel-Crafts reactions using such catalysts are named for Charles Friedel and James Crafts, who first used them in 1877. The acidic halides used as Friedel-Crafts catalysts are also known as *Lewis acids.*

Frosting. A light-scattering surface resembling fine crystals. See also CHALKING, BLOOM, HAZE. (ASTM D 883-75a).

Frost Line. In the extrusion of lay-flat film, a ring-shaped zone of frosty appearance located at the point where the film reaches its final diameter, caused by a reduction in film temperature below the softening temperature of the resin.

Frothing. A technique for applying urethane foam in which blowing agents or small air bubbles are introduced under pressure into the liquid mixture of foam ingredients.

Frozen Strains. (residual strains) Strains which remain in an article after it has been shaped and cooled to its final form, due to a nonequilibrium configuration of the polymer molecules. Such strains result when cooling is carried below a certain temperature before stresses of a molding or forming operation have been fully relieved.

Full-Flighted Screw. An extruder screw in which the flights extend over the entire length of the screw.

Fumed Silica. (pyrogenic silica) An exceptionally pure form of silicon dioxide made by reacting silicon tetrachloride in an oxy-hydrogen flame. Individual particles of fumed silica, ranging from .007 to .05 micron, tend to link together by a combination of fusion and hydrogen bonding to form chain-like aggregates with high surface areas which retard the flow of liquids in which they are dispersed. This characteristic makes fumed silica useful as a thickening agent imparting thixotropy to liquid resins which are normally Newtonian, for example polyesters. Fumed silica is also used in dry molding powders to make them free flowing and easier to disperse with colorants. Improved electrical properties, prevention of blocking, and reduction of plasticizer migration are other benefits said to result from the addition of pyrogenic silica to vinyl compounds.

Functionality. The ability of a molecule or group to form covalent bonds with another molecule or group in a chemical reaction. Compounds may be monofunctional, difunctional, trifunctional or polyfunctional depending on the number of functional groups participating in a reaction.

Fungicides. Agents incorporated in plastic compounds to control fungus growth, usually by killing the organisms. Most plastics with a few exceptions, notably some of the cellulosics, are inherently resistant to fungus attack; however, many plasticizers are highly susceptible to attack. Examples of fungicides used in plastics are copper-8-quinolinate, mercury compounds, and Captan. Agents which retard fungal growth without killed the organisms are called *fungistats*. See also BIOCIDES.

Funginertness. (fungus resistance) Not susceptible to the formation of fungus growths.

Fungistats. Agents incorporated in plastics compounds to control fungus growth without killing the fungi. See also BIOCIDES.

Furan Prepregs. Developments in latent catalysts announced in 1972 broadened the uses of furan resins to include prepregs for laminates, which overcame the difficulties previously experienced with the wet lay-up process. These composites possess good heat and chemical resistance, excellent surface hardness and fire resistance.

Furan Resins. (furfuryl resins) Dark colored thermosetting resins obtained primarily by the condensation polymerization of furfuryl alcohol in the presence of strong acids, sometimes in combination with formaldehyde or furfuryldehyde. The term also includes resins made by condensing phenol with furfuryl alcohol or furfuryl, and furfuryl-ketone polymers. The resins are available as liquids ranging from low viscosity fluids to thick heavy syrups, which cure to highly cross-linked, brittle substances. They are used for impregnating cured plaster structures, as binders for foundry sand cores, for binding high explosives, and as wood adhesives. The cured resins exhibit good resistance to chemicals and solvents. Improved resin/catalyst systems have made the older furan systems obsolete and enabled the use of fire-retardant furans in hand lay-up, spray-up and filament winding operations, competing with polyesters.

Furfural. C_4H_3OCHO (ant oil, fural, 2-furaldehyde, furfuraldehyde, pyromucic aldehyde) A liquid obtained by distilling acid-digested corn cobs, oat hulls, rice hulls or cottonseed hulls. It is colorless when first distilled, but darkens on exposure to air. Furfural is used as a solvent, and in the production of furans and tetrahydrofuran compounds. See also FURAN RESINS.

Furnace Black. A type of CARBON BLACK, which see, made in a refractory-lined furnace.

Fusion. With respect to vinyl plastisols and organosols, fusion is the state attained in the course of heating when all of the resin particles have dissolved in the plasticizers present, so that upon cooling a homogeneous solid solution results. Should not be confused with CURE, which see. In the 1940's, especially in Europe, the terms *gelation, gelatinization,* and *fluxing* were widely used for the state now known as fusion. *Curing* was also used, although improperly.

Fusion Temperature. In vinyl dispersions, the temperature at which fusion occurs. The *optimum fusion temperature* is that at which maximum physical properties are obtained before any thermal degradation is encountered. The term *minimum fusion temperature* is sometimes employed for that at which a substantial degree of physical properties is attained, usually on the order of 75% of ultimate properties. Also called *fluxing temperature* and, in some European literature, *gelling temperature.*

Fuzz. An accumulation of short broken filaments collected from passing glass strands, yarns or rovings over a contact point. The fuzz may be weighed and used as an inverse measure of abrasion resistance.

Gamma-. A prefix denoting the position of a group of atoms or a radical in the main group of a compound.

Gamma Rays. (Nuclear X-Rays) Quanta of electromagnetic energy emitted from radioactive substances. They are similar to but of much higher energy than ordinary X-rays. Gamma rays are highly penetrating, being capable of passing through several centimeters of lead.

Gamma Transition. See GLASS TRANSITION.

Gap. In filament winding, an unintentional space between two windings that should lay next to each other.

Gardner-Holt Bubble Viscometer. See AIR BUBBLE VISCOMETER.

Gas Black. See CARBON BLACK.

Gas Chromatography. A method of analysis by which the specimen is vaporized and introduced into a stream of carrier gas (usually helium), which stream is then pushed through a chromatographic column which separates it into its constituent parts. These fractions pass through the column at characteristic rates, and are detected as they emerge in a time sequence by a device such as a thermal conductivity cell. The detecting cell responses are recorded on a strip chart, from which the components can be identified both qualitatively and quantitatively. See also CHROMATOGRAPHY.

Gas-Liquid Chromatography. A variation of gas chromatography process in which the chromatographic column is packed with a finely divided solid impregnated with a non-volatile organic liquid. The sample to be analyzed is injected into the inlet of the column where it is quickly and completely vaporized. The gas stream carries it into the packed section, where the vapors contact the impregnated solids. The non-volatile liquid phase tends to condense the vapors from the gas stream, and the moving gas phase tends to evaporate the condensed sample vapors. The vapors of each compound present in the sample in effect spend a characteristic fraction of time in the condensed phase, and the remainder in the mobile gas phase. Each chemical species will tend to migrate at its own characteristic rate and will be separated from other species by the time they emerge from the column. The detector senses the emergence of each sample component and provides an electrical signal to the recorder proportional to the concentration of each component in the emergent stream.

Gas-Phase Polymerization. A polymerization process developed by Union Carbide for high-density polyethylene, particularly for a grade preferable for making paper-like films. Purified gaseous ethylene and a highly active chromium-containing catalyst in dry powder form are fed continuously into a fluidized-bed reactor. The resin is formed directly in powder form, avoiding gels, discoloration and contamination problems associated with conventional polymerization processes.

Gas Transmission Rate. (gas permeability) The rate at which a given gas will diffuse through a stated area of a specimen at a standard pressure and temperature. The result is usually expressed as the volume or weight of gas per 24 hours per 100 square inches of membrane. Plastics vary widely in this property, ranging from zero permeability to most gases for polyvinyl alcohol, to fairly high rates in the case of polycarbonate and the polyolefins.

Gas Welding. See HOT GAS WELDING.

Gate. In injection and transfer molding, the channel through which the molten resin flows from the runner into the cavity. It may be of the same cross section as the runner, but more often is restricted to 1/8″ or less. When the gate diameter is very small, e.g. 0.02″, it is known as a *pinpoint gate*. A *submarine gate* is shaped to carry the material down below the parting line of the mold and into the cavity at a point just below its edge. Other types of gates are the *fan gate* (a relatively wide and thin edge gate) and the *tab gate* (one which extends the runner system into the molded part). The term *gate* is also used for the portion of the plastic molding formed by the gate orifice.

Gate Blush. (gate splay) A blemish or disturbance in the gate area of an injection molded article. It occurs when the melt fractures in leaving the gate due to relaxation of elastic forces.

Gauge Band. A term used in the packaging film industry for a thickness irregularity found in rolls of film. A thick area in a film will produce a raised ring in a finished roll. Conversely, a thin area will result in a soft ring. Problems with gauge irregularities are more severe with thinner films because it is more likely that they will tear during stretching or burn through during the shrink cycle.

Gauss. In the old cgs system, the unit of magnetic induction (flux density). it is equal to 1 maxwell per cm². Its value is such that if a conductor 1 cm long moves through a magnetic field at a velocity of 1 cm, in an induction mutually perpendicular, the induced emf is one abvolt. In the new SI system, the gauss is to be expressed as 1.000 000*E-04 tesla (T).

Gay-Lussac's Law. See CHARLES' LAW.

Gel. (n) (1) A semisolid system consisting of a network of solid aggregates in which a liquid is held. (2) The initial jelly-like solid phase that develops during the formation of a resin from a liquid. Note: Both types of gel have very low strengths and do not flow like a liquid. They are soft, flexible and will rupture under their own weight unless supported externally. (ASTM D 883-65T). (3) A defect in plastic film such as polyethylene or PVC characterized by hard, glassy particles which appear in an otherwise clear, transparent film. Such defects are believed to be cross-linked particles which fail to flow out uniformly with the melt because of their network structure. They can be also introduced into a compound via an additive such as impact modifier or a processing aid. (4) In physical chemistry, a semi-rigid colloidal dispersion of a solid with a liquid or gas. (5) For a special meaning of the term *gel* in connection with vinyl plastisols and organosols, see GELATION.

Gelation. The formation of a GEL, which see. With regard to vinyl plastisols and organosols, gelation is the change of state from the liquid to the solid condition that occurs in the course of heating and/or ageing, when the plasticizer has been absorbed by the resin to an extent resulting in a dry but weak and crumbly mass. Within normal proportions of resin and plasticizer, this state is attained when the resin particles have become so swollen by diffusion of plasticizer into them that they touch each other. As heating progresses the swollen particles begin to weld together, resulting in some degree of strength. However, the state of gelation is considered to exist up to the point at which useful degrees of physical properties are attained, such as the CLEAR POINT, which see. Much confusion has existed regarding the meaning of this term because in the early years of the art, especially in Europe, *gelation* was used in place of the term *fusion* now employed in the United States.

Gelation Time. With reference to thermosetting resins, the interval of time between introduction of the catalyst and the formation of a gel.

Gel Coat. In reinforced plastics, a thin outer layer of resin, sometimes containing pigment, applied to give the structure its surface gloss and finish. It also serves as a barrier to liquids and ultraviolet radiation. The gel coat is usually the first to be applied to the mold in the lay-up process (after the mold release agent), becoming permanently bonded to the successive layers of reinforcement and resin.

Gel Effect. See AUTOACCELERATION.

Gelling Agents. See THICKENING AGENTS.

Gel Permeation Chromatography (GPC). A recently developed column chromatography technique employing as the stationary phase a swollen gel made by polymerizing and cross-linking styrene in the presence of a diluent which is a non-solvent for the styrene polymer. The polymer to be analyzed is introduced at the top of the column and then is eluted with a solvent. The polymer molecules diffuse through the gel at rates depending on their molecular size. As they emerge from the bottom of the column they are detected by a differential refractometer coupled to a recording chart, on which a molecular size distribution curve is plotted. This method yields results more rapidly than other chromatographic processes. See also CHROMATOGRAPHY.

Gel Point. The stage at which a liquid begins to exhibit pseudo-elastic properties. Note: This stage may be conveniently observed from the inflection point on a viscosity-time plot. See GEL. (ASTM D 883-75a).

Geodesic. Pertaining to the shortest distance between two points on a surface.

Geodesic Isotensoid. A filamentary structure in which there exists a constant stress in any given filament at all points in its path.

Geodesic-Isotensoid Contour. In filament-wound reinforced plastics pressure vessels, a dome contour in which the filaments are placed on geodesic paths so that the filaments will exhibit uniform tensions throughout their length under pressure loading.

Geodesic Ovaloid. A contour for end domes, the fibers forming a geodesic line on the surface of revolution. The forces exerted by the filaments are proportioned to meet hoop and meridional stresses at any point.

Geometric Metamerism. See METAMERISM.

Germicidal Lamp Test. A rapid screening method for evaluating the relative resistance of vinyl compounds to discoloration and degradation upon exposure to light and weather. The specimens are placed approximately three inches below a germicidal lamp with a special output of 2537 Angstroms in wave length. The materials are rated according to the degree of discoloration and plasticizer spewing after a specified interval of time, e.g. 24 or 48 hours.

Gibbs Indophenol Test. A test for detecting the presence of free phenol in phenolic parts after curing, as an indication of the completeness of cure. A few drops of dibromoquinone chloroimide reagent are added to an aqueous extract of the resin which has been rendered slightly alkaline. A bright blue color indicates the presence of phenols.

Giga-. (G) The SI-approved prefix for a multiplication factor to replace 10^9.

Glass Fiber Reinforcements. A family of reinforcing materials for reinforced plastics based on single filaments of glass ranging in diameter from 0.00012″ to 0.00075″. Single filaments are produced by mechanically drawing molten glass streams, then these filaments are usually gathered into bundles called *strands* or *rovings*. The strands may be used in continuous form for filament winding; chopped into short lengths for incorporation in molding compounds or use in "spray-up" processes; or formed into fabrics and mats of various types for use in hand lay-up, matched-die molding and other laminating processes. The glass fibers are usually coated with a material known as a COUPLING AGENT, which see, which serves to promote adhesion of the glass to the specific resin being used. Glass fiber reinforcements are classified according to their properties. "E" glass is electrical-grade glass, the most common general-purpose type. "C" glass is the chemical grade, and "S" glass denotes high tensile strength. Glass fibers coated with nickel by the electron beam deposition process are used in molding compounds for producing electrically conductive articles.

Glass Finish. See COUPLING AGENT.

Glass Flake. A filler produced by blowing molten type E glass into a very thin tube, then smashing the tube into small fragments. The flakes pack closely in thermosetting resin systems, producing strong products with good moisture resistance.

Glassine. Thin transparent paper treated with ureaformaldehyde resin, used for packaging.

Glass-Reinforced Plastics. See REINFORCED PLASTICS.

Glass Spheres. Solid glass spheres of diameters ranging from 5 to 5000 microns are used as fillers and/or reinforcements in both thermosetting and thermoplastic compounds. The size used most frequently is less than 325 mesh, with an average sphere diameter of 30 microns. The spheres are available with various silane coupling agent coatings to improve bonding between the polymer and the glass. They improve physical properties, assist flow and mold filling, and reduce costs of materials, processing and end products.

Glass Stress. In a filament wound part, usually a pressure vessel, the stress calculated by using the load and the cross sectional area of the reinforcement only.

Glass Transition. (gamma transition, second order transition, rubber transition) A reversible change that occurs in an amorphous polymer when it is heated to a certain temperature range, characterized by a rather sudden transition from a hard, glassy or brittle condition to a flexible or elastomeric condition. Other properties such as coefficient of thermal expansion, specific heat and density usually undergo changes at the same time. The transition occurs when the polymer molecular chains, normally coiled, tangled and motionless at temperatures below the glass transition range, become free to rotate and slip past each other. This temperature varies widely among polymers; for example, the glass transition temperature for polystyrene is in the neighborhood of 100 °C, and that of a 75/25 copolymer of butadiene and styrene is near −50 °C.

Glass Transition Temperature. (Tg) The approximate midpoint of the temperature range over which glass transition takes place. The Tg can be determined readily only by observing the temperature at which a significant change takes place in a specific electrical, mechanical, or other physical property. Moreover, the observed temperature can vary significantly depending on the specific property chosen for observation and on details of the experimental technique (for example, rate of heating, frequency). Therefore, the observed Tg should be considered only an estimate. The most reliable estimates are normally obtained from the loss peak observed in dynamic mechanical tests or from dialatometric data. (ASTM D 883-75a).

Glitter. (flitter, spangles) A family of decorativve pigments comprising light-reflective flakes of sizes large enough so that each separate flake produces a visible sparkle or reflection. They are incorporated into the plastic material during compounding.

Glossmeter. An instrument for measuring mar resistance of plastics, described in ASTM D673. Light from a standard source is directed at a 45° angle at the abraided specimen, and the reflected light is measured by a photoelectric cell and a galvanometer.

Glycerol. $CH_2OHCHOHCH_2OH$ (1,2,3-Propanetriol, glycerin, glycerine, glycyl alcohol) The term *glycerol* applies to the pure product; *glycerin* applies to commercial products containing at least 95% of glycerol; and *glycerine* is an improper spelling. Glycerol is a colorless, viscous liquid derived from soap manufacture as a by-product, or, more recently, synthesized from propylene and sugar. Its uses in the plastics industry include the manufacture of alkyd resins (esters of glycerol and phthalic anhydride); the plasticization of cellophane; and the production of urethane polymers.

Glycerol Diacetate. See DIACETIN.

Glycerol Ether Acetate. $C_3H_5(OCH_2CH_2)_n(OCOCH_3)_3$ A plasticizer for cellulosics and polyvinyl acetate.

Glycerol Monoacetate. $CH_3COOCH_2CHOHCH_2OH$ A plasticizer for cellulose acetate, cellulose nitrate, and vinyl resins.

Glycerol Mono-Lactate Triacetate. A plasticizer for cellulose acetate, imparting resistance to gasoline.

Glycerol Monolaurate. $C_{11}H_{23}COOCH_2CHOHCH_2OH$ (glyceryl monolaurate, lauryl glycerin) A plasticizer for cellulosic and vinyl resins and polystyrene.

Glycerol Mono-Oleate. $(C_{17}H_{33})COOCH_2CHOHCH_2OH$ A yellow oil, approved by FDA as a plasticizer for food packaging.

Glycerol Monoricinoleate. $C_{17}H_{32}(OH)COOCH_2CHOHCH_2OH$ A plasticizer for cellulose nitrate, ethyl cellulose and polyvinyl butyral.

Glycerol Phthalic Anhydride. An alkyd resin made by modifying glycerol phthalate with an equal portion of oil, fatty acid and natural or synthetic resin. it is used in varnishes, lacquers and enamels.

Glycerol Triacetate. (triacetin) $CH_3COOCH_2CH(OCOCH_3)CH_2OCOCH_3$ A plasticizer for cellulosic resins, PMMA and polyvinyl acetate. It has been approved by FDA for food contact use.

Glyceryl Tribenzoate. (tribenzoin) $C_3H_5(OOCC_6H_5)_3$ A colorless crystalline solid, used as a solid plasticizer for rigid PVC.

Glycerol Tributyrate. $C_3H_5(OCOC_3H_7)_3$ A plasticizer for cellulose acetate, cellulose acetate butyrate and cellulose nitrate.

Glycerol Tripropionate. $C_3H_5(OCOC_2H_5)_3$ (glyceryl tripropionate) A plasticizer for cellulosic plastics and polyvinyl acetate.

Glyceryl Diacetate. See DIACETIN.

Glyceryl Tri-(12-acetoxystearate). $C_3H_5(OOCC_{17}H_{34}OCOCH_3)_3$ A plasticizer for cellulosic and vinyl resins.

Glyceryl Tri-(12-acetylricinoleate). $C_3H_5(OOCC_{17}H_{32}OCOCH_3)_3$ A plasticizer for cellulose nitrate, ethyl cellulose and PVC.

Glycidol. CH_2OCHCH_2OH (2,3-epoxy-1-propanol) A stabilizer for vinyl resins.

gamma-Glycidoxypropyltrimethoxysilane. $CH_2 \cdot CH\text{-}CH_2\text{-}O(CH_2)_3Si(OCH_3)_3$ A silane coupling agent used in reinforced thermosetting and thermoplastic resins.

Glycidyl Ester Resins. A family of epoxide resins derived from the condensation of epichlorohydrin with polycarboxylic acids, first available in commercial quantities in 1968. The preferred curing agents for these resins are anhydrides, polycarboxylic acids, aromatic amines and phenolics. Advantages claimed for glycidyl ester resins are electrical tracking resistance, high physical strength, high modulus combined with toughness, good weatherability, low viscosity, long pot life and high reactivity at moderately elevated temperatures. Limitations are poorer properties at temperatures over 100°C, higher shrinkage, poor alkali resistance, and inability to cure at room temperature.

Glycidyl Ether Resins. See EPOXY RESINS.

Glycol. A compound having hydroxyl groups on adjacent carbon atoms, for example ethylene glycol, $HO\text{-}CH_2\text{-}CH_2\text{-}OH$. The term glycol is also used for dihydric alcohols, that is, alcohols containing two hydroxyl groups.

Glycol Phthalates. A type of thermoplastic polyester used mainly for fibers. See POLYESTERS, SATURATED.

Glyptal Resins. See ALKYD MOLDING COMPOUNDS.

Gold Leaf Stamping. See HOT STAMPING.

Goniophotometry. A procedure for evaluating the manner in which materials geometrically redistribute light, described in ASTM E 166-60T and E 167-60T.

GP. Abbreviation for *general purpose,* sometimes used to denote types of resins and molding compound suitable for a wide range of applications.

GPC. Abbreviation for GEL PERMEATION CHROMATOGRAPHY, which see.

GR-1. See BUTYL RUBBER.

Gradient Tube Density Determination. A convenient method for measuring densities of very small samples, often used in the plastics industry. A vertical glass tube (the gradient tube) is filled with a heterogeneous mixture of two or more liquids, the density of the mixture varying linearly or in other known fashion with the height. A drop or small particle of the specimen is introduced in the tube and falls to a position of equilibrium which indicates its density by comparison with positions of known standard samples.

Graft Copolymer. See GRAFT POLYMER.

Graft Polymer. A polymer comprising molecules in which the main backbone chain of atoms has attached to it at various points side chains containing different atoms or groups from those in the main chain. The main chain may be a copolymer or may be derived from a single monomer.

Grain Alcohol. See ETHYL ALCOHOL.

Gram Atom. (gram atomic weight) The mass of a substance in grams numerically equal to the atomic weight of the substance.

Gram Equivalent. The weight of a substance displacing or otherwise reacting with 1.008 grams of hydrogen or combining with one-half of a gram atomic weight (8.00 grams of oxygen).

Granular Polymerization. See SUSPENSION POLYMERIZATION.

Granular Structure. Nonuniform appearance of finished plastic material due to retention of, or incomplete fusion of, particles of composition either within the mass or on the surface; or to the presence of coarse filler particles.

Granulates. Molding compounds in the form of spheres or small cylindrical pellets.

Granulators. (Scrap grinders) Machines for cutting waste material such as sprues, runners, excess parison material and reject parts into particles which can be reused. The most common form of granulator comprises a series of two or more rotating knives passing in close proximity to stationary knives, and a screen through which particles of the desired size are discharged.

Graphite. (plumbago, black lead) A crystalline form of carbon with atoms arranged hexagonally, characterized by a soft, greasy feel. It occurs naturally, and is also produced synthetically by heating petroleum coke or other organic materials under controlled atmospheric conditions. In powder form, graphite is used as a lubricating filler for nylon and fluorocarbon resins. Pyrolytic graphite fibers, made by decomposing organic filaments at high temperatures in controlled atmospheres, have recently been developed as reinforcements for high-performance applications. Graphite WHISKERS, which see, are also used as high-performance reinforcements.

Grasshopper. A stiff bunch of parallel strands in a fibrous mat.

Gravure Coating. (engraved roll coating) A roller coating process in which the amount of coating applied to the web is metered by the depth of an all-over engraved pattern in the application roll. This process is frequently modified by interposing a resilient offset roll between the engraved roll and the web.

Gravure Printing. The depressions in an engraved printing cylinder or plate are filled with ink, the excess on raised portions being wiped off by a doctor blade. Ink remaining in the depressions is deposited on the plastic film as it passes between the gravure roll and a resilient back-up roll.

Gray. The SI unit of absorbed dose, defined as the energy imparted by ionizing radiation to a mass of matter corresponding to one joule per kilogram.

Grex Number. The weight in grams of 10 kilometers of a yarn or fiber. See also CUT, DENIER, TEX.

Grid. (n) Channel-shaped mold-supporting members.

Grinding-Type Resin. A vinyl resin which requires grinding to effect dispersion in plastisols or organosols.

Grit Blasting. A mold finishing process in which abrasive particles are blasted onto the mold surfaces in order to produce a roughened surface. This process is often used on molds for blow molding in order to assist air escape, and on molds for other processes in order to produce a desired texture.

GRP. Abbreviation for glass reinforced plastic. See REINFORCED PLASTIC.

GR-S. Abbreviation for "Government Rubber-Styrene," a copolymer of 75 parts of butadiene with 25 parts of styrene, also known as Buna S and SRB. See also STYRENE BUTADIENE RUBBERS.

Guide Eye. In filament winding, a metal or ceramic loop (eye) through which the fiber passes when directed from the creel to the mandrel.

Guide-Pin Bushing. (dowel-pin bushing) A guiding bushing through which the leader pin moves.

Guide Pins. (dowel pins) In compression, transfer and injection molding, hardened steel pins that maintain proper alignment of the mold halves as they open and close.

Guignet's Green. See CHROME OXIDE GREEN.

Gums. Although rarely used in the plastics industry, this term is used in other industries to include materials that can be dissolved in water to produce viscous or mucilaginous solutions. Thus, water soluble polymers such as PVP, PVA, ethylene oxides, polyacrylic acid and polyacrylamide are regarded as gums.

Guncotton. A highly explosive form of CELLULOSE NITRATE, which see, made by digesting clean cotton in a mixture of 1 part nitric acid and 3 parts sulfuric acid.

Gunk. Colloquial term for a premixed charge for premix molding which contains all of the ingredients for molding, usually chopped roving, resin, pigment, filler and catalyst.

Gussets. The tucks in the sides of bags made from plastic film, which permit bags to assume a rectangular form when opened. Gussets may be formed in the tubular LAY FLAT FILM (which see) from which bags are made.

Gutta-Percha. A rubber-like polymeric substance extracted from the milky sap of leaves and bark of certain trees belonging to the family *Sapotaceae,* such as Palaquim, Payena and Dichopsis plants native to Malaya, Borneo and Sumatra. It has the same empirical formula as natural rubber, but is the *trans* isomer of polyisoprene whereas natural rubber is the *cis* isomer. Gutta Percha is a tough, horny substance at room temperature, but becomes soft and tacky upon warming to 100 °C. It is used in compounds for golf ball covers, electrical insulation, cutlery handles, machinery belting, and as a stiffening agent in natural rubber.

Gutta Percha, Synthetic. See POLYISOPRENE.

Halides. Binary compounds of the halogen family of elements which comprises astatine, bromine, chlorine, fluorine and iodine. The term is sometimes incorporated in the names of plastics containing one of these elements, for example polyvinyl chloride is sometimes referred to as a polyvinyl halide.

Halocarbon Plastics. A term listed by ASTM for polymers containing only carbon and one or more halogens. The primary members are the FLUOROCARBON RESINS, which see.

Halogens. The elements fluorine, chlorine, bromine, iodine, and astatine.

Hand. The softness of a piece of film, fabric or coated fabric as judged by the touch of a person.

Hand Lay-up Molding. See LAY-UP MOLDING.

Hand Mold. A mold that is removed from the press after each shot for removal of the molded article; generally used only for short runs and experimental moldings.

Hardeners. See CURING AGENTS.

Hard Fibers. Fibers produced from leaves.

Hardness. The resistance of a plastic material to compression, indentation and scratching. See BARCOL HARDNESS, BRINELL HARDNESS, INDENTATION HARDNESS, MOHS VALUE, ROCKWELL HARDNESS, SCRATCH HARDNESS, VICKERS HARDNESS, KNOOP MICROHARDNESS TEST, DUROMETER.

Haze. The cloudy or turbid aspect of appearance of an otherwise transparent specimen caused by light scattered from within the specimen or from its surfaces. Note: For the purpose of ASTM Method D 1003, Test for Haze and Luminous Transmittance of Transparent Plastics, haze is the percentage of transmitted light which, in passing through the specimen, deviates from the incident beam through forward scatter more than 2.5 degrees on the average. (ASTM D 883-75a).

HDPE. Abbreviation for high density polyethylene. See POLYETHYLENES.

Head. In a blow molding operation, the end section of the apparatus in which the melt becomes transformed into a hollow parison.

Head-to-Head Polymers. Polymers in which the monomeric units are alternately reversed as in the following polymer of the monomer $CH_2 = CHR$:

$$CH_2{-}\underset{\underset{R}{|}}{CH}{-}\underset{\underset{R}{|}}{CH}{-}CH_2CH_2{-}\underset{\underset{R}{|}}{CH}{-}\underset{\underset{R}{|}}{CH}$$

Head-to-Tail Polymers. Polymers in which the monomeric units regularly repeat as in the following polymer of the monomer $CH_2 = CHR$:

$$CH_2{-}\underset{\underset{R}{|}}{CH}{-}CH_2{-}\underset{\underset{R}{|}}{CH}$$

Heat Distortion Point. (deflection temperature under load, tensile heat distortion temperature) The temperature at which a standard test bar (ASTM D648) deflects .010 inches under a stated load of either 66 or 264 psi.

Heat Equivalent of Fusion. The quantity of heat necessary to change one gram of a solid substance to its liquid state with no change in the temperature of the substance.

Heated Tool Welding. See HOT PLATE WELDING.

Heater-Adapter. That part of an extrusion die around which the heating medium is held.

Heater Bands. Electrical heating units shaped to fit extruder barrels, injection molding cylinders and the like, for heating the plastic material to the desired temperature.

Heat Forming. See THERMOFORMING.

Heating Cylinder. (heating chamber) In injection molding, that part of the machine in which the cold molding compound is heated to the molten condition before injection.

Heat of Combustion. The amount of heat evolved by the combustion of one gram molecular weight of the substance.

Heat Mark. An extremely shallow depression or groove in the surface of a plastic visible because of a sharply defined rim or a roughened surface. See also *shrink-mark*. (ASTM D 883-75a).

Heat Sealing. The process of joining two or more thermoplastic films or sheets by heating areas in contact with each other to the temperature at which fusion occurs, usually aided by pressure. When the heat is applied by dies or rotating wheels maintained at a constant temperature, the process is called *thermal sealing*. In *impulse* sealing, heat is applied by resistance elements which are applied to the work when relatively cool, then rapidly are heated. Simultaneous sealing and cutting can be performed by this method. *Dielectric sealing* is accomplished by inducing heat within the films by means of radio frequency waves. When heating is performed by means of ultrasonic vibrations, the process is called *ultrasonic sealing*. See also WELDING.

Heat Sink. A device for the absorption or transfer of heat away from a critical part or element.

Heat Stability. The resistance to change in color or other properties as a result of heat encountered by a plastic compound or article in either processing or end use. Such resistance may be enhanced by the incorporation of a stabilizer.

Heat Transfer. The movement of energy in the form of heat from one body to another body which is colder. The three methods by which heat can be transferred are radiation, conduction and convection. Radiation heating is effected when heat passes from the emitting body to the receiving body through a medium which is not warmed. Conduction heating is the transfer of heat from one zone to a colder zone in either a single homogeneous substances or two substances in contact with each other. Convection is the transfer of heat by the flow of a fluid, either a gas or a liquid, and either by natural convection currents caused by differences in density or by forced movement caused by pumping.

Heavy Spar. See BARIUM SULPHATE.

Hecto-. (h) The SI-approved prefix for a multiplication factor to replace 10^2.

Hehner Number. (Hehner value) A number expressing the percentage of water-insoluble fatty acids in an oil or fat.

Helical Screw Feeders. Devices for conveying and metering dry materials, comprising a tube containing a screw, fed from a supply hopper.

Helical Winding. A winding in which the filament or band advances along a helical path, not necessarily at a constant angle except in the case of a cylindrical article.

Henry. The inductance of a closed circuit in which an electromotive force of one volt is produced when the electric current in the circuit varies uniformly at a rate of one ampere per second.

Henry's Law. The mass of a slightly soluble gas that dissolves in a definite mass of a liquid at a given temperature is very nearly directly proportional to the partial pressure of that gas. This law is valid only for gases that do not unite chemically with the solvent.

n-Heptyl n-Decyl Phthalate. A general-purpose plasticizer for PVC and several other thermoplastics with better volatility and low-temperature performance than DOP. It is excellent for use in vinyl plastisols.

n-Heptyl n-Nonyl Adipate. A plasticizer for PVC similar to DOA but with better low temperature performance and lower volatility. It is also approved for food-contact use.

n-Heptyl n-Nonyl Trimellitate. A plasticizer similar to TOTM, with better low-temperature performance and lower volatility.

Hertz. (Hz) Cycles per second of electromagnetic waves, named after the German scientist who first investigated radio telegraphy. The term and its multiple Megahertz (MHz, one million cycles per second) are used most frequently for waves in the radio frequency range, but there is a growing trend toward its use for low frequency electrical waves. In the new SI system of units the Hertz is not limited to electrical waves, but is defined as the frequency of a periodic phenomenon of which the period is one second.

Heterocyclic Compounds. Compounds containing molecules whose atoms are arranged in a ring, the ring containing two or more chemical elements.

Heteropolymer. A copolymer formed by HETEROPOLYMERIZATION, which see.

Heteropolymerization. A special case of additive copolymerization which involves the combination of two dissimilar unsaturated organic monomers.

Hevea Rubber. See RUBBER, NATURAL.

Hexabromobiphenyl. A flame retardant suitable for use in thermosetting resins and thermoplastics such as ABS, nylon, polycarbonates, polyolefins, PVC, PPO and SAN. It is insoluble in water, heat-stable, and furnishes a high bromine content in the end product.

Hexachloroethane. Cl_3CCCl_3 (carbon trichloride, carbon hexachloride, perchloroethane) A substitute for camphor in celluloid manufacture.

Hexachlorophene. $(C_6HCl_3OH)_2CH_2$ A white, essentially odorless, free flowing powder widely used as a bacteriostat in many thermoplastics including vinyls, polyolefins, acrylics and polystyrene.

Hexafluoroacetone Sesquehydrate. A solvent cement, active at room temperatures, for bonding acetal resin articles to themselves and to other polymers such as nylon, ABS, styrene-acrylonitrile, polyester, cellulosics, and natural or synthetic rubber.

Hexahydrophenol. See CYCLOHEXANOL.

Hexahydrophthalic Anhydride. $C_6H_{10}(CO)_2O$ (HHPPA) A curing agent for epoxy resins, and an intermediate for the production of alkyd resins for coatings.

Hexamethylene. See CYCLOHEXANE.

Hexamethylene Adipamides. (Nylon 6/6) The type of nylon made by condensing hexamethylenediamine with an adipic acid.

Hexamethylenediamine. $H_2N(CH_2)_6NH_2$ (1,6-diaminohexane) A colorless solid in leaflet form, which when condensed with an adipic acid forms Nylon 6/6.

1,6-Hexamethylene Diisocyanate. (HDI) $OCH(CH_2)_6NCO$ A colorless liquid, the first aliphatic diisocyanate to be used commercially in the production of urethanes. When used with certain metal catalysts it produces urethane polymers with good resistance to discoloration, hydrolysis and heat degradation.

Hexamethylenetetramine. $(CH_2)_6N_4$ (methenamine, ammonioformaldehyde) (HMTA) The reaction product of ammonia and formaldehyde, used as a basic catalyst and accelerator for phenolic and urea resins, and a solid catalytic-type curing agent for epoxy resins.

Hexamethylphosphoric Triamide. $[N(CH_3)_2]_3PO$ A pale water-soluble liquid used as an ultraviolet absorber in PVC compounds.

Hexane. C_6H_{14} A straight-chain hydrocarbon, derived from petroleum or natural gas. Commercial grades contain other hydrocarbons such as cyclohexane, methyl-cyclopentane and benzene. Hexanes are used as catalyst carrying solvents in the polymerization of olefins and elastomers.

Hexanedioic Acid. See ADIPIC ACID.

Hexyl. The straight-chain radical C_6H_{13} of the hexane series.

Hexyl Acetate. $CH_3COOC_6H_{13}$ A solvent for cellulose nitrate, CAB resins, PVAc, polystyrene, phenolics, alkyds, and coumarone-indene resins.

Hexyl Methacrylate. $C_6H_{13}OOCC(CH_3):CH_2$ One of the monomers used for the production of acrylic resins.

n-Hexyl, n-Octyl, n-Decyl Phthalate. (NODP) See n-OCTYL, n-DECYL PHTHALATE.

n-Hexyl Trimellitate. A permanent plasticizer for vinyls and PVC/ABS blends characterized by resistance to fogging, extraction and migration.

H.F. Abbreviation for HIGH FREQUENCY, which see.

H.F. Preheating. See DIELECTRIC HEATING.

High Density Polyethylene. This term is generally considered to include polyethylenes ranging in density from about .94 to .96 and over. Whereas the molecules in low density polyethylene are branched and linked in random fashion, those in the higher density polyethylenes are linked in longer chains with fewer side branches, resulting in more rigid material with greater strength, hardness, chemical resistance and higher softening temperature. See also POLYETHYLENES.

High Frequency. The electrical frequency range approximately between 3 and 200 mc per second, employed in plastics welding and sealing operations. Frequencies of 30 mc and below are most often employed.

High-Frequency Heating. (H. F. HEATING). The heating of materials by dielectric loss in a high-frequency electrostatic field. The material is exposed between electrodes, and by absorption of energy from the electrical field is heated quickly and uniformly throughout.

High Frequency Welding. A method of welding thermoplastic articles in which the surfaces to be joined are heated by contact with electrodes of a high frequency electrical generator.

High-Load Melt Index. The rate of flow of a molten resin through an orifice of 0.0825 inch diameter when subjected to a force of 21,600 grams at 190 °C. See also MELT INDEX.

High Polymer. A polymer with molecules of high molecular weight, sometimes arbitrarily designated as greater than 10,000. All materials commonly regarded as plastics are high polymers, but not all high polymers are plastics. See POLYMER.

High Pressure Laminates. Laminates molded and cured at pressures not lower than 1000 p.s.i. and more commonly in the range of 1200 to 2000 p.s.i. See also LAMINATE.

High Pressure Molding. A molding or laminating process in which the pressure used is greater than 200 p.s.i. (ASTM D 883-75a).

High Pressure Powder Molding. Some polymers in powder form can be molded by high-pressure compaction at room temperature followed by heating to complete curing or polymerization reactions. The process is limited to polymers that do not release vapors when heated, and is most successful with semi-crystalline polymers which can be post-heated for a sufficient time at a temperature within the crystalline endotherm of the polymer. Examples of such polymers are PPO, PP and PTFE.

High Pressure Spot. In reinforced plastics, an area containing very little resin, usually due to an excess of reinforcing material.

High Temperature Plasticizers. Plasticizers which impart higher than normal resistance to high temperatures to plastics compounds in which they are incorporated. An example is di-tridecyl phthalate, which permits vinyl compounds to be used in temperatures up to 136 °C.

Hindered Isocyanate. See ISOCYANATE GENERATOR.

Hob. (n) A master model of hardened steel, which is pressed into a block of softer metal to form a mold cavity.

Hobbing. (v) A process of forming a mold by forcing a hob of the shape desired into a soft metal block.

Hog. A machine for reducing particle size or grinding scrap, similar to a granulator but more heavily constructed, equipped with more cutting knives, and using forced air to urge the material through the perforated screen.

Hold-Down Groove. A small groove cut into the side wall of the molding surface to assist in holding the molded article in that member while the mold opens.

Holder Block. See CHASE.

Holomicroscopy. A three-dimensional photographic process utilizing laser beams and time differential interferometry. Part of the laser beam is split and directed through the sample under observation, the other part of the beam on another precise path eventually interfering with the first part. The beam that passes through the sample suffers a phase change, a slowing down, and the amount of slowing down can be seen in the hologram. The process is a valuable polymer research tool, enabling the observation of molecular crystals.

Homogenizer. A machine used to break up agglomerates and disperse elemental particles in fluids, consisting of a positive-displacement pump capable of attaining very high pressures, an orifice through which the material is pumped, and an impact ring on which the fluid stream impinges. Homogenizers are used in the preparation of monomer emulsions, coating compounds, and for dispersing pigments and plasticizers into resins. See also COLLOID MILL.

Homologous Series. A series of organic compounds that are identical except that each successive member has one more CH_2 group in its molecule than the preceding member. An example is the series *methanol* (CH_3OH), *ethanol* (C_2H_5OH), *propanol* (C_3H_7OH), and *butanol* (C_4H_9OH).

Homologous Temperature. The ratio of the absolute temperature of a material to its absolute melting temperature.

Homopolymer. A polymer resulting from the polymerization of a single monomer; a polymer consisting substantially of a single type of repeating unit.

Honeycomb. Manufactured product consisting of sheet metal or resin-impregnated sheet material (paper, fibrous glass, etc.) which has been formed into hexagonal-shaped cells. Used as core materials for sandwich constructions.

Hookean Elasticity. (ideal elasticity) The type of elasticity in which the deformation or strain of the material is proportional to the applied stress, in accordance with HOOKE'S LAW, which see.

Hooke's Law. With the elastic limit of any body the ratio of the stress to the strain produced is constant. This law may be expressed by the equation

$$T = E \frac{L - L_o}{L_o}$$

in which T is the imposed tensile stress, E is the constant of proportionality (Young's modulus or modulus of elasticity), L_o is the original length of the specimen, and L is the final length of the specimen.

Hoop Stress. The circumferential stress in a material of cylindrical form subjected to internal or external pressure.

Hopper. In extrusion or injection molding, the container holding a supply of molding material to be fed to the screw or ram. The hopper may be intermittently filled or continuously fed. (See HOPPER LOADER). Feeding from the hopper to the screw or ram may be by gravity, or aided by flow control devices such as disc feeders, rotary vane feeders, vibratory feeders, helical screw feeders or belt feeders.

Hopper Dryer. A combination feeding and drying device for extrusion and injection molding of thermoplastics. Hot air flows upward through the hopper containing the feed pellets. The efficiency of the apparatus can be improved by passing the hot air through a dessicant to reduce its humidity and increase its resin drying ability.

Hopper Loader. (hopper filler) A device for automatically feeding molding powders to hoppers of extruders, injection molding machines and the like. The functions of drying and blending colors with the molding powders are also sometimes accomplished by loaders. There are two general types of hopper loaders: mechanical and pneumatic. The mechanical systems use a rotating screw in a tube, or a conveyor belt on which are fastened small containers which dump their contents into the hopper. Pneumatic loading systems employ positive pressure from a blower, or negative pressure from a vacuum source, to convey the material through a tube.

Hotbench Test. A method of determining gelation properties of plastisols, employing a temperature gradient plate on which a film is cast.

Hot Gas Welding. A welding process used for plastics similar to that used for metals, except that open flame is not used because of excessive temperature and oxidation effects. Welding guns for plastics contain an electrically or gas heated chamber through which a gas, usually dry air or nitrogen, is passed. The heated gas is directed at the joint to be welded, while a rod of the same material as the thermoplastic being welded is applied to the heated area.

Hot-Leaf Stamping. See HOT STAMPING.

Hot Manifold Mold. An injection mold equipped with an internally heated torpedo located in the center of the melt stream in the manifold and nozzle system. This type of mold was developed for thermally sensitive resins such as acetals, which present temperature control problems with externally heated runners and nozzles because of extreme temperature striations.

Hot Melts. Thermoplastic compounds which are normally solid at room temperatures but become sufficiently fluid when heated to be pourable or spreadable. They are used as adhesives and coatings.

Hot Plate Welding. (hot tool welding) Two plastic surfaces to be joined are first held lightly against a heated metal surface, which may be coated with PTFE to prevent sticking, until the surface layers are melted. The surfaces are then quickly brought together and held under light pressure until cool. See also WELDING, THERMOBANDE WELDING.

Hot Runner Mold. (insulated runner mold) An injection mold for thermoplastics, in which the runners and sometimes the secondary sprues are insulated from the chilled cavities so that they remain hot during the entire cycle. The material in the center of the mold remains fluid so that it is not ejected with the molded part, thus avoiding the need for handling and reprocessing scrap normally generated from runners and sprues.

Hot-Short. (adj.) Inelastic, nonstretchable and easily broken in tension, when hot.

Hot Stamping. (roll leaf stamping, gold leaf stamping) A method of marking plastics in which a special pigmented, dyed or metallized foil is pressed against the plastic article by means of a heated die, welding selected areas of the foil to the article. The term also includes the process of impressing inked type into the material when the type is heated.

Hot-Tip Gate Molding. An injection molding process used for large-area thin articles molded in a single cavity. In conventional molding of multiple-cavity parts, sprues are needed to connect the mold filling orifice with runners leading to each cavity gate. In the hot-tip gate molding process the sprues are eliminated and material is injected directly from the mold-filling orifice through a heated bushing which serves as the sprue and gate for the individual cavity. Advantages are faster molding cycles, lower material waste, fewer post-molding operations, and reduced sink marks and flow lines.

Hot Wire Cutter. A device used for splitting certain types of plastic foam blocks into smaller pieces. Electrically heated wires slowly melt through the blocks as they are fed into the wires by gravity or conveyor.

HT-1. A type of nylon made from phenylenediamine and iso- or terephthalic acid, with good high temperature properties.

Hull. Dark speck of foreign matter which appears to be in the fabric of fabric-base laminated sheet. (ASTM D 883-65T).

Humectants. Agents which have a pronounced effect on the ability of moisture to adhere to a substance. They are sometimes used in anti-static coatings for plastics.

Humidity Blush. See BLUSHING.

Hydantoin Epoxy Resins. See EPOXY RESINS.

Hydrated Alumina. See ALUMINA TRIHYDRATE.

Hydroabietyl Alcohol. $C_{19}H_{31}CH_2OH$ A derivative of abietic acid, used as a plasticizer for cellulose nitrate, ethyl cellulose and PVC.

Hydrocarbon Plastics. Plastics based on resins made by the polymerization of monomers composed of carbon and hydrogen only. (ASTM D 883-75a). In the plastics industry, hydrocarbon resins are considered to be those thermoplastic resins of low molecular weight made from relatively impure monomers derived from coal-tar fractions, cracked petroleum distillates, and turpentine. The family includes COUMARONE-INDENE RESINS, which see; cyclopentadiene resins; petroleum resins; terpene resins; and many others of little commercial importance. Having little strength, the hydrocarbon resins are rarely used alone. Their primary applications are as binders in asphalt flooring, processing aids in elastomers and polyolefins, and coating additives.

Hydrogels. Three dimensional networks of hydrophilic polymers, generally covalently or ionically cross-linked. The most widely used polymer is poly(hydroxyethyl methacrylate), especially in medical applications such as implants as well as blood bags and syringes, etc.

Hydrogenated Methyl Abietate. $C_{19}H_{31}COOCH_3$ A derivative of abietic acid, used as a plasticizer for cellulose nitrate, ethyl cellulose, acrylic and vinyl resins, and polystyrene.

Hydrogenation. The combination of hydrogen with another substance, usually an unsaturated organic compound. The process usually requires elevated temperatures, high pressures and catalysts.

Hydrogen Equivalent. The number of replaceable hydrogen atoms in one molecule of a substance, or the number of atoms of hydrogen with which one molecule could react.

Hydrolysis. A double decomposition reaction between water and another substance in which water is split into its ions and reacts to form a weak acid or base or both.

Hydrophilic. Having an affinity for water.

Hydrophobic. Having a disaffinity for water; not capable of uniting or mixing with water.

Hydroquinone. $C_6H_4(OH)_2$ (quinol, hydroquinol, paradihydroxybenzene) A white crystalline material derived from aniline, used along with many of its derivatives as an inhibitor in unsaturated polyester resins and in monomers such as vinyl acetate. The hydroquinones are relatively colorless, and require trace amounts of oxygen in the polyester resin in order to be activated.

Hydroquinone Di(Beta-Hydroxyethyl) Ether. (HQEE) A white solid material used as a reactant in the preparation of polyesters, polyolefins and polyurethanes. When used as a chain extender in urethane prepolymers, HQEE increases the high temperature resistance of parts molded from the prepolymer up to 300°F.

Hydrosol. (1) In physical chemistry, a colloidal suspension in water. (2) In the plastics industry, a suspension of resin such as PVC or nylon in water, not necessarily of colloidal nature. See also LATEX.

2(2'-Hydroxy 3', 5 Ditertiarybutylphenyl) 7-Chloro-Benzotriazole. An off-white, non-toxic crystalline powder with high thermal stability, used as an ultraviolet absorber for polyolefins, PVC, polyurethanes, polyamides and polyesters.

Hydroxyethyl Acetamide. See n-ACETYL ETHANOLAMINE.

Hydroxyethyl Cellulose. A water-soluble, film-forming substance formed by reacting alkali cellulose with ethylene oxide. It is also used as a stabilizer for vinyl polymerization.

Hydroxyethylmethyl Methacrylate. A hydrophilic acrylic polymer that is rigid when dry but when saturated with water becomes a soft, clear material. It was introduced by Hydron Laboratories under the trademark Hydron, under license to patents held by the Czechoslovak Academy of Science. Early applications were soft contact lenses, masonry coatings and biomedical applications.

Hydroxyl Value. A measure of hydroxyl (univalent OH) groups in an organic material. In plasticizers, the hydroxyl value includes OH groups present in any free unesterified alcohol as well as those of the plasticizer molecule itself. In some plasticizers, large hydroxy values are indicative of the possibility of the plasticizer becoming incompatible on aging. In urethane technology, hydroxyl number is an important factor in the selection of polyols to achieve desired characteristics of elastomers and foams.

2-Hydroxy-4-Methoxy-Benzophenone. A U.V. absorber for many thermoplastics.

2(2′-Hydroxy 5′-Methylphenyl) Benzotriazole. An off-white, non-toxic crystalline powder with high thermal stability, used as an ultraviolet absorber for polystyrene, acrylics, PVC, polyesters and polycarbonate resins.

2-Hydroxy-4-Methoxy-5-Sulfobenzophenone. A U.V. absorber for thermoplastics.

2-Hydroxy-4-n-Octoxybenzophenone. $C_{21}H_{26}O_3$ A pale yellow powder useful as a UV light absorber for PVC and several other plastics. It is compatible with highly plasticized vinyls, and has a very low order of toxicity.

Hydroxypropylglycerin. A pale straw-colored liquid used as a plasticizer for cellulosic resins and as an intermediate for alkyd and polyester resins.

$$\overset{CH_3}{\underset{|}{}}\qquad\overset{CH_3}{\underset{|}{}}$$

Hydroxypropyl Methacrylate. (HPMA) $CH_2 = CCOO\text{-}CH_2CHOH$ A reactive monomer copolymerizable with a wide variety of acrylic and vinylic monomers, used for thermosetting resins and surface coatings.

Hygroscopic. Having the tendency to absorb moisture from the air. Some resins are hygroscopic, thus requiring drying before molding.

Hysteresis. (electrical) The magnetization of a body of iron or steel due to a magnetic field which is made to vary through a cycle of values, and the magnetization of the metal lags behind that of the field.

Hysteresis. (mechanical) The cyclic noncoincidence of the elastic loading and the unloading curves under cyclic stressing. The area of the resulting elliptical hysteresis loop is equal to the heat that is generated in the system..

Hysteresis Loop. The closed area between two curves on a graph plotting results of a changing force, first with ascending values, then with descending values. Such hysteresis loops are present in viscosity curves of thixotropic and dilatant fluids such as plastisols, and in stress-strain curves of tensile tests.

Hytrel. Du Pont's trademark for a family of copolyester elastomers. Typical reactants from which the elastomers are derived are terephthalic acid, polytetramethylene glycol, and 1,4-butanediol. Grades of Hytrel are marketed in powder and pellet form for molding and extrusion by techniques used for conventional thermoplastics. The products are highly resilient, have good resistance to flex fatigue at low and high temperatures, and are resistant to oils and chemicals. Other grades termed segmented copolyesters are excellent modifiers for polyvinyl chloride resins, imparting good abrasion resistance, impact resistance, improved processability and fungus resistance.

Hz. Abbreviation for HERTZ, the term which has replaced cycles per second in the new metric system.

Immediate Set. The deformation found by measurement immediately after the removal of the load causing the deformation.

Immiscible. Incapable of mixing. Thus, two fluids which are not mutually soluble, e.g. water and oil, are described as immiscible.

Impact Adhesive. See CONTACT ADHESIVE.

Impact Modifier. A general term for any additive, usually an elastomer or plastic of different type, incorporated in a plastic compound to improve the impact resistance of finished articles.

Impact Resistance. The relative susceptibility of plastic articles to fracture under stresses applied at high speeds. A widely used impact test (ASTM D 256-73, found in Part 35 of the 1976 Book of ASTM Standards) employs the Izod pendulum striker swung from a fixed height to strike a specimen in the form of a notched bar mounted as a cantilever beam. The Charpy type of tester uses a specimen in the form of a beam supported at both ends. Impact strength is also measured by the DROP WEIGHT TEST, FREE FALLING DART TEST, and TENSILE IMPACT TEST, all of which see. However, these tests are all affected by complex variables, and they are often accompanied by trial-and-error processes of subjecting actual manufactured pieces to conditions stimulating end-use conditions; for example dropping filled blown plastic bottles on hard floors. See also BRITTLENESS TEMPERATURE.

Impact Strength. (1) The ability of a material to withstand shock loading. (2) The work done in fracturing, under shock loading, a specified test specimen in a specified manner. See IMPACT RESISTANCE, IZOD IMPACT STRENGTH.

Impregnation. The process of thoroughly soaking a material of a porous nature with a resin. When webs or shapes of reinforcing fibers are impregnated with a thermosetting resin advanced to the B-stage, or with a thermoplastic, and such webs are intended for subsequent shaping or laminating, the masses are called PREPREGS, which see. The main difference between impregnation and ENCAPSULATION, which see, is that in encapsulation there is formed an outer protective coating with little or no penetration of the resin into the article, whereas in impregnation there is little or no outer protective coating.

Impulse Sealing. (thermal impulse sealing) The process of joining thermoplastic sheets or films by pressing them between elements equipped to provide a pulse of intense thermal energy to the sealing area for a very short time, followed immediately by cooling. The heating element may be a length of thin resistance wire such as nichrome, or an RF heated metal bar which is cored for water cooling. See also HEAT SEALING.

Inching. A reduction in the rate of mold closing travel just before the mating mold surfaces touch each other.

Inclusion. A foreign or impurity phase in a solid.

Inclusion Complexes. See ADDUCTS.

Indene. $C_6H_4CH_2CHCH$ A colorless liquid derived from the fraction of coal tar which boils at 176 to 182 °C. It is used in the production of COUMARONE-INDENE RESINS, which see.

Indentation Hardness. The hardness of a material as determined by either the size of an indentation made by an indenting tool under a fixed load, or the load necessary to produce penetration of the indenter to a predetermined depth. The test usually employed for plastics is by means of a durometer such as the Shore instrument, comprising a spring-loaded indentor point projecting through a hole in a presser foot, and a device to indicate the distance this point projects beyond the face of the foot. The scale readings range from 0 (for 0.100″ penetration) to 100 (for zero penetration). The Shore A instrument employs a sharp indentor point with a load of 822 grams. In the Shore D instrument, used for very hard plastics, the point is blunt and the load is 10 pounds.

Index of Refraction. (refractive index) The ratio of the velocity of light in a vacuum to its velocity in a transparent specimen. It is expressed as the ratio of the sine of the angle of incidence to the sine of the angle of refraction. The index of refraction of a substance usually varies with the wave length of the refracted light. The ASTM test for Index of Refraction of transparent organic plastics is given in D 542-50, published in Part 35 of the 1976 Book of ASTM Standards.

India Rubber. See RUBBER, NATURAL.

Indicia. Any markings such as symbols, lettering, small pictures, etc. applied to a plastic article.

Indophenol Test. See GIBBS INDOPHENOL TEST.

Inductance. The change in a magnetic field due to the variation of current in a conducting circuit causing an induced counter electromotive force in the circuit itself. The Henry is that inductance in which an induced electromotive force of one volt is produced when the inducing current is changed at the rate of one ampere per second.

Induction Heating. A method of heating electrically-conductive materials, usually metallic parts, by placing the part or material in a high-frequency electromagnetic field generated by passing an alternating electric current through a primary coil. The alternating magnetic field induces electromotive forces in the work which generate heat. Plastics, being poor conductors, cannot be heated directly by induction heating but the process is used indirectly in welding of plastics and in heating extruder barrels.

Induction Welding. A method of welding thermoplastic materials by placing a conductive metal insert on the interface of two sections to be joined, applying pressure to hold the sections together, heating the metallic insert by means of a high frequency generator until the surrounding plastic material is softened and welded together, then cooling the joint.

Inert Additive. A material added to a plastic compound, such as a filler, which may alter the properties of the finished article but which does not react chemically with any other constituents of the composition.

Infra-Red Heating. A heating process used occasionally in plastics processing, employing lamps or heating elements which emit invisible radiation of wavelengths below the limit of the visible spectrum at the red end. Such radiation is more penetrating than visible light rays.

Infra-Red Polymerization Index. (IRPI) A number representing the degree of cure of phenolic resins, defined as the ratio of absorbances of absorption peaks at 12.2 and 9.8 microns. The background absorbance at each wavelength is subtracted from the peak absorbance, and the index is obtained by dividing the difference at 12.2 μ, by the difference at 9.8 μ. The test is often used in conjunction with the MARQUARDT INDEX, which see.

Infrared Spectrophotometry. A technique used to identify organic substances such as plastics. All chemical compounds have characteristic intramolecular vibratory motions and can absorb incident radiant energy if such energy is sufficient to increase the vibrational motions of the atoms. In the case of organic molecules, vibrational motions of the substituent groups within the molecules coincide with the electromagnetic frequencies of the infrared region. An infrared spectrometer directs infrared radiation through a film or layer of sample, and measures and records the relative amount of energy absorbed by the sample as a function of the wavelength or frequency of the infrared radiation. The chart produced is compared with correlation charts for known substances to identify the sample.

Infusible. Not capable of melting when heated, as are all thermosetting resins.

Inherent Viscosity. (logarithmic viscosity number) In dilute solution viscosity measurements, inherent viscosity is the ratio of the natural logarithm of the relative viscosity to the concentration of the polymer in grams per 100 ml of solvent. See also DILUTE SOLUTION VISCOSITY.

Inhibitor. A substance capable of retarding or stopping an undesired chemical reaction. Inhibitors are used in certain monomers and resins to prolong storage life. When used to retard degradation of plastics by heat and/or light, inhibitors function as STABILIZERS, which see.

Initial Modulus. See MODULUS OF ELASTICITY, YOUNG'S MODULUS.

Initial Viscosity. A term used in the vinyl plastisol industry to express the viscosity taken immediately after the plastisol has been mixed. The viscosity of a plastisol normally increases at a declining rate after mixing.

Initiator. An agent which causes a chemical reaction to commence and which enters into the reaction to become part of the resultant compound. Initiators differ from catalysts in that catalysts do not combine chemically with the reactants. Initiators are used in many polymerization reactions, especially in emulsion polymerization processes. Initiators most commonly used in the polymerization of monomers and resins having ethylenic unsaturation (-C = C-) are the organic peroxides.

Injection Blow Molding. A blow molding process in which the parison is formed over a mandrel by injection molding, after which the mandrel and parison are shifted to a blow mold where the remainder of the cycle is completed. While the part is being blown, cooled and ejected, another parison is being injection molded. Advantages of the process are that a completely finished part is formed requiring no post finishing operations, closer tolerances are possible, and parison wall thicknesses can be varied at desired areas.

Injection Mold. A mold used in the process of INJECTION MOLDING, which see. The mold usually comprises two sections held together by a clamping device with sufficient strength to withstand the pressure of the molten plastic when injected, and is provided with channels for heating, cooling and venting.

Injection Molding. The method of forming objects from granular or powdered plastics, most often of the thermoplastic type, in which the material is fed from a hopper to a heated chamber in which it is softened, after which a ram or screw forces the material into a mold. Pressure is maintained until the mass has hardened sufficiently for removal from the mold. In a variation called *flow molding* additional molten material is forced into the mold during cooling of the initial charge to overcome shrinkage. Machines employing screws for forcing the material through the plasticating zone (for the process called *screw plasticating injection molding*) are either single stage or double stage. In single stage machines, plastication and injection are done in the the same cylinder, the injection pressure being attained by a forward motion of the screw. This process is called *reciprocating screw injection molding*. In double stage machines, the material is plasticated by the screw and then is delivered to an accumulator chamber through a check valve, from which it is injected into the mold by a piston. This process is called *screw-piston injection molding*. See also REACTION INJECTION MOLDING, HOT-TIP GATE MOLDING, TWO-SHOT INJECTION MOLDING.

Injection Molding pressure. The pressure applied to the cross-sectional area of the material cylinder, expressed in pounds per square inch. (ASTM D 883-65T).

Injection Nozzle. A hardened steel nozzle serving to conduct the molten plastic material emerging from an injection cylinder into the mold. It usually terminates in a spherical tip which fits into a recessed fitting in the mold called a sprue bushing, and is equipped with separate heater bands for temperature control.

Injection Ram. The ram which applies pressure to the plunger in the processes of injection- and transfer-molding.

Injection Stamping. A term proposed for a modification of the injection molding process wherein first the plastic melt is injected under relatively low pressure into a mold which is vented at this stage, then after the cavity is filled additional clamping pressure is applied to completely close the mold and compress or "stamp" the molded shape. Molds are designed so that even in the venting position no material exudes from the land areas. Advantages of the process are said to be a reduction in injection time and pressure, and shorter cycle because after the filling operation the injection screw begins to plasticate the next shot.

Inlay Printing. See VALLEY PRINTING.

In-Mold Decorating. The process of applying labels or decorations to plastic articles simultaneously with the molding operation by which they are formed. Two basic methods, each with many variations, are in use. The first employs a pre-printed label of plastic film, paper or cloth which is positioned in the mold prior to molding. During the molding cycle, the label or its printed image fuses to and becomes an integral part of the article. In the second basic method, the image is printed directly onto the mold surface with wet or dry ink, or applied to the mold by an offset process. In-mold decorating processes can be used in injection molding, blow molding and casting operations.

Inorganic Pigments. Pigments derived from naturally occurring minerals or synthesized from inorganic substances. They are always opaque as opposed to organic pigments and dyes, and are usually resistant to heat and light. Examples of those used in plastics are titanium dioxide, iron oxides, ultramarines, lead chromates, and cadmium compounds.

Inorganic Polymers. An inorganic polymer within the scope of the plastics industry may be defined as a polymer without carbon in its backbone and of a degree of polymerization sufficient for the polymer to exhibit considerable mechanical strength, plastic or elastomeric properties, and the capability of being formed into shapes by processes employed for plastics. Organic-group side chains may be present. SILICONES (which see) are the most commercially important of the inorganic polymers.

Insert. (n) An article of metal or other material which is incorporated in a plastic molding either by pressing the insert into the finished molding or by placing the insert in the cavity so that it becomes an integral part of the molding. An example is an internally threaded bushing used for attaching the molded article to another article.

Insert Molding. See DOUBLE-SHOT MOLDING.

"In Situ" Foaming. The technique of depositing a foamable plastic (prior to foaming) into the place where it is intended that foaming shall take place. An example is the placing of foamable plastics into cavity brickwork to provide insulation. After being positioned, the liquid mix foams to fill the cavity. See also CELLULAR PLASTIC.

Instant Set Polymer. (ISP) A modified polyurethane material developed by Dow Chemical Co. The isocyanate component of ISP contains a proprietary organic modifier which is soluble in the liquid phase but insoluble in the solid phase. During the rection the modifier precipitates out of solution and is trapped in spherical droplets of about 0.5 microns in diameter. These cells act as a heat sink to retard heat build-up and thermal degradation, and also contribute to good machinability and resistance to fatigue and stress cracking. ISP can be used to cast very thick parts in short molding cycles, with mechanical properties comparable to those of engineering plastics such as nylons, acetals, and ABS resins.

Insulated Runner Mold. See HOT RUNNER MOLD.

Insulation Resistance. The electrical resistance between two conductors or systems of conductors separated only by an insulating material. The electrical resistance of a plastic insulating material is determined in the same manner as that of a metallic conductor, that is by dividing the voltage applied to two electrodes in contact with or embedded in the specimen by the current flowing between the electrodes.

Integral Skin Molding. A method of producing urethane foam articles with substantially non-porous integral skins in one operation. Whereas normal urethane castings are foamed by carbon dioxide formed by reacting an excess of isocyanate with water present in the reaction mix (see URETHANE FOAMS), integral skin foams are expanded by gases generated by the volatilization of a solvent such as trichlorotrifluoroethane. Molds for integral skin molding must be heat conductive so that a layer of the reaction mix next to the mold will remain at a temperature below the boiling point of the blowing agent while the interior is being foamed. The reaction mixture, typically consisting of a polyol, an isocyanate and the solvent-type blowing agent, is introduced rapidly into a closed mold through a single port located near the geometric center of the mold. The mold is generally preheated to between 100° and 150°F, but the remaining heat necessary to foam and cure the mass is provided by exothermic reaction of the components. After the mass has gelled, the skin should be pierced to equalize pressure and prevent shrinkage. The solvent blowing agent can be removed from the finished article by allowing it to stand at ambient conditions for about 24 hours, or by oven drying for about 30 minutes at 250°F. In some instances, integral skin moldings have replaced composites of vinyl-covered urethane foam articles and cast vinyl skins filled with urethane foam, such as automotive arm rests, crash pads and instrument panel covers.

Intensive Mixers. Mixers for dry-blending resins such as PVC with plasticizers and other additives, comprising a propeller-like impeller rotating at high speed in the bottom of a stationary can, continuously recirculating the materials between closely spaced stationary and rotating pins.

Interface. The junction point or surface between two different media.

Interfacial Polymerization. A polymerization reaction that occurs at or near to the interfacial boundary of two immiscible solutions. A simple example is the often-performed demonstration of making nylon thred from a beaker container a lower layer of a solution of sebacyl chloride in carbon tetrachloride and an upper layer of hexamethylene diamine in aqueous solution. A pair of tweezers is gently lowered through the top layer, closed on the interfacial layer of polymer, then drawn upward to pull with it a continuous strand of nylon.

Interlaminar Strength. The strength of the adhesive bond between adjacent layers of a laminated material.

Interlayer. An intermediate sheet in a laminate.

Internal Lubricant. A lubricant which is incorporated in a plastic compound or resin prior to processing, as opposed to one which is applied to the mold or die. Examples of internal lubricants are waxes, fatty acids, fatty acid amines, and metallic stearates such as zinc, calcium, magnesium, lead and lithium stearate. The lubricants reduce friction between polymers and metal surfaces, improve flow characteristics, and enhance knitting and wetting properties of compounds. They are used primarily in rigid and flexible PVC, high molecular weight polyethylene, polystyrene, ABS, melamine and phenolic resins.

Internal Mixers. Mixing machines using the principle of cylindrical containers in which the materials are deformed by rotating blades or rotors. The containers and rotors may be cored for heating or cooling to control the batch temperature. These mixers are used extensively in the compounding of rubbers and plastics, and have the inherent advantage of keeping dust and fume hazards to a minimum.

Internal Plasticizer. An agent incorporated in a resin during its polymerization, as opposed to a plasticizer added to the resin during compounding.

Internal Stabilizer. An agent incorporated in a resin during its polymerization, as opposed to a stabilizer added to the resin during compounding.

Interpolymer. A particular type of COPOLYMER, which see, in which the two monomer units are so intimately distributed in the polymer molecule that the substance is essentially homogeneous in chemical composition. An interpolymer is sometimes called a true copolymer.

Intrinsic Viscosity. In dilute solution viscosity measurements, intrinsic viscosity is the limit of the reduced and inherent viscosities as the concentration of the polymeric solute approaches zero and represents the capacity of the polymer to increase viscosity. Interactions between solvent and polymer molecules have the effect of yielding different intrinsic viscosities for the same polymer in various solvents. The IUPAC term for intrinsic viscosity is *limiting viscosity number.* To determine the intrinsic viscosity of a polymer from dilute solution viscosity data, the reduced and inherent viscosities of solutions of various concentrations are determined at constant temperature and these values are plotted against the respective concentrations. The two lines thus obtained converge upon a point of zero concentration of the solute which represents the intrinsic viscosity of the polymer in that solvent of the temperature of the determination. (ASTM D 1243-60).

Introfaction. The change in fluidity and wetting properties of an impregnating material, produced by the addition of an INTROFIER, which see.

Introfier. A chemical which will convert a colloidal solution into a molecular one, by changing the wetting properties and fluidity of the solution.

Intumescence. The foaming and swelling of a plastic when exposed to high surface temperatures or flames. It has particular reference to ablative urethanes used on rocket nose cones, and to INTUMESCENT COATINGS, which see.

Intumescent Coatings. Coatings which when exposed to flame or intense heat decompose and bubble into a foam which protects the substrate and prevents the flame from spreading. Such coatings are used, for example, on reinforced plastic building panels. Examples of intumescent coatings are magnesium oxychloride cement used on urethane foams, and certain epoxy coatings used on polyester panels.

Inventory. (n) In injection molding or extrusion, the amount of plastic contained in the heating cylinder or barrel.

Iodine Value. (Iodine number) The number of grams of iodine that 100 grams of an unsaturated compound will absorb in a given time under arbitrary conditions. It is used to indicate the residual unsaturation in epoxy plasticizers; a low iodine value implies a high degree of saturation. See also OXIRANE VALUE.

Ion. An atom, molecule or radical which has become electrically charged by virtue of having either gained or lost an electron. When an electron is gained the ion is negatively charged and is called an *anion.* A positively charged ion is called a *cation.*

Ion Exchange. A reversible interchange of ions between a solid phase and a liquid phase in which there is no permanent change in the structure of the solid phase. In a typical application, water softening, an ION EXCHANGE RESIN (which see) extracts insoluble calcium ions from the water, replacing them with equivalent amounts of soluble sodium ions. Subsequently, the calcium-loaded resin may be treated with sodium chloride (regenerated) to bring it back to the sodium form ready for another cycle of operation.

Ion Exchange Resins. Small granular or bead-like resins consisting of two principal parts: a resinous matrix serving as a structural portion, and an ion-active group serving as the functional portion. The functional group may be acidic or basic. The resin matrix most often used is a copolymer of styrene and divinylbenzene. Acidic ion exchangers are made by sulfonating the resin beads with, for example, sulfuric acid, chlorosulfonic acid or sulfur trioxide. The basic materials often contain quaternary ammonium groups fixed to the resin beads. Acidic ion exchange resins are used for softening water. Complete deionization of water is accomplished by use of both acidic and basic resins, in sequence or in mixed beds. Ion exchange resins are also used for other chemical processes such as electrodialysis.

Ionic. Pertaining to an atom, radical or molecule which is capable of being electrically charged, either negatively (anionic) or positively (cationic) or both (amphoteric).

Ionic Initiators. Substances providing either carbonium ions (cationic) or carbanions (anionic) which attack the reactive double bonds of vinyl monomers and add on, regenerating the ion species on the propagating chain.

Ionic Polymerization. (cationic polymerization, anionic polymerization) A polymerization process conducted in the presence of electrically charged ions which become attacked to carboxylic groups on carbon atoms in the polymer chain. The carboxylic groups are produced along the polymer chain by copolymerization, providing the anionic portion of the ionic cross links. The cationic portion is provided by metallic ions added to the polymerization mixture. The electrostatic (ionic) forces binding the chains together are much stronger than the covalent bonds between the molecules in conventional polymers. Some polymers produced by cationic polymerization are polyisobutylene, butyl rubber, polyvinyl ethers and coumarone-indene resins. A typical anionic polymerization product is polybutadiene, prepared with an alkali metal catalyst.

Ionic Urethanes. Urethanes containing electrical charges in their backbones or side chains. Cationic urethanes (those containing positive electrical charges) can be formed by reacting diisocyanates with diols containing tertiary nitrogen to yield urethanes which are then treated by a quaternization reaction which forms positive charges in the macromolecular backbone or in the side chains. Applications of such ionic urethanes are aqueous dispersions with outstanding film-forming ability, polyelectrolyte complexes which can be cast from solutions to form coatings on films or tubings used in medical applications, and electrically conducting elastomers.

Ionitriding. See NITRIDING.

Ionization Foaming. The process of foaming polyethylene by exposing it to ionizing radiation which evolves hydrogen from the polymer, causing it to foam.

Ionization Potential. The work required to remove a given electron from its atomic orbit and place it at rest at an infinite distance. The customary unit of ionization potential is electron volts (ev), one ev being equal to 23,053 calories per mole.

Ionomer. (1) The product of an IONIC POLYMERIZATION, which see. (2) See IONOMER RESINS.

Ionomer Resins. Modified polymers obtained by heating and pressing certain polymers containing carboxylic groups in the presence of metallic ions. Examples of the original type are Du Pont's "Surlyn A" (polyethylene containing metallic ions), and copolymers of ethylene and a vinyl monomer with an acid group such as methacrylic or carboxylic acid. Such ionomer resins have low density, high transparency, toughness, resilience, and resistance to greases and solvents. They are FDA-approved for food use. A newer type of ionomer, announced in 1968, is obtained by mixing a finely powdered resin such as an acrylic polymer with powdered metallic oxides and heating the mixture under high pressure to 575 °F. Ionomer resins can be processed by extrusion, blow molding, injection molding, and rotational molding. Major applications to date have been in the fields of skin and blister packaging, bottles, golf ball covers, shoe soles, auto bumper guards, and laminated bags for food and drug items.

Ion Plating. A process for deposition of metal of dielectric films onto plastic substrates with a highly adherent bond. The process is performed in a tank similar to a vacuum metallizing tank. A negative charge is developed on a metal bias plate located behind the plastic substrate. Next, the plating material is converted to a plasma of positive ions by filament or r-f heating. At this point a phenomenon known as Crooke's dark space appears, surrounding the entire surface of the substrate, and constituting a large electrical potential difference between the ions and the charged plastic surface. This causes the ions of the plating material to strike the plastic with high kinetic energy and form strong bonds.

IR. Abbreviation for isoprene rubber (British Standards Institution), the cis-1,4-type of POLYISOPRENE, which see.

Iron Oxide Pigments. Heat-stable pigments ranging from blacks through yellows are obtained from various iron oxides. Reds are formed from ferric oxide (Fe_2O_3), yellows from hydrated ferric oxide, and blacks from ferroferric oxide (Fe_3O_4).

Irradiation. The subjection of a material to radiant energy for the purpose of producing a desired effect or of determining the effect of the radiant energy on the material. Thermosetting resins such as unsaturated polyesters, acrylics, acrylic-modified polyesters and acrylic-modified epoxies can be cured rapidly at room temperature and without catalysts by exposure to high-energy ionizing radiation. Such radiation also forms free radicals in thermoplastics such as polyolefins, PVC, acrylics, nylon, polystyrene, polyvinylidene fluoride, silicone resins, polyethylene terephthalate and acetal resins, which combine to link the molecular chains together in a three-dimensional network. This cross-linking in thermoplastics imparts higher density, increased softening points, lower dielectric loss and improved chemical resistance. Radiation sources most widely used in the plastics industry are electron accelerators and isotopes such as Cobalt 60.

Irregular Block. A block that can not be described by only one species of constitutional repeating unit in a single sequential arrangement. (IUPAC).

Irregular Polymer. A polymer whose molecules can not be described by one one species of constitutional unit in a single sequential arrangement. (IUPAC).

Isano Oil. A fatty oil extracted from an African tree of the same name, used as a flame retardant for acrylic resins. When heated to 200°C it polymerizes rapidly and may explode.

Iso-. The strict meaning of this prefix according to chemical nomenclature is "one methyl group on the next-to-last carbon atom, and no other branches." In the plasticizer field, the prefix is used to denote an isomer of a compound, most specifically an isomer having a single, simple branching at the end of the straight chain.

I.S.O. Abbreviation for the International Standards Organization. Its standards pertaining to plastics can be obtained from U.S.A. Standards Institute, 10 East 40th Street, New York, N.Y. 10016.

Isoamyl Acetate. Rectified AMYL ACETATE, which see.

Isoamyl Butyrate. $C_5H_{11}OOCC_3H_7$ A colorless liquid derived by treating isoamyl alcohol with butyric acid, used as a solvent and plasticizer for cellulose acetate.

Isoamyl Salicylate. $C_6H_4OHCOOC_5H_{11}$ (amyl salicylate, orchidae) A colorless liquid used as a plasticizer.

Isobars. Chemical elements of the same atomic mass but of different atomic numbers. The sum of their nucleons is the same but there are more protons in one than in the other.

Isobutene. $(CH_3)_2C:CH_2$ (isobutylene, 2-methylpropene) A colorless, volatile liquid derived from petroleum, easily polymerized to form polybutene.

Isobutyl Acetate. $CH_3COOCH_2CH(CH_3)_2$ A colorless liquid with a mild fruity odor, used as a solvent for cellulosic plastics and lacquers. Its properties are similar to those of butyl acetate, except that the evaporation rate is higher.

Isobutyl Isobutyrate. $(CH_3)_2CHCOOCH_2CH(CH_3)_2$ A colorless solvent with a fruity odor, slow evaporation rate, good flow and leveling characteristics and good blush resistance. It is used as a solvent for nitrocellulose and several vinyl resins.

Isocyanate Foams. See URETHANE FOAMS.

Isocyanate Generator. (hindered isocyanate) A mixture of an isocyanate, a phenol and a polyester which remains stable at room temperature. When heated to 160°, the phenol and isocyanate components dissociate and react with the polyester to form a polyurethane resin.

Isocyanate Plastic. A plastic based on polymers made by the polycondensation of organic isocyanates with other compounds. Reaction of isocyanates with hydroxyl-containing compounds produces polyurethanes having the urethane group -NH-CO-O-. Reaction of isocyanates with amino-containing compounds produces polyureas having the urea group -NH-CO-NH. (ISO). See also POLYURETHANES, URETHANE FOAMS.

Isocyanates. Compounds containing the isocyanate group, —NCO, attached to an organic radical or hydrogen. Isocyanates containing just one —NCO group (monoisocyanates) have limited uses in the plastics industry. The term is often used with reference to compounds contianing two —NCO groups (diisocyanates) or many such groups (polyisocyanates). However, in the case of a trimer compound containing three NCO groups in a six-membered ring the term *isocyanurate* is used. See also DI-ISOCYANATES.

Isocyanurate. A trimer of an isocyanate, formed by the catalytic cyclization of three isocyanate molecular groups to a six-membered ring.

Isocyanurate Foams. Foams prepared from unmodified isocyanurates have excellent flame resistance but are brittle and of little commercial value. However, isocyanurate foams modified with epoxides, polyimides or (most commonly) urethane groups and polyols possess flame resistance far superior to that of conventional urethane foams and can be processed into a variety of foam products suitable for insulation.

Isocyanurate Plastic. A plastic based on isocyanate polymers in which trimerization of the isocyanates incorporates six-membered isocyanurate ring groups in a chain. Note: In commercial polyisocyanurate cellular plastics, 10 to 30% of the available isocyanate groups are reacted with polyols to introduce urethane groups into the chain. (ISO).

Isodecyl Octyl Phthalate. A primary plasticizer for vinyls, cellulose nitrate, polystyrene and ethyl cellulose. In vinyls, its performance is said to be superior to that of DOP.

Isodecyl Diphenyl Phosphate. A plasticizer for vinyls and many other thermoplastics. It has good flame retardance, low temperature performance, and is a good plasticizer for plastisols.

Isodecyl Octyl Adipate. A light-colored, oily liquid used as a plasticizer for vinyls.

Isoindolinones. A relatively new family of pigments, available in three bright yellows and a red. They have good lightfastness, heat stability and bleed resistance.

Isomers. From the Greek *iso* (the same, equal, alike) and *meros* (part or portion), isomers are substances comprising molecules which contain the same number and kind of atoms but which differ in structure, so that they form materials with wide differences in properties. Isomeric polymers are formed by the polymerization of isomeric monomers which link together in different ways.

Isomorphism. A state of crystallization characterized by a similar arrangement of geometrically similar structural units. See also POLYMORPHISM.

Isooctyl Adipate. See DI-ISOOCTYL ADIPATE.

Isooctyl Palmitate. $C_8H_{17}OOCC_{15}H_{31}$ A plasticizer for polystyrene and cellulosic plastics.

Isophorone. $COCHO(CH_3)CH_2C(CH_3)_2CH_2$ A powerful solvent for vinyl and cellulosic resins, with moderate power to dissolve nearly all common thermosetting and thermoplastic resins.

Isophorone Diisocyanate. (IPDI) An isocyanate used in the production of urethane elastomers and foams. It is less volatile than TDI, and thus easier to maintain at low levels in worker airspace.

Isoprene. $CH_2:C(CH_3)CH:CH_2$ (3-methyl-1,3-butadiene; 2-methyl-1,3-butadiene) A colorless, volatile liquid derived from propylene or from coal gases or tars, chemically similar to the natural rubber molecule. Its polymer of the cis-1,4-type of polyisoprene is chemistry's nearest approach to synthesizing rubber equivalent to the natural material.

Isoprene Rubber. The cis-1,4-type of POLYISOPRENE, which see.

Isopropyl Acetate. $CH_3COOCH(CH_3)_2$ A colorless, aromatic liquid used as a solvent for cellulose nitrate, ethyl cellulose, PVAc, PMMA, polystyrene, and certain alkyd and phenolic resins.

Isopropylbenzene. (isopropylbenzol) See CUMENE.

para, para'-Isopropylidenediphenol. See BISPHENOL A.

Isopropyl Myristate. $C_{13}H_{27}COOCH(CH_3)_2$ A plasticizer for CAB, cellulose nitrate and ethyl cellulose.

Isopropyl Oleate. $C_{17}H_{33}COOCH(CH_3)_2$ A plasticizer for cellulose nitrate, ethyl cellulose and polystyrene, and, with partial compatibility, vinyl resins.

Isopropyl Palmitate. $C_{15}H_{31}COOCH(CH_3)_2$ A plasticizer for cellulose nitrate and ethyl cellulose.

Isotactic. (1) Derived from the Greek words *iso,* meaning "the same" and *tatto,* meaning "to put in order," the term isotactic is sometimes used to denote a polymer structure in which monomer units attached to a polymer backbone are identical on one side and/or on the other side of the backbone. See also SYNDIOTACTIC. (2) Pertaining to a type of polymeric molecular structure containing a sequence of regularly spaced asymmetric atoms arranged in like configuration in a polymer chain. (ASTM D 883-75a). Materials containing isotactic molecules may exist in highly crystalline form because of the high degree of order that may be imparted to such structures. See also STEREO-SPECIFIC.

Isotactic Polymer. A regular polymer whose molecules can be described by one one species of configurational base unit (having chiral or prochiral atoms in the main chain) in a single sequential arrangement. Note: In an isotactic polymer molecule the configurational repeating unit is identical with the configurational base unit. (IUPAC).

Isotactic Pressing. The process of pressing a powder under a gas or liquid so that pressure is transmitted equally in all directions. Such a process is sometimes used in sinter molding.

Isotopes. In chemistry, two or more nuclides having the same atomic number, hence constituting the same element, but differing in mass number. Isotopes of a given element have the same number of nuclear protons but differing numbers of neutrons. Naturally occurring chemical elments are usually mixtures of isotopes so that the observed atomic weights are average values.

Isotropic Laminate. A laminate in which the strength properties are equal in all directions.

Isotropy. The tendency of a material to react the same regardless of the direction of measurement of a property.

Itaconic Acid. $CH_2:C(COOH)CH_2COOH$ A white, crystalline powder usually obtained by oxidative fermentation of sucrose or glucose with Aspergillus terreus. It is capable of polymerization alone, or as a comonomer with acrylic acid, acrylonitrile, styrene, methyl methacrylate and vinylidene chloride. It is used as an additive in acrylic resins to increase their adhesion to cellulose. By polycondensation of itaconic acid with diols, polyesters are obtained which contain methylene side groups.

IUPAC. Abbreviation for International Union of Pure and Applied Chemistry.

IVE. Abbreviation for VINYL ISOBUTYL ETHER, which see.

Izod Impact Strength. A measure of impact strength (described in ASTM D 256) determined by the difference in energy of a swinging pendulum before and after it breaks a notched specimen held vertically as a cantilever beam. The pendulum is released from a vertical height of two feet, and the vertical height to which it returns after breaking the specimen is used to calculate the energy lost. See also IMPACT RESISTANCE, DROP-WEIGHT TEST.

Jacquet Indicator. A tachometer used for measuring very slow linear speeds such as those encountered in plastics extrusions.

Jar Mill. A small version of a BALL MILL, which see, utilizing a portable jar of porcelain or metal rather than a fixed cylinder for holding the material to be ground and the grinding media.

Jet Molding. (offset molding) A modification of the injection molding process designed for thermosets. An elongated nozzle or "jet" is attached to the front of the molding cylinder, provided with a high intensity heating element and rapid cooling means. It is also necessary to control cylinder temperatures carefully to prevent premature hardening of the resin.

Jet Spinning. For most purposes, similar to melt spinning. Hot gas jet spinning uses a directed blast or jet of hot gas to "pull" molten polymer from a die lip and extend it into fine fibers.

Jetting. Turbulent flow of resin from an undersize gate or thin section into a thicker mold section, as opposed to laminar flow of material progressing radially from a gate to the extremities of the cavity.

Jig. A device for holding component parts of an assembly during a manufacturing operation, or for holding other tools; a clamping device used to secure a bonded assembly until the adhesive has set.

Jig Welding. The welding of thermoplastic materials between suitably shaped jigs. Heat may be applied to the material by heating the jigs, or by any other appropriate means.

Joggles. (keys) A term sometimes employed for matching inserts for exact positioning of a multi-piece mold.

Joining. The process of assembling plastic parts by means of mechanical fastening devices such as rivets, screws, clamps, etc. See also FABRICATE.

Joule. (1) A unit of work, or force acting against resistance to produce motion in a body, equal to 1×10^7 ergs. (An erg is the metric term for a force of one dyne acting through a distance of one centimeter.) (2) The *International joule,* a unit of electrical energy, is the work expended per second by a current of one International Ampere flowing through one International Ohm. (3) The new SI definition for Joule (J) as a unit of energy is the work done when the point of application of a force of one newton is displaced a distance of one meter in the direction of the force. Conversion factors for two familiar units are: 1 Btu = 1055.056 Joules, and 1 foot pound = 1.355818 Joules. The decimal *kilojoule* may be used for small units such as the calorie. (4.1868 kj = 1 calorie.)

Jute. A fiber obtained from the stems of several species of the plant *Corchorus* found mainly in India and Pakistan. It is used in the form of fiber, yarn and fabric for reinforcing phenolic and polyester resins.

Kalrez. DuPont's tradename for fluoroelastomer, a terpolymer of tetrafluoroethylene, perfluorovinylmethylether, and a small amount of a cross-linkable monomer. The elastomer combines the properties of Viton with the thermal stability, chemical resistance and electrical characteristics of Teflon fluorocarbon resins.

Kaolin. (china clay, bolus alba) A variety of CLAY, which see, consisting essentially of the minerals *kaolinite, dickite* and *nacrite* (all $Al_2O_3 \cdot 2SiO_2 \cdot 2H_2O$). The name kaolin comes from the Chinese *kao-ling,* meaning high hill, the name of the mountain in China which yielded the first kaolin sent to Europe. Deposits exist also in England and the southeastern states of the U.S. Kaolins and china clays are used as inexpensive fillers in many plastics.

Karl Fischer Reagent. A solution of iodine, sulfur dioxide and pyridine in methanol or methyl "Cellosolve." It is used for determining the water content of plastic resins, etc.

Kelvin.. (K) The absolute or thermodynamic temperature scale proposed by Lord Kelvin, and recommended by SI for universal international use. Although the temperature interval of one degree Kelvin is exactly equal to one degree on the Celsius scale (formerly called Centigrade), The Kelvin scale starts with zero at absolute zero ($-273.15\,°C$, the temperature at which a molecule of a perfect gas has no kinetic energy). Thus, the temperature of melting ice, $0\,°C$ is $273.15\,°K$, and the temperature of boiling water, $100\,°C$ is $373.15\,°K$.

Keratin. The protein derived from feathers, hair, hoofs, horns, etc. of animals by calcination. It is sometimes used as a filler in plastics, particularly urea formaldehyde molding compounds in which it reduces brittleness and permits drilling and tapping.

Ketohexamethylene. See CYCLOHEXANONE.

Ketones. A group of organic compounds containing a carbonyl (C = O, keto) group bound to two carbon atoms. The lower ketones are widely used as solvents for vinyl and cellulosic resins, and as intermediates in the production of resins. The most important members of the family are diacetone alcohol, diisobutyl ketone, 1,2-diketones, 1,3-diketones, 1,4-diketones, ethyl n-butyl ketone, mesityl oxide, methyl n-amyl ketone, methyl ethyl ketone, and methyl isobutyl ketone.

Keys. See JOGGLES.

K-Factor. A term sometimes used for thermal insulation value or coefficient of thermal conductivity. See THERMAL CONDUCTIVITY.

Kieselguhr. See DIATOMITE.

Kilo-. (k) The SI-approved prefix for a multiplication factor to replace 10^3.

Kinematic Viscosity. (kinetic viscosity) The absolute (dynamic) viscosity of a fluid divided by the density of the fluid. The c.g.s. unit of kinematic viscosity is the Stoke.

Kirksite. An alloy of aluminum and zinc, easily castable at relatively low temperatures, often used for molds for blow molding. Its high thermal conductivity aids in accelerating cooling cycles.

Kiss-Roll Coating. A coating process by which very thin plastics coatings can be applied to substrates such as paper. A roll immersed in the coating fluid transfers a layer of coating to a second roll from which a portion of the layer is transferred to the substrate. See also ROLLER COATING.

Kling Test. A method for determining the relative degree of fusion of flexible vinyl sheets, coated fabrics and thin sections of cast or molded parts, by immersing the folded specimen in a solvent and noting the time in which disintegration commences. Typical solvent systems are based on MEK, THF, ethyl acetate and carbon tetrachloride. The preferred solvent system is one that will initiate degradation within 5 to 10 minutes on a fully fused specimen.

Kneaders. Mixers with a pair of intermeshing blades, often in the shape of the letter Z, used for working plastic masses of semi-dry or rubbery consistency.

Knife Coating. See SPREAD COATING, AIR KNIFE COATING.

Knit Lines. See WELD MARK.

Knockout. Any part or mechanism of a mold used to eject the molded article.

Knockout Bar. A bar or plate in a knockout frame used to back up a row or rows of knockout pins.

Knockout Pin. (ejector pin) A pin that ejects a molded article from the mold. It is usually activated automatically as the mold opens.

Knockout-Pin Plate. See EJECTOR PLATE.

Knoop Microhardness Test. A test employing a diamond pyramid indentor. The Knoop hardness number (KHN) is obtained by the formula

$$KHN = 14.2 \; P/l^2$$

wherein P is the load and l is the length of indentation.

Knot Tenacity. (knot strength) The tenacity in grams per denier of a yarn where an overhand knot is put into the filament or yarn being pulled to show up sensitivity to compressive or shearing force.

Knuckle Area. In reinforced plastics, the area of transition between sections of different geometry in a filament-wound part. (ASTM D 883-75a).

Kohinoor Test. A test for SCRATCH HARDNESS, which see, employing a series of pencils of different hardnesses.

Kraton^{TM} Thermoplastic Elastomers. Shell's elastomers with a block copolymer structure, combining rigid polystyrene end blocks with elastomeric center blocks of polybutadiene, polyisoprene, or (in the case of Kraton G) EPR. Unlike chemically crosslinked elastomers, Kraton materials are true thermoplastics and can be processed on high speed equipment.

Krebs Unit. A measurement of viscosity used with reference to the STORMER VISCOMETER, which see. The Krebs Unit is the weight in grams which will rotate the viscometer paddle 100 revolutions in 30 seconds.

K-Value. (1) A number calculated from dilute solutiopn viscosity measurements of a polymer, used to denote degree of polymerization or molecular size. The formula is:

$$\frac{\log (N_s/N_0)}{c} = \frac{75 \; K^2}{1 + 1.5 \; Kc} + K$$

$$\text{where } N_s = \text{viscosity of the solution}$$
$$N_0 = \text{viscosity of the solvent}$$
$$c = \text{concentration in grams per ml}$$

Utilization of K-Value as a resin specification originated in Europe and is still used there, but is rarely used in the U.S. today. (2) The term K-Value (K-factor) is also used for coefficient of thermal conductivity. See THERMAL CONDUCTIVITY.

Lacquer. A solution of a film-forming natural or synthetic resin in a volatile solvent, with or without a color pigment, which when applied to a surface forms an adherent film that hardens solely by evaporation of the solvent. The dried film has the same properties as the resins used in making the lacquer. Cellulosic, alkyd and vinyl resins are most often used in lacquers.

Lactams. Cyclic amides obtained by removing one molecule of water from an amino acid. An example is CAPROLACTAM, which see.

Lactic Acid. $CH_3CHOHCOOH$ (milk acid, alpha-hydroxypropionic acid) A colorless or yellowish liquid with several applications in plastics. Reacted with glycerine it forms an alkyd resin. It is a catalyst for vinyl polymerizations, and an additive for phenolic casting resins.

Ladder Polymers. (double stranded polymers) Polymers comprising chains made up of fused rings. Examples are cyclized (acid-treated) rubber and polybutadiene.

Lake. A type of organic pigment prepared from water-soluble acid dyes, precipitated on an inert substrate by means of a metallic salt, tannin or other reagent. Lakes were used in plastics at one time, but have been replaced by more permanent pigments.

Lambert. The unit of brightness equal to 1/pi candle per square centimeter. In the new SI it is defined as 3,183 009 E + 04 candela per meter^2.

Lamellar Structures. Platelike single crystals which exist in some crystalline polymers.

Laminar Flow. (lamular flow) The movement of one layer of fluid past another layer with no transfer of matter from one to the other. The friction between two such layers is called VISCOSITY COEFFICIENT, which see. Laminar flow of thermoplastic resins in a mold is achieved by solidification of the layer in contact with the mold surface, thus providing an insulating tube through which material flows to fill the remainder of the cavity. This type of flow is essential in order to duplicate the mold surface.

Laminate. (n) A product made by bonding together two or more layers of material or materials. The term most usually applies to preformed layers joined by adhesives or by heat and pressure. However, some authors apply the term to composites of plastic films with other films, foil and paper even though they have been made by spread coating or by extrusion coating. In the reinforced plastics industry, the term refers mainly to superimposed layers of resin-impregnated or resin coated fabrics or fibrous reinforcements which have been bonded together, usually by heat and pressure, to form a single piece. When the bonding pressure is at least 1000 psi, the product is called a *high pressure laminate*. Products pressed at pressures under 1000 psi are called *low pressure laminates*; and those made with little or no pressure, such as hand lay-ups, are sometimes called *contact pressure laminates*. The term *parallel laminate* refers to a laminate in which all layers are oriented approximately parallel with respect to the grain or strongest dimension in tension. In *cross laminates,* one or some of the plies are perpendicular to the grain. The term laminate is sometimes also used to include composites of resins and fibers which are not in distinct layers, such as filament wound structures and spray-ups. Resins most widely used in laminates are the epoxies, phenolics, polyesters, melamine-formaldehydes and silicones. The base materials include fabrics of cotton cloth, asbestos, glass fiber, and nylon. See also REINFORCED PLASTICS, COMPOSITE LAMINATES, DECORATIVE BOARDS.

Lamp Black. A type of CARBON BLACK, which see, used as a pigment.

Land. (n) (1) The horizontal bearing surface of a semipositive or flash mold by which excess material escapes. (See CUT-OFF). (2) The bearing surface along the top of the flights of a screw in a screw extruder. (3) The surface of an extrusion die parallel to the direction of melt flow. (4) The mating surfaces of any mold, adjacent to the cavity depressions, which prevent the escape of material.

Land Area. The area of surfaces of a mold which contact each other when the mold is closed, measured in a direction perpendicular to the direction of application of closing pressure.

Landed Force. A force with a shoulder which seats on the land in a landed positive mold.

Lap. In filament winding, the amount of overlay between successive windings, usually intended to minimize gapping.

Lap Joint. A joint made by placing one surface to be joined partly over another surface and bonding the overlapping portions.

Lap Winding. A variation of the FILAMENT WINDING (which see) process, consisting of convolutely winding a resin-impregnated tape onto a mandrel of the desired configuration. The process is used for making large chemical and heat-resistant, conical or hemispherical parts such as ablative heat shields for ballistic re-entry bodies.

Laser. A term coined from the understored letters of "Light amplication by stimulated emission of radiation." Laser rays are formed by means of a synthetic ruby rod encircled by a flash lamp, the rod having a mirrored reflective surface at each end, but with less reflective material at one (the emitting) end. When the lamp is flashed, electrons of the atoms of the ruby rod are elevated to a higher orbit for about 1 millionth of a second. As they return, excess energy is released. This energy is reflected back and forth between the mirrors until it has attained enough force to break through the weaker mirror at the emitting end of the rod. This energy or light emerging from the rod is concentrated in a slim beam, which is further condensed through lenses and directed to the workpiece where it is put to work in drilling, cutting or welding operations. The laser was first used for perforating and welding plastics. More recently, low-power lasers have been used for highly accurate and rapid inspection systems for gaging thickness and detecting physical flaws in plastics. The newest development is a medium-power "CO_2 laser" preferred for machining of plastics because, unlike other industrial lasers, they produce a light wavelength of 10.6 microns which is completely absorbed by plastics. The CO_2 laser consists of a glass or quartz tube under mild vacuum conducting a mixture of gases such as helium, nitrogen and carbon dioxide along its longitudinal axis. A high voltage electrical discharge passed between electrodes at opposite ends of the tube excites the gas mixture to the extent causing emission of light, which is then amplified by reflection between two opposing mirrors as in the original system. A beam delivery system at the end of the partially reflecting mirror forms the desired shape and size of beam for the machining job to be performed.

Latch Plate. A plate used for retaining a removable mold core of relatively large diameter, or for holding insert-carrying pins on the upper part of a mold. Release of the pins or core is effected by moving the latch plate.

Latent Heat of Fusion. See HEAT EQUIVALENT OF FUSION.

Latent Heat of Vaporization. The quantity of heat necessary to change one gram of liquid to its vapor state without change of temperature, measured as calories per gram in the cgs system. The value is converted to joules in the SI.

Latent Solvent. An organic liquid which has little or no solvent effect on a particular resin until it is activated by either heat or admixture with a true solvent.

Latex. A stable dispersion of a polymeric substance in an essentially aqueous medium. The term latex was originally coined from a Greek word by botanists early in the 19th Century to describe the milky fluid in certain plants such as milk weeds, which fluid exudes when the plants are cut and which coagulates on exposure to air. Later, the rubber industry applied the term to the exudate of rubber trees. Today, the meaning has been expanded to include aqueous dispersions of all high polymers, natural and synthetic. Those of interest to the plastics industry are based mainly on styrene-butadiene copolymers, polystyrene, and vinyl polymers and copolymers. See also HYDROSOL.

Latices. Plural of latex, preferred over the sometimes used "latexes."

Lattice Pattern. In filament winding, a pattern with a fixed arrangement of open voids producing a basket-weave effect.

Lauryl Methacrylate. $CH_2:C(CH_3)COO(CH_2)_{11}CH_3$ One of the monomers used in the production of acrylic resins.

Lay. (n) The length of twist produced by stranding singly or in groups, such as fibers or rovings; or the angle that such filaments make with the axis of the strand during a stranding operation. The length of twist of a filament is usually measured as the distance parallel to the axis of the strand between successive turns of the filament. (ASTM D 883-75a). The term is also used in the packaging of glass fibers for the spacing of the roving lands on the package expressed as the number of bands per inch.

Lay-Flat Film. Film which is extruded as a tube, usually blown, cooled, then gathered by rollers and wound up in flattened form.

Lay-Up. (n) A fibrous reinforcing material with or without a resinous impregnation, ready for positioning in a mold. The reinforcement is usually cut and fitted to mold contours. See also PREPREG.

Lay-Up Molding. (hand lay-up molding) A method of forming reinforced plastics articles comprising the steps of placing a web of the reinforcement, which may or may not be preimpregnated with a resin, in a mold or over a form and applying fluid resin to impregnate and/or coat the reinforcement, followed by curing of the resin. When little or no pressure is used in the curing process, the process is sometimes called CONTACT PRESSURE MOLDING. When pressure is applied during curing, the process is often named after the means of applying pressure, such as autoclave molding or BAG MOLDING, which see. A related process is SPRAY-UP, which see.

LDPE. Abbreviation for low density polyethylene. See POLYETHYLENES.

L/D Ratio. In an extruder screw, the ratio of the screw length to the screw diameter. The Extrusion PAG of SPE has established definitions for two types of L/D ratios of extruders. *Total L/D ratio* is the distance from the rear edge of the feed opening to the forward end of the barrel bore divided by the bore diameter and expressed as a ratio wherein the diameter is reduced to one such as 15:1 or 20:1. *Effective (enclosed) L/D ratio* is defined as the distance from the forward edge of the feed opening to the forward end of the barrel bore and expressed as a ratio wherein the diameter is reduced to one, such as 15:1 or 20:1.

Leaching. The process of extraction of a component from a mixture by treating the mixture with a solvent which will dissolve the component but has no effect on the remaining portions of the mixture.

Lead Carbonate. See BASIC LEAD CARBONATE.

Lead Chrome Pigments. A series of inorganic pigments including yellows, oranges and greens, used in PVC, polyolefins, cellulosics, acrylics and polystyrene.

Leader Pin. See GUIDE PINS.

Leader Pin Bushing. See GUIDE PIN BUSHING.

Lead Phosphite, Dibasic. See DIBASIC LEAD PHOSPHITE.

Lead Salicylate. $Pb(C_6H_4OHCOO)_2 \cdot H_2O$ A white crystalline material used as a heat stabilizer.

Lead Stabilizers. A family of stabilizers which are highly effective as heat stabilizers, but are limited to use in applications where toxicity, sulfur staining and lack of clarity can be tolerated. Examples of lead stabilizers are:

> Basic lead carbonate
> Basic lead sulfate complexes
> Basic silicate of white lead
> Co-precipitated lead silicate and silica gel
> Dibasic lead maleate
> Dibasic lead phosphite
> Dibasic lead phthalate
> Dibasic lead salicylate
> Dibasic lead silicate-sulphate
> Dibasic lead stearate
> Lead chlorosilicate complexes
> Lead stearate
> Monohydrous tribasic lead sulphate

Lead Stearate. $Pb(C_{18}H_{35}O_2)_2$ A white powder used as a vinyl resin stabilizer and lubricant in extrusion compounds, and in phonograph record compounds.

Leakage Flow. In the metering section of an extruder screw, leakage flow is the backward flow of material through the clearance between the screw flight lands and the barrel wall. It is usually a very small negative component of total material flow. See also DRAG FLOW.

Leathercloth. A term sometimes used, especially in Europe, for plastic-coated fabric with a leather-like texture.

Lemon Yellows. Pigments containing up to 40% lead sulfate in the rhombic form. The higher the rhombic content, the greener the shade.

Let-Go. An area in laminated glass over which an initial adhesion between interlay and glass has been lost. (ASTM D 883-75a).

Let-Off. (pay-off) A device used in coating by calendering or extrusion to suspend a coil or reel from which the material to be coated is fed to the coating machine.,

Letterpress Printing. The process used for paper is adapted to plastics by the use of special inks and transfer rolls, and possibly a modification of press speed. Flexible printing plates are usually employed, such as vinyl or rubber.

Lewis Acid Catalysts. See FRIEDEL-CRAFTS CATALYSTS.

Lexan. Registered trademark of the General Electric Co. for thermoplastic carbonate-linked polymers produced by reacting bisphenol A and phosgene. For other types of polycarbonates see POLYCARBONATE RESINS.

Lift. The complete set of moldings produced in one cycle of a molding press.

Light-Resistance. (light fastness, colorfastness) The ability of a plastic material to resist fading, darkening or degradation upon exposure to sunlight or ultraviolet light. Nearly all plastics tend to change color under these conditions, due to characteristics of the polymeric material and/or pigments incorporated therein. Tests for light resistance are made by exposing specimens to natural sunlight or to artificial light sources such as the carbon arc, mercury lamp, germicidal lamp, or xenon arc lamp.

Light Scattering. In a dilute polymer solution, light rays are scattered and diminished in intensity by a number of factors including fluctuations in molecular orientation of the polymer solute. Observations of the intensity of light scattered at various angles provide the basis for an important method of measuring molecular weights of high polymers.

Light Stabilizer. An agent added to a plastic compound to improve its resistance to light. See also STABILIZER, ULTRAVIOLET STABILIZERS.

Light Transmittance. See LUMINOUS TRANSMITTANCE.

Lignin. The major non-carbohydrate constituent of wood and woody plants, functioning in nature as a binder to hold the matrix of cellulose fibers together. Lignins are obtained commercially from by-products of coniferous wood, for example by treating wood flour with a derivative of lignosulfonic acid. They are used as extenders in phenolic resins, and sometimes as reactants in the production of phenol-formaldehyde resins.

Lignin Plastics. Plastics based on resins made by the treatment of lignin with heat or by reaction of lignin with chemicals or synthetic resins, the lignin being in greatest amount by weight. (ASTM D 883-75a).

Lignic Resin. A resin made by heating lignin or by reaction of lignin with chemicals or resins, the lignin being in greatest amount by mass. (ISO).

Ligroin. (petroleum ether, benzine) A saturated petroleum fraction boiling in the range 58-275 °F, used as a solvent. The term benzine is outmoded due to confusion with benzene, and should not be used.

Lime. See CALCIUM OXIDE.

Limestone. See CALCIUM CARBONATE.

Limiting Viscosity Number. The IUPAC term for INSTRINSIC VISCOSITY, which see.

Linear Expansion. See COEFFICIENT OF THERMAL EXPANSION.

Linear Polymer. A polymer in which the molecules form long chains without branches or crosslinked structures. The molecular chains of a linear polymer may be intertwined, but the forces tending to hold the molecules together are physical rather than chemical and thus can be weakened by energy applied in the form of heat. Such linear polymers are thermoplastics.

Linear Unsaturated Polyesters. See POLYESTERS, UNSATURATED..

Liner. A continuous, usually flexible coating on the inside surface of a filament wound pressure vessel, used to protect the laminate from chemical attack or to prevent leakage under stress.

Linters. Short fibers that adhere to the cotton seed after ginning. Used in rayon manufacture, as fillers for plastics, and as a base for the manufacture of cellulosic plastics.

Liquid Colorants. In the compounding of plastisols and organosols it has always been common practice to use liquid or pasty dispersions of colorants, which are easily stirred into the compounds. In recent years, similar liquid dispersions and metering systems have been developed for adding the colorants directly onto the screw of an extruder at the base of the feed hopper. The term *liquid colorants* is being used for such dispersions designed for adding to dry molding or extrusion compounds. The advantages of liquid colorants are better pigment dispersion, higher let-down ratios, savings in handling and pollution control, and less build-up on the screw and screen pack.

Liquid Injection Molding. (LIM) A process of injection molding thermosetting resins in which the uncured resin components are mixed, metered and injected at relatively low pressures through nozzles into mold cavities, the curing or polymerization taking place in the mold cavities. The process is most widely used with resins that cure by addition polymerization such as polyesters, epoxies, silicones, alkyds, DAP, and (occasionally) urethanes. However, the term REACTION INJECTION MOLDING is most often used when urethane reactants are employed. The term *liquid resin molding* has also been used for LIM or RIM.

Liquid Reaction Molding. (LRM) Synonymous with REACTION INJECTION MOLDING, which see.

Lithium Stearate. $LiC_{18}H_{35}O_2$ A white crystalline material used as a lubricant in plastics.

Lithopone. (lithophone, Orr's white, Charlton white, zinc baryta white, zinc sulfide white) A composite pigment obtained by mixing solutions of barium sulfate and zinc sulfide, filtering, washing and drying the precipitate. It is used as a filler and pigment in plastics.

Loading Board. See LOADING TRAY.

Loading Space. Space provided in a compression mold or in the pot used with a transfer mold to accommodate the molding material before it is compressed.

Loading Tray. (charging tray, loading board) A device for charging measured amounts of molding compound simultaneously into each cavity of a multiple cavity mold, comprising a compartment tray with a sliding bottom.

Locating Ring. A ring which serves to align the nozzle of an injection mold cylinder with the entrance of the sprue bushing and the mold to the machine platen.

Locking Pressure. The amount of pressure applied to an injection or transfer mold to keep it closed during molding.

Locking Ring. A slotted plate in an injection or transfer mold which locks the parts of the mold together and prevents it from opening while the material is being injected.

Logarithmic Viscosity Number. The IUPAC term for INHERENT VISCOSITY, which see.

Longos. A colloquial term used in the filament winding industry to describe low-angle helical or longitudinal windings.

Loop Test. A simple test for evaluating the compatibility of vinyl resin plasticizers based on the fact that a material under compressive stress will exude plasticizer more rapidly. A specimen in sheet form is folded double, forming a loop with internal radius equal to the sheet thickness. The inside of the loop is examined at intervals over a period of a week or more for signs of exudation of plasticizer.

Loose Punch. A male portion of a mold constructed so that it remains attached to the molding when the press is opened, to be removed from the part after demolding.

Loss Angle. The anti-tangent of the electrical dissipation factor. See DIELECTRIC LOSS ANGLE.

Loss Factor. The product of the power factor and the dielectric constant of a dielectric material.

Loss Modulus. A damping term describing the dissipation of energy into heat when a material is deformed. The imaginary portion of the complex modulus. The product of the storage modulus and the tangent of the loss angle.

Low Density Polyethylene. This term is generally considered to include polyethylenes ranging in density from about .915 to .925. In low density polyethylenes, the ethylene molecules are linked in random fashion, with the main chains having side branches. This branching prevents the formation of a closely knit pattern, resulting in material that is relatively soft, flexible and tough, and which will withstand moderate heat. See also HIGH DENSITY POLYETHYLENE, POLYETHYLENES.

Low Pressure Injection Molding. A term sometimes used for the process of injecting a fluid material such as vinyl plastisol into a closed mold by means of a grease gun or similar low-pressure apparatus.

Low Pressure Laminates. Various definitions place the upper limit of pressure for this term at from 1000 pounds per square inch down to pressures obtained by mere contact of the plies. According to ASTM the upper limit for a low pressure laminate is 200 psi. The Decorative Board Section of NEMA has recommended abandonment of the term low pressure laminate in favor of *decorative board* in the case of "a product resulting from the impregnation or coating of a decorative web of cloth, paper, or other carrying media with a thermosetting resin and consolidation of one or more of these webs with a cellulosic substrate under heat and pressure of less than 500 pounds per square inch." This includes all boards that were formerly called low pressure melamines and polyester laminates, but does not include vinyls. See also CONTACT PRESSURE MOLDING, LAMINATE.

Low Pressure Molding. Molding or laminating in which the pressure used is 200 psi or less (ASTM D 883-75a).

Low Pressure Resins. See CONTACT PRESSURE RESINS.

Low Temperature Flexibility. All plastics which are flexible at room temperature become less flexible as they are cooled, finally becoming brittle at some temperature. This property is often measured by torsional tests over a wide range of temperatures, from which apparent moduli of elasticity are calculated. See also CLASH-BERG POINT.

L-Sealer. A heat sealing device used in packaging, which seals a length of flat folded film on the edge opposite the folded edge and simultaneously seals a strip across the width at 90° from the edge seal. The article to be packaged is inserted between the two layers of folded film prior to sealing. When it is desired to sever the continuous length of sealed compartments into individual packages, a heated wire or knife is incorporated between two sealing bars forming the bottom of the "L." These bars then form the top seal of the filled bag and the bottom seal of the next bag to be filled.

Lubricant. A substance which when interposed between parts or particles tends to make surfaces slippery, reduce friction, and prevent sticking between the lubricated surfaces. Lubricants are added to plastics to (1) assist flow in calendering, molding and extrusion by lubricating the metal surfaces in contact with the plastic; (2) assist in knitting and wetting of the resin in mixing and milling operations; and (3) impart lubricity to finished products. Among the lubricants most commonly used are fatty acid soaps, metallic soaps or salts of fatty acids (such as calcium or barium stearate), paraffin waxes, hydrocarbon oils, fatty alcohols, low molecular weight polyethylenes, synthetic waxes of the fatty amide and ester type, and certain silicones. Graphite, molybdenum disulfide and fluorocarbon polymers are used to impart lubricity to finished articles made of acetals, nylon and polycarbonate. Lubricity is also contributed by certain plasticizers, stabilizers and pigment dispersions.

Lubricant Bloom. Irregular, cloudy, greasy exudation on the surface of a plastic. (ASTM D 883-65T). Such effects can also be caused by exudation of plasticizers, stabilizers and other additives, so the term *lubricant bloom* should be used only when the exudation is caused by a lubricant contained in the plastic compound or applied to it during processing.

"Lucite." Du Pont's trademark for a group of methacrylate ester polymers, including PMMA and several other acrylics.

Lumen. The SI unit of LUMINOUS FLUX, which see.

Luminescent Pigments. Pigments which produce striking effects in darkness or light. See PHOSPHORESCENT PIGMENTS, FLUORESCENT PIGMENTS.

Luminous Flux. (1) The total visible energy emitted by a source per unit time. (2) The new SI unit of luminous flux is the lumen, defined as the luminous flux emitted in a solid angle of one steradian (the solid angle which encloses a surface on the sphere equivalent to the square of the radius) by a point source having a uniform intensity of one candela.

Luminous Transmittance. The ratio of transmitted light to incident light.

Lux. (lx) The SI unit of illuminance defined as the illuminance produced by a luminous flux of one lumen uniformly distributed over a surface of one square meter.

Lyophilic. (adj.) Descriptive of the ability of a substance to form colloidal suspensions easily. Such ability when the suspending medium in water is called *hydrophilic*. PVC plastisol is an example of a lyophilic suspension.

Lyophobic. (adj.) Descriptive of a material in the colloidal state which lacks affinity for the suspending medium. Such lyophobic dispersions require the presence of a stabilization agent to prevent their coagulation or settling out.

Machine Shot Capacity. See SHOT CAPACITY.

Machining of Plastics. Machining operations commonly used for metals are applicable to plastics, with slight variations in tooling and speeds. Among such operations are blanking, boring, drilling, grinding, milling, planing, punching, routing, sanding, sawing, shaping, tapping, threading and turning.

Macromolecular. Pertaining to a substance consisting of macromolecules.

Macromolecule. The large ("giant") molecules which make up the high polymers. Each macromolecule may contain hundreds of thousands of atoms.

Magnesium Carbonate. $MgCO_3$ (magnesium carbonate, precipitated; magnesia alba) A white powder of low density, used as a filler or modifier in phenolic resins.

Magnesium Glycerophosphate. $MgPO_4 \cdot C_3H_5(OH)_2$ A colorless powder derived by the action of glycerophosphoric acid on magnesium hydroxide, used as a stabilizer for plastics.

Magnesium Hydroxide. $Mg(OH)_2$ Used as a thickening agent for polyester resins. Its action is slow or medium compared to that of magnesium oxide.

Magnesium Oxide. MgO (magnesia, periclase) A white powder, used as a filler, and as a thickening agent in polyester resins. Magnesium oxide does not occur naturally, but is made by calcining magnesium hydroxide or magnesium salts.

Magnesium Oxychloride Cement. (Sorel cement) A mixture of magnesium chloride and magnesium oxide that reacts with water to form a solid mass, presumed to be magnesium oxychloride. It has been found to be useful as an intumescent coating for urethane foams, and other materials such as ABS, styrenes, nylons, acetals, polyesters and silicones.

Magnesium Phosphate, Dibasic. $MgHPO_4 \cdot H_2O$ A white, crystalline powder derived by reacting orthophosphoric acid with magnesium oxide, used as a non-toxic stabilizer for plastics.

Magnesium Phosphate, Monobasic. $MgH_4(PO_4)_2 \cdot 2H_2o$ A white, hygroscopic, crystalline powder derived by reacting orthophosphoric acid with magnesium hydroxide. It is used as a flame retardant and stabilizer for plastics.

Magnesium Phosphate, Tribasic. $Mg_3(PO_4)_2 \cdot 8H_2O$ or $\cdot 5H_2O$ A fine, soft white powder derived by reacting magnesium oxide and phosphoric acid at high temperatures, used as a non-toxic stabilizer.

Magnesium Stearate. $Mg(C_{18}H_{35}O_2)_2$ A white, soft powder used as a lubricant and stabilizer.

Maleic Anhydride. $(COCH)_2O$ (2,5-furandione) Colorless needles obtained by passing a mixture of benzene and air over a heated vanadium oxide catalyst. It has many applications in plastics, including the production of alkyd, polyester and vinyl copolymer resins, and as a curing agent for thermosetting resins such as phenolics and ureas. About 55% of the maleic anhydride produced in the U.S. is used in the production of unsaturated polyester resins, to which it imparts fast-curing and high-strength characteristics.

MAN. Abbreviation for METHACRYLONITRILE, which see.

Mandrel. (1) The core around which paper, fabric, or resin-impregnated fibrous glass is wound to form pipes or tubes. (2) In extrusion (also called a *pin*), an insert in the flow channel of a die which converts the melt section from the solid to some type of hollow or annular cross section. The outer surface of the mandrel guides the flow of the inner surface of the plastic melt as it leaves the discharge end of the die.

Manifold. A pipe or channel with several inlets or outlets. With reference to blow molding, extrusion and injection molding equipment, a manifold is a piping or distribution system which takes the single channel flow output of the extruder or injection cylinder and divides it to feed several blow molding heads or injection nozzles. Manifolds are also used in cooling and heat distribution systems. Types of manifolds used in extrusion are named according to shape, as follows: (a) *Tear Drop* — the cross-section is streamlined in tear-drop shape as it leads to the die land area. (b) *Fish Tail* — the flow channel leading to the die land area is trapezoidal in shape and resembles a fish tail. (c) *T-Shape* — a center feed die where the cross section of the die manifold is circular as it leads to the die land area. (d) *Coat Hanger* — a center feed die in which the flow channel resembles a coat hanger in shape.

Man-Made Fiber. A class name for various fibers (including filaments) produced from fiber-forming substances which may be: (1) polymers synthesized from chemical compounds, e.g. acrylic, nylon, polyester, polyurethane, polyethylene, and polyvinyl fibers; (2) modified or transformed natural polymers, e.g. alginic and cellulose-based fibers such as acetates and rayons; or (3) mineral, e.g. glass. The term man-made usually refers to all chemically produced fibers to distinguish them from truly natural fibers such as cotton, wool, silk, flax, etc.

Mannich Reaction. The condensation of ammonia or primary or secondary amine with formaldehyde and a compound containing at least one hydrogen atom of pronounced activity. The active hydrogen is replaced by an amino-methyl or substituted amino-methyl or substituted amino-methyl group. This type of reaction has been employed in the production of "Mannich Polyols" for use in urethane foams.

Marble. See CALCIUM CARBONATE.

Marquardt Index. In an infrared absorption curve study of the cure advancement of phenolic resins, the marquardt index is the numerical difference in the percent transmission between the absorption peaks at 12.2 microns and 13.3 microns. As resin cure progresses, the intensity of the 13.3μ absorption changes more rapidly than that of the 12.2μ peak; thus the marquardt index number decreases as the cure advances. See also INFRA-RED POLYMERIZATION INDEX.

Mar Resistance. The resistance of glossy plastic surfaces to abrasive action. It is measured (ASTM D673) by abrading a specimen to a series of degrees, then measuring the gloss of these abraded spots with a GLOSSMETER (which see) and comparing the results with an unabraded area of the specimen.

Mask. A stencil used for spray painting plastics, consisting of a relatively thin base sheet shaped to fit the part to be painted with openings for areas to be painted. Masks for irregularly shaped articles are often made by electroforming a thin shell over a part, then cutting openings in the desired areas. Masks for spherical articles such as playballs can be made from spun metal hemispheres, and those for flat articles from flat metal sheets.

Mass. The unit of quantity of matter. The term is often confused with *weight,* which in commercial and everyday use is used in lieu of the true technical meaning of *mass.* When an object is weighed on a balance with arms, the units of mass and weight are identical because the force of gravity is cancelled out. In the SI system, the term kilogram is restricted to the unit of mass. See also *weight.*

Mass Dyeing. See SPIN DYEING.

Mass Polyermization. See BULK POLYMERIZATION.

Mass Spectrometry. An analytical process in which molecules of the specimen are exposed to a positive ion beam and are identified by measurements of deflection and intensities of the beam. The process is used for both qualitative and quantitative analyses. Until recent years the process was largely supplemented by GAS CHROMATOGRAPHY, which see. However, a more recent modification of the mass spectrometry process called *pulsed positive negative ion chemical mass spectrometry* generates both positive and negative ions simultaneously from a sample and separates them of the basis of their charges and masses. The spectra from these charges provide much more structural information about the sample than would be obtained from spectra from either type of ion alone.

Masterbatch. A term used primarily in the rubber industry for rubber compounds containing high percentages of pigment or other additives, to be added in relatively small amounts to batches during compounding. The term is sometimes used in the plastics industry for COLOR CONCENTRATE, which see.

Mastic. (mastic gum) A solid resinous material obtained from the tree *Pistacia lentiscus,* used in adhesives and lacquers.

Mat. A fabric or felt of glass or other reinforcing material cut to the contour of a mold, for use in reinforced plastics processes such as matched-die molding and hand lay-up or contact pressure molding. The mat is usually impregnated with resin just before or during the molding process.

Matched Metal Die Molding. (matched die molding, matched metal molding) The process of forming shaped articles of reinforced plastics by pressing mats or preforms between matching male and female mold sections. For simple shapes without compound curves or deep draws, mats cut from rolls or sheets of compacted glass fiber or other reinforcement may be used. For the more intricate shapes, "preforms" are made by depositing cut fibers mixed with a resin binder on a screen shaped approximately to the contours of the finished article. Fibers are deposited on the screen by spraying, flotation on a water slurry, or by a suction process from a rotating plenum chamber. The mat or preform is placed on one half of the mold, then a measured quantity of resin is poured on and spread in a controlled pattern. The mold is then closed and subjected to heat and pressure in the range of 150 to 400 psi until the resin has cured. The process is used for making boat hulls, automotive parts, furniture seats, and a wide variety of shaped panels for other applications. In a variation called *prepreg molding,* the fibrous mat is preimpregnated with resin, fillers, pigments and other additives so that it is ready for molding without further treatment.

Matched Mold Thermoforming. A sheet thermoforming process in which the heated plastic sheet is shaped between male and female portions of a matched mold. The molds may be of metal or inexpensive materials such as plaster, wood, epoxy resin, etc., and must be vented to permit the escape of air as the mold closes. See also THERMOFORMING.

Material Well. Space provided in a compression or transfer mold to care for bulk factor, that is to provide for the difference in volume between the loose molding powder and the final molding.

Matrix. (1) A mold. (2) The resinous phase of a reinforced plastic.

MBS. Abbreviation for methacrylate-butadiene-styrene resins. These are mixtures of PMMA and butadiene-styrene copolymers, formulated in a variety of types with markedly different characteristics according to their composition and molecular weight. MBS resins can be processed by extrusion, injection molding and compression molding.

Mc. Abbreviation for megacycles, or one million cycles. The term often refers to Mc per second even though "per second" is not stipulated.

MDI. Abbreviation for diphenylmethane-4,4'-diisocyanate. See DI-ISOCYNATES.

Measling. The appearance of spots or stars under the surface of the resin portion of an epoxy-glass fiber laminate.

Mechancial Equivalent of Heat. The quantity of energy which, when transformed into heat, is equivalent to a unit quantity of heat. In the old cgs system such units are a calorie at 20 °C, or 4.18×10^7 ergs. In the new SI, the stipulated unit is the joule, the conversion factor at 20 °C being 4.181 90 E + 00 joules.

Mechanically Foamed Plastic. A cellular plastic in which the cells have been produced by gases introduced by physical means. See also CELLULAR PLASTIC.

Mechanical Property. The observed manifestation of a function which expresses the relations between a strain and the stress having induced it. Properties of plastics which are classified as mechanical include abrasion resistance, creep, ductility, friction resistance, elasticity, hardness, impact resistance, stiffness and strength.

Medium Yellows. Pigments based on essentially pure monoclinic lead chromate.

Mega-. (M) The SI-approved prefix for a multiplication factor to replace 10^6.

Megahertz. (MHz) Equivalent to one million cycles (megacycles or mc) per second.

MEK. Abbreviation for METHYL ETHYL KETONE, which see.

MEKP. Abbreviation for METHYL ETHYL KETONE PEROXIDE, which see.

Melamine. $(C_3N_3)(NH_2)_3$ (2,4,6-triamino-s-triazine) A crystalline material derived from cyanuric acid, of which it is the triamide. Its main use is for MELAMINE FORMALDEHYDE RESINS, which see.

Melamine-Formaldehyde Resins. (melamine resins) Thermosetting resins of the AMINO RESIN (which see) family, made by reacting melamine with formaldehyde. The lower molecular weight, uncured melamine resins are water soluble syrups, used for impregnating paper, laminating, etc. The higher molecular weight resins are powders, widely used for plastic tableware.

Melamine/Phenolic Resins. Mixtures of melamine and phenolic resins which combine the dimensional stability and ease of molding of phenolics with the ease of coloring characteristic of melamines.

Melt. (n) A normally solid thermoplastic material which has been heated to a molten condition.

Melt Extractor. Usually refers to a type of injection machine torpedo, but could refer to any type of device which is placed in a plasticating system for the purpose of separating fully plasticated melt from partially molten pellets and material. It thus insures a fully plasticated discharge of melt from the plasticating system.

Melt Flow Index. See MELT INDEX.

Melt Fracture. A phenomenon sometimes encountered in extrusion, characterized by irregularities in the extrudate ranging from slight surface ripples to gross annular distortions in the entire cross section. For a given set of standard processing conditions and die geometry, there is a critical shear rate for a specific compound below which melt fracture does not occur and above which it will occur. Theories attribute the phenomenon to non-uniform or irregular elastic strains in the material at the die entrance, or to alternate sticking and slippage of material as it progresses from die entry to the die land. Corrective measures include raising the melt temperature, changing to a resin of higher melt index, and decreasing the die entry angle.

Melt Index. The amount, in grams, of a thermoplastic resin which can be forced through an orifice of 0.0825 inch diameter when subjected to a force of 2160 grams in ten minutes at 190 °C. The test is performed by an extrusion rheometer described in ASTM D 1238. It is most widely used in classifying polyethylene resins, but is sometimes used for evaluating acrylics, ABS, polystyrene and nylon. Melt index values for commercial polyethylenes range from 0.1 to about 20. Those of low melt index have high molecular weights, and are used mainly for heavy duty applications such as pipe. The high melt index polymers are of low molecular weights, used for extrusion and molding of flexible products.

Melt Instability. An instability in the melt flow through a die starting at the land of the die. It leads to the same surface irregularities on the finished part as melt fracture.

Melt Spinning. See SPINNING.

Melt Strength. The strength of the plastic while in the molten state. This is a pertinent factor in extrusion, blow molding and drawing of molten resin from a die. It is also important when a plastic film is reheated for shrink packaging.

Melt Zone. The zone of an extruder barrel in which the material has been plasticized by heat and pressure.

Memory. The tendency of a plastic article to revert in dimension to a size previously existing at some stage in its manufacture. For example, a film which has been oriented by hot stretching at the proper temperature will, upon reheating, tend to revert to its size before orientation due to its "memory." See also ORIENTATION.

Mer. Derived from the Greek word *meros,* meaning a part or unit, a mer is the repeating structural unit of any polymer. One mer is a monomer, two mers form a dimer, three mers a trimer, four mers a tetramer, and so forth. A great many mers form a polymer.

Mercuric Chloride. $HgCl_2$ (corrosive sublimate, mercury bichloride) White crystals, used as a polymerization catalyst for PVC.

Mesityl Oxide. $CH_3COCH::C(CH_3)_2$ (4-methyl-3-penten-2-one) An oily, colorless liquid used as a powerful solvent for cellulosic and vinyl resins, and as an intermediate in the production of plasticizers.

Metallic Fiber. Generic name for a manufactured fiber composed of metal, plastic-coated metal or metal-coated plastic. (Federal Trade Commission). An example is aluminum fiber covered with cellulose acetate butyrate.

Metallic Flake Pigments. Flake-shaped particles of either aluminum or copper and its alloys which reflect light specularly when incorporated in a plastic substance or coating vehicle with their reflecting surfaces approximately parallel. The aluminum pigments reflect very strongly throughout the visible spectrum, producing brilliant blue-white highlights. The copper-based pigments, called gold bronzes but actually brasses, range from the characteristic red of copper to a progressively more yellow hue as the zinc content rises.

Metallic Soaps. See SOAPS, METALLIC.

Metallizing. A term covering all processes by which plastics are coated with metal. The most commonly used processes are described under ELECTROPLATING, VACUUM METALLIZING and SILVER SPRAY PROCESS. Other methods are spraying with metallic pigments, chemical reduction, gas plating and vapor pyrolysis.

Metamer. From the Greek *meta* (change, transposition, transfer) and *meros* (part or portion), the term metamer was once used in chemistry for a specific kind of isomer having to do with positional differences in molecules of the same composition. The term ISOMER, which see, is now used in the same sense.

Metamerism. A term sometimes used in the color industry for the phenomenon exhibited by two surfaces which appear to be of the same color when viewed under one light source (e.g. daylight), but which do not match in color when viewed uner a different light source (e.g. incandescent lamp). The term *geometric metamerism* has been proposed for a change in appearance of a colored surface when the viewing angle is changed.

Metastable. (adj.) An unstable condition of a plastic evidenced by changes of physical properties not caused by changes in composition or in environment. Note: Metastable refers, for example, to the temporarily more flexible condition of some plastics after molding. No physical tests should be made while the plastic is in a metastable condition unless data regarding this condition are desired. (ASTM D 883-65T).

Meter. This unit is now defined by the SI as the length of exactly 1 650 763.73 wavelengths of the radiation in vacuum corresponding to the unperturbed transition between the levels $2p_{10}$ and $5 d_5$ of the atom of Krypton 86, the orange-line. It has been defined previously as 1 553 164.13 wave lengths of the red cadmium line measured in dry air at 15 °C and normal pressure. It is equal to approximately 39.37 inches.

Metering Screw. An extrusion screw which has a shallow constant depth, and constant pitch section over, usually, the last three to four flights. The function of the metering zone is to control the pressure and temperature of the plastic melt.

Metering Zone. The final zone of an extruder barrel, in which the melt is advanced at a uniform rate to the breaker plate or die.

Methacrylate Esters. Esters of methacrylic acid having the formula $CH_2:C(CH_3)COOR$, wherein R is usually methyl, ethyl, isobutyl or n-butyl-octyl. They are polymerizable to acrylic resins.

Methacrylate Plastics. See ACRYLIC RESINS.

Methacrylic Acid. $CH_2:C(CH_3)COOH$ (alpha-methacrylic acid, 2-methylpropen-2-oic acid) A colorless liquid prepared by the acid hydrolysis of acetone, from which is derived all of the methacrylate compounds. Most important of these compounds are the esters, especially methyl methacrylate.

Methacrylonitrile. (MAN, alpha methyl acrylonitrile) A recently developed vinyl monomer containing the nitrile group. Its homopolymers are true thermoplastics with good mechanical strength and high resistance to solvents, acids and alkalis. Modified properties can be obtained through blending, grafting, or copolymerization with other monomers such as styrene and methyl methacrylate. MAN is also used as a replacement for acrylonitrile in the preparation of nitrile elastomers.

gamma-Methacryloxypropyltrimethoxy Silane. $CH_2:C-C-O(CH_2)_3Si(OCH_3O_3)$ A silane coupling agent used in reinforced polyesters, epoxies, and many thermoplastics.

$$\overset{CH_3}{\underset{}{}} \overset{O}{\underset{}{}}$$

Methanal. See FORMALDEHYDE.

Methanol. See METHYL ALCOHOL.

Methoxyethyl Acetoxystearate. $C_{17}H_{34}(OCOCH_3)COOCH_2CH_2OCH_3$ A plasticizer for vinyl and cellulosic resins.

Methoxyethyl Acetyl Ricinoleate. $C_{17}H_{32}(OCOCH_3)COOCH_2CH_2OCH_3$ A plasticizer for vinyl and cellulosic plastics.

Methoxyethyl Ricinoleate. $C_{17}H_{32}(OH)COOCH_2CH_2OCH_3$ A plasticizer for cellulosic and vinyl resins.

Methoxyethyl Stearate. $C_{17}H_{35}COOCH_2CHOHCH_3$ (1,2-propylene glycol monostearate) A solvent and plasticizer for cellulosic plastics.

Methyl Abietate. $C_{19}H_{29}COOCH_3$ A derivative of abietic acid, used as a plasticizer for cellulose nitrate, ethyl cellulose, acrylic and vinyl resins and polystyrene.

Methyl Acetate. CH_3COOCH_3 A colorless, volatile liquid with a fragrant odor, used as a solvent for cellulose nitrate, acetyl cellulose and cellulose esters.

Methyl Acetylricinoleate. $C_{17}H_{32}(OCOCH_3)COOCH_3$ A plasticizer for some vinyl resins and polystyrene.

Methyl Acrylate. $CH_2:CHCOOCH_3$ A colorless, volatile liquid, used as a monomer for acrylic ressins.

Methyl Alcohol. CH_3OH (methanol) A colorless, highly toxic liquid usually obtained by synthesis from hydrogen and carbon monoxide. It is sometimes called wood alcohol, but true wood alcohol is obtained from wood distillates and contains contaminants in addition to methyl alcohol. Methyl alcohol is used as an intermediate in the production of formaldehyde, phenolic, urea, melamine, and acetal resins; and as a solvent for cellulose nitrate, ethyl cellulose, PVAc and polyvinyl butyral.

Methyl n-Amyl Ketone. $CH_3CO(CH_2)_4CH_3$ (2-heptanone) A high boiling ketone solvent with a fruity odor, used in synthetic resin finishes. It is especially useful in lacquers and finishes for roll coating where improved blush resistance is required.

Methylbenzene. (methylbenzol) See TOLUENE.

Methyl Butadiene. See ISOPRENE.

Methyl Butyl Ketone. $CH_3COC_4H_9$ (MBK, propylacetone) A solvent for vinyl and many other resins, often used in conjunction with MEK to control the drying rate of lacquers. A higher content of MBK slows the drying rate.

Methylbutynol. $HC:CCOH(CH_3)_2$ A viscosity stabilizer and solvent for some nylons.

Methyl Butyrate. $CH_3CH_2CH_2COOCH_3$ A solvent for ethyl cellulose and cellulose nitrate.

Methyl-2-Cyanoacrylate. An adhesive used for bonding plastics such as CAB, cellulosics, nylon, polyesters, polyolefins, polystyrene and polyurethanes to each other or to other materials such as wood, metal, rubber, glass or leather.

2,2'-Methylenebis (6-tert-Butyl-4-Ether Phenol). An antioxidant for ABS packaging, appliances, pipe and automotive items, marketed by American Cynamid as Cyanox 425.

2,2'-Methylenebis (6-tert-Butyl-4-Methyl-Phenol). A phenolic-type antioxidant marketed by American Cyanamid as Cyanox 2246, used in polyolefins and ABS.

4,4'-Methylenebis (Cyclohexylisocyanate). ($H_{12}MDI$) A diisocyanate used in the production of urethane elastomers and foams.

Methylene Chloride. CH_2Cl_2 (methylene dichloride, dichloromethane) A colorless, heavy non-flammable liquid used as a solvent for cellulose triacetate and vinyl resins, a solvent in the polymerization of polycarbonate resins, and as a reactant for certain phenolic resins. It is widely used as a paint stripper. A more recent use for methylene chloride is as a chemically inert blowing agent in conjunction with water in urethane foams to control foam density.

Methylene Group. The radical $—CH_2—$ or $=CH_2$, existing only in combination.

Methyl Ethyl Ketone. $CH_3COC_2H_5$ (MEK, ethyl methyl ketone, 2-butanone) A colorless, flammable liquid with an acetone-like odor. One of the most widely used solvents for several thermoplastics including cellulose acetate butyrate, ethyl cellulose, acrylic resins and vinyl copolymers.

Methyl Ethyl Ketone Peroxide. (MEKP, MEK Peroxide) A curing agent for polyester resins. In some systems, curing can be performed at room temperature with MEKP.

Methyl Glucoside. $CH_2OHCH(CHOH)_3CHOOCH_3$ A plasticizer for alkyd, amino and phenolic resins. It is also used as a polyol for urethane foam production.

Methyl Group. The radical $CH_3—$, existing only in combination.

Methyl Hexyl Ketone. $CH_3COC_6H_{13}$ (2-octanone) A colorless liquid with a pleasant odor, used as a solvent for epoxy resin coatings.

Methyl Isoamyl Ketone. $CH_3COC_2H_4CH(CH_3)_2$ (5-methyl-2-hexanone) A colorless liquid with a pleasant odor, used as a solvent for cellulose nitrate, cellulose acetate butyrate, acrylic resins and certain vinyl copolymers. It has a high solvent power and low evaporation rate, making it useful as a retarder solvent which promotes flow-out and reduces blushing.

Methyl Isobutyl Ketone. $(CH_3)_2CHCH_2COCH_3$ (hexone, 2-methyl-4-pentanone, MIBK) A solvent with a moderate evaporation rate, used with cellulosic, vinyl, alkyd, acrylic, phenolic, coumarone-indene resins; and polystyrene.

Methyl Isopropenyl Ketone. $CH_3COC(CH_3):CH_2$ A flammable liquid used as a copolymerizable monomer.

Methyl Lactate. $CH_3CHOHCOOCH_3$ A solvent for several cellulosic plastics.

Methyl Methacrylate. $CH_2CCH_3COOCH_3$ A colorless, volatile liquid derived from acetone cyanohydrin, methanol and dilute sulphuric acid, and used in the production of ACRYLIC RESINS, which see.

n-Methyl Morpholine. $CH_2CH_2OCH_2CH_2NCH_3$ A colorless liquid used as a catalyst in urethane foams.

Methyl Myristate. $CH_3(CH_2)_{12}COOCH_3$ (methyl tetradecanoate) The methyl ester of myristic acid, with applications in stabilizers and plasticizers.

Methyl Oleate. $C_{17}H_{33}COOCH_3$ A plasticizer for cellulose nitrate, ethyl cellulose, polystyrene, and, with limited compatibility, vinyl resins.

Methylol Urea. $H_2NCONHCH_2OH$ Colorless crystals derived from the combination of urea and formaldehyde, the first stage in the production of urea-formaldehyde resins.

Methyl Palmitate. $CH_3(CH_2)_{14}COOCH_3$ The methyl ester of palmitic acid, with applications in stabilizers and plasticizers.

Methyl Pentachlorostearate. $Cl_5C_{17}H_{30}COOCH_3$ A plasticizer for polystyrene, PMMA, cellulosics and vinyl resins.

Methylpentene Resin. See POLY(4-METHYL-PENTENE-1).

Methyl Phthalyl Ethyl Glycolate. $CH_3OCOC_6H_4COOCH_2COOC_2H_5$ A plasticizer for several thermoplastics including PVC, polystyrene, cellulosics and phenolics.

Methyl Propionate. $CH_3CH_2COOCH_3$ A solvent for cellulose nitrate.

n-Methyl-2-Pyrrolidone. $CH_3NCH_2CH_2CH_2CO$ A solvent with a low order of inhalation toxicity, good thermal and chemical stability, and a high flash point. It is capable of dissolving resins such as polyamide-imides, epoxies, urethanes, nylon and PVC. PVC fiber-spinning solutions often contain the material as a solvent.

Methyl Ricinoleate. $CH_3(CH_2)_5CH(OH)CH_2CH:CH(CH_2)_7COOCH_3$ A plasticizer for cellulose acetate butyrate, cellulose nitrate, polyvinyl acetate and polystyrene.

alpha-Methylstyrene. $C_6H_5C(CH_3):CH_2$ A colorless liquid, easily polymerizable by heat or catalysts.

MF. Abbreviation for melamine-formaldehyde. See MELAMINE FORMALDEHYDE RESINS.

MFC. Abbreviation for *multifunctional concentrate.* See COLOR CONCENTRATE.

M-Glass. Glass with a high content of beryllia, designed especially for high modulus of elasticity.

MHz. Abbreviation for MegaHertz, equivalent to Mc, which see.

Mica. One of a series of crystalline silicate minerals characterized physically by a perfect basal cleavage, consisting essentially of orthosilicates of aluminum and potassium. They occur naturally, mainly as the minerals *muscovite* (white mica), *phlogopite* (amber mica), and *biotite*; and are also synthesized from potassium fluorosilicate and alumina. The micas are used as fillers in thermosetting resins, imparting good electrical properties and heat resistance. A new grade of mica called "HAR" or "high aspect ratio" mica has been developed. 3 to 5 micron thick flakes with aspect ratios as high as 200 can be processed, although the optimum maximum ratio is around 70. The larger flakes increase flexural modulus and strength, have lower moisture content, and increase heat distortion temperature.

Micelle. A colloidal particle formed by the reversible aggregation of dissolved molecules. Micelles may be in the shape of spheres, cylinders or platelets. Soaps, detergents and other emulsifying agents used in emulsion polymerization contain micelles generally composed of from 50 to 100 molecules of emulsifier, within which the polymerization reaction is initiated.

Micro-. (μ) The SI-approved prefix for a multiplication factor to replace 10^{-6}.

Microballoons. This term has been used for (1) tiny plastic spheres used to reduce evaporation of liquids such as oils, by floating a layer of spheres on the surface of liquids in storage tanks; and for (2) MICROSPHERES, which see.

Microbial Degradation. See BIODEGRADATION, PINK STAINING.

Microcrystalline. (adj.) Minutely crystalline; composed of crystals of microscopic size. Microcrystalline particles are sometimes defined as those having maximum dimensions of one micron.

Microcrystalline Silicate. A derivative of chrysotile asbestos, consisting of tiny rod-shaped particles of hydrated magnesium silicate. The particles have hydroxyl groups on their surfaces which bond with hydrogen-bonding sites of molecules of a fluid in which they are incorporated. They are also used as viscosity building agents in unsaturated polyester and other resins.

Microencapsulation. The process of encasing a small solid particle or a discrete amount of liquid or gas in a capsule. The term applies to capsules ranging in diameter from a few microns to about 500 microns. The capsule is usually of a synthetic plastic, although waxes, glass and metals are also used. Methods used for forming polymeric capsules fall into three broad classes: phase separation, interfacial reaction, and physical methods. Phase separation methods include COACERVATION (which see), applying meltable dispersions, and spray-drying of a suspension of the material in a vaporizable solvent. Interfacial reaction methods include interfacial polymerization, in-situ polymerization, and chemical vapor deposition. The physical methods include fluidized bed coating processes, spray coating, electrostatic coating methods, and extrusion. Typical examples of microencapsulations are "carbonless" carbon paper, slow-release drugs and fertilizers, and battery separators.

Microgels. Small particles of cross-linked polymer of very high molecular weight and containing closed loops. Microgels may be present in trace amounts due to impurities in monomers, and exert an influence on polymer properties and molecular weight studies.

Micron. This term and its abbreviation, the Greek symbol μ, were dropped by action of General Conference on Weights and Measures on October 13, 1967. The symbol "μ" is to be used solely as the abbreviation for the prefix *micro*, standing for exponential multiplication factor 10^{-6}. The micron now should be stated as μM (micro-meters).

Microporous. Having pores of microscopic dimensions. Some plastic films and fabric coatings are rendered microporous in order to permit "breathing," while retaining waterproofness.

Microspheres. Tiny, hollow spheres of glass or plastic material which are used as fillers to impart low density to plastics. Such plastics are called *syntactic foams*. Prior to 1969, the microspheres in use were made of glass, phenolic resin or epoxy resin. A new type of microsphere introduced in 1969 is made of a saran-type copolymer of vinylidene chloride and acrylonitrile, containing a blowing agent activated by moderate heat. The unblown spheres range from 4 to 20 microns in diameter and have a bulk density of about 45 pounds per cubic foot. The spheres can be preblown by heating in an oven or in steam prior to incorporation in a plastic, or they may be incorporated in the plastic prior to blowing and expanded in situ by a subsequent processing step. Glass-reinforced polyesters containing about 3% of saran microspheres have higher strength-to-weight ratios, higher impact strengths, improved resistance to crazing, and more "wood-like" characteristics than foams of equivalent density made with glass or phenolic microspheres. They may be nailed or screwed. See also GLASS SPHERES.

Micro-Wave Heating. A heating process similar to dielectric heating, but using frequencies in the 10^9 to 10^{10} cps (radar) range. The Federal Communications Commission has allocated the specific frequencies 915, 2450 and 5850 MHz for industrial use. Microwave ovens similar to those used in restaurants and households for rapidly cooking foods have been used experimentally for preheating molding powders, vacuum bag curing, autoclave molding, and nylon overwrap curing. Plastic films coated with water-containing materials such as polyvinylidene chloride can be dried rapidly and economically by microwave energy such as is used for rapid heating of foods. Line speeds above 1000 feet per minute have been attained with polyethylene film, by means of a microwave cabinet only 8 feet long.

Migration. The transfer of a constituent of a plastic compound to another contacting substance.

Migration of Plasticizer. In plasticized thermoplastics or elastomers, the movement of molecules of plasticizer from their site when the article was initially formed to the surface layer of the article where the plasticizer appears as a greasy or oily layer and may be rubbed off or dissolved. The phenomenon occurs most often in vinyl compounds containing an incompatible plasticizer.

Mil. A unit of length equal to .001 inch, often used for specifying diameters of wires and glass fibers.

Mill. (n) In the plastics industry, the term mill is generally taken to refer to a roll mill such as a two-roll mill used in compounding and mixing compounds. The dictionary definition would include all mechanical devices for transforming raw materials into a condition ready for use. See ROLL MILLS.

Milled Fibers. Small modules of glass filaments produced by hammer milling continuous glass strands. They are useful as anticrazing reinforcing fillers for adhesives.

Milli-. (m). The SI-approved prefix for a multiplication factor to replace 10^{-3}.

Mineral Black. Black pigments made by grinding and/or heating slate, shale or coal.

Mineral Spirits. An aliphatic hydrocarbon fraction of petroleum evolved in the distillation range of about $150°$ to $200°C$. An example is "V.M. & P.," used as a diluent in organosols.

Mixers. Devices used to intimately intermingle two or more materials to some defined state of uniformity. Types used in the plastics industry are:

BALL MILLS	INTENSIVE MIXERS
BANBURY MIXERS	INTERNAL MIXERS
CHANGE CAN MIXERS	MILLS
CENTRIFUGAL IMPACT MIXERS	MOTIONLESS MIXERS
COLLOID MILLS	PROPELLER MIXERS
CONICAL DRY BLENDERS	RIBBON BLENDERS
DISC AND CONE AGITATORS	ROLL MILLS
DRUM TUMBLERS	SAND MILLS
HOMOGENIZERS	TUMBLING AGITATORS
KNEADERS	VIBRATORY MILLS

Mixture. A combination of two or more substances intermingled with no constant percentage composition, in which each component retains its essential original properties.

MMA. Abbreviation for METHYL METHACRYLATE, which see.

Mn. Abbreviation for *number-average molecular weight*, the total weight of all molecules divided by the total number of molecules. Also, the chemical symbol for the element manganese.

MOCA. Du Pont's trademark for methylene-bis-ortho-chloroaniline, $CH_2(C_6H_4ClNH_2)_2$. Sold in the form of tan-colored, coarsely ground lumps, MOCA was widely used as a curing agent for urethane rubbers and epoxy resins prior to alleged findings that it is a carcinogen. Although as of mid-1976 the OSHA restrictions on the use of MOCA had not become effective due to industry court actions, many urethane formulators switched to MDI and other systems rather than face any future MOCA restrictions.

Modacrylic Fiber. A manufactured fiber in which the fiber forming substance is any long chain synthetic polymer composed of less than 85% but at least 35% by weight of acrylonitrile units. (FTC Definition).

Modified Resins. Synthetic resins modified by the incorporation of natural resins, elastomers or oils, which alter the processing characteristics or physical properties of the basic resins.

Modulus. Derived from the Latin word meaning "small measure," modulus is a number which expresses a measure of some property of a material, e.g. modulus of elasticity.

Modulus At 300%. The tensile stress necessary to elongate a specimen to 300% of its original length. Although other elongations may be used, 300% is the one most often employed for rubbers and flexible plastics.

Modulus In Compression. The ratio of compressive stress to strain within elastic limits of the material.

Modulus In Flexure. The ratio of the flexure stress to strain, within elastic limits of the material.

Modulus In Shear. The ratio of shear stress to strain within elastic limits of the material.

Modulus Of Elasticity. (elastic modulus, Young's modulus) The ratio of stress (nominal) to corresponding strain below the proportional limit of a material. It is expressed in force per unit area, usually pounds per square inch. (ASTM D 638-72). The strain may be a change in length (Young's modulus); a twist or shear (modulus of rigidity or modulus of torsion); or a change in volume (bulk modulus). In the SI, all types of moduli of elasticity are to be reported in pascals, the conversion factor being one psi = 6.894 757 E + 03 pascals.

Modulus Of Resilience. The energy that can be absorbed per unit volume without creating a permanent distortion. It can be calculated by integrating the stress-strain curve from zero to the elastic limit and dividing by the original volume of the specimen.

Modulus of Rigidity. See FLEXURAL MODULUS.

Mohs Value. A measure of hardness based on a scale established in 1822 by Frederick Mohs, giving a relative ranking of minerals in the order in which one will scratch another.

Mohs Scale No.	Standard Mineral
1	talc
2	gypsum
3	calcite
4	flourite
5	apatite
6	orthoclase
7	quartz
8	topaz
9	corundum
10	diamond

See also SCRATCH HARDNESS.

Moiety. A constituent present in a material to some indefinite extent.

Moil. A term rarely used for flash remaining on a molded plastic item.

Moisture Absorption. The pick-up of water vapor by a material upon exposure to an atmosphere. Should not be confused with WATER ABSORPTION, which see, which relates to water acquired by immersion of the specimen in water.

Moisture Adsorption. The pick-up of moisture from the air by a material on its surface only.

Moisture Vapor Transmission. The rate at which water vapor permeates through a plastic film or wall at a specified temperature and relative humidity.

Molal Solution. A solution which contains one mole of the solute per 1000 grams of the solvent.

Molar Solution. A solution which contains one mole or gram molecular weight of the solute per liter of solution.

Mold. (n) A hollow form or matrix into which a plastic material is placed and which imparts to the material its final shape as a finished article.

Mold. (v) To impart shape to a plastic mass by means of a confining cavity or matrix. The term *molding* is usually employed for processes using dry thermoplastic compounds, e.g. injection, transfer or compression molding. The verb "to cast" is preferable for processes employing fluid dispersions.

Mold Base. An assembly of ground flat steel plates, usually containing dowel pins, bushings and other components of injection or compression molds excepting the cavities and cores.

Mold Efficiency. In an multimold blowing system, the percentage of the total turn-around time of the mold actually required for forming, cooling and ejection of the part.

Molding. For molding processes see the following:

BAG MOLDING	FLOW MOLDING
BLOW MOLDING	HIGH PRESSURE MOLDING
CLAMSHELL MOLDING	INJECTION BLOW MOLDING
COINING	INJECTION MOLDING
COLD FORMING	INJECTION STAMPING
COLD HEADING	INTEGRAL SKIN MOLDING
COLD MOLDING	JET MOLDING
COLD PRESSING	LIQUID INJECTION MOLDING
COLD RUNNER INJECTION MOLDING	LOW PRESSURE INJECTION
COMPOSITE MOLDING	MOLDING
COMPRESSION MOLDING	LOW PRESSURE MOLDING
CONTACT PRESSURE MOLDING	MATCHED DIE MOLDING
DIP FORMING	ONE-SHOT MOLDING
DOUBLE-SHOT MOLDING	OUTSERT MOLDING
FILM BLOWING	PLATFORM BLOWING

PLUNGER MOLDING
POWDER MOLDING
PULP MOLDING
REACTION INJECTION MOLDING
RECIPROCATING SCREW INJECTION
  MOLDING
ROTARY MOLDING
ROTATIONAL INJECTION MOLDING
ROTATIONAL MOLDING
ROTOMOLDING
RUBBER PLUNGER MOLDING

RUNNERLESS INJECTION MOLDING
SHAW POT
SINTER MOLDING
SLUG MOLDING
SLUSH MOLDING
SOLID PHASE FORMING
STEAM MOLDING
STRUCTURAL FOAM MOLDING
TRANSFER MOLDING
WARM FORGING

Molding Compounds. Granules or pellets of polymers or resins, usually mixed with additives such as plasticizers, stabilizers, fillers and colorants, ready for processing by extrusion, molding, calendering or other forming processes into finished or semi-finished products. The term includes DRY BLENDS and MOLDING POWDERS, both of which see.

Molding Cycle. (1) The period of time occupied by the complete sequence of operations on a molding press requisite for the production of one set of moldings. (2) The operations necessary to produce a set of moldings without reference to the time taken.

Molding, High-Pressure. Molding or laminating in which the pressure used is greater than 200 psi. (ASTM D 883-65T).

Molding Index. A test for determining the molding index of thermosetting molding powder according to ASTM D 731-57T comprises molding the specimen compound in a standard flash-type cup mold under prescribed conditions, and expressing the molding index as the total minimum force in pounds required to close the mold.

Molding, Low Pressure. Molding or laminating in which the pressure used is 200 psi or less. (ASTM D 883-65T).

Molding Powders. This term usually denotes pellets or granules of molding compounds made by blending a resin with all of the necessary compounding ingredients, heating the mixture to a homogeneous mass, cooling, and grinding or pelletizing to form a raw material ready for processing by extrusion, molding, calendering or other forming processes. See also DRY BLENDS.

Molding Pressure. The pressure applied to the ram of an injection machine or press to force the softened plastic completely to fill the mold cavities. It is expressed in pounds per square inch of cross sectional area of the material in the pot or cylinder.

Molding Pressure, Compression. See COMPRESSION MOLDING PRESSURE.

Molding Pressure, Injection. See INJECTION MOLDING PRESSURE.

Molding Pressure, Transfer. See TRANSFER MOLDING PRESSURE.

Molding Shrinkage. (Mold shrinkage, Shrinkage, Contraction) The decrease in dimensions, expressed in inches per inch, between a molding and the mold cavity in which it was molded, both the mold and the molding being at normal room temperature when measured.

Mold Lubricant. See PARTING AGENT.

Mold-Mat. A relatively new term for a prepreg containing a chemical thickening agent. Mold-mats may be heated until formable (about 120°F), then compression molded or stamped to shape by dies. By means of high-energy radiation, cure can be effected very rapidly.

Mold Release. See PARTING AGENT.

Mold Seam. Line on a molded or laminated piece, differing in color or appearance from the general surface, caused by the parting line of the mold. (ASTM D 883-65T).

Mold Wiper. In injection molding, a device which enters between the opened mold halves during the ejection cycle, engages the molded piece and lifts or shoves it from the mold. The wiper movement is interlocked with the mold closing mechanism to prevent closing of the mold until the wiper is retracted.

Mole. (1) A mass numerically equal to the molecular weight of a substance. It is most often expressed as the gram molecular weight, e.g. the weight of one mole expressed in grams. (2) In the new SI system, the mole is defined as the amount of substance of a system which contains as many elementary entities as there are atoms in 0.012 kilogram of carbon-12. Note: When the mole is used, the elementary entities must be specified and may be atoms, molecules, ions, electrons, other particles, or specified groups of such particles.

Molecular Orientation. See ORIENTATION.

Molecular Sieves. Porous mineral particles, for example of zeolite, with the ability to absorb molecules of other materials. They have been used as carriers for blowing agents in plastics, so that upon heating the blowing agents are released at the desired rate.

Molecular Volume. The volume occupied by one mole; numerically equal to the molecular weight divided by the density.

Molecular Weight. The sum of the atomic weights of all atoms in a molecule. In most non-polymeric materials the molecular weight is a fixed constant value. In high polymers, the molecular weights of individual molecules vary widely so that they must be expressed as averages. Average molecular weights of polymers may be expressed as *number-average molecular* weight (M_n) or *weight-average molecular weight* (M_w). Methods for determining molecular weights include measurements of osmotic pressure, light scattering, sedimentation equilibrium, dilute solution viscosity, freezing points, vapor pressure; and analyses of ultracentrifugation and spectroscopy. See also MOLECULAR WEIGHT DISTRIBUTION.

Molecular Weight Distribution. The relative amounts of polymers of different molecular weights that comprise a given specimen of a polymer. Two samples of the same polymer with the same weight-average molecular weight may perform quite differently in processing because they have different molecular weight distributions. Two basic groups of methods are used for measuring molecular weight distributions. Fractionation methods, which actually divide the specimen into fractions of various molecular weight ranges, include fraction precipitation and fraction solution (these two are the most widely used); chromatography; liquid-liquid partition; ultracentrifugation; zone refining; and thermogravimetric diffusion. After fractionation by any of these methods, the weight of each fraction is plotted against the average molecular weight of each fraction to obtain a curve of the distribution. Methods for estimating molecular weight distribution without fractionation include light scattering studies, electron microscopy, dilute solution viscosity measurements, gel permeation chromatography, ultracentrifugation and diffusion. The ratio of the weight-average molecular weight to the number-average molecular weight also gives an indication of the distribution.

Molecule. The smallest unit quantity of matter which can exist by itself and retain all of the properties of the original substance.

Molybdate Orange Pigments. Solid solutions of lead chromate, lead molybdate and lead sulfate, used as dark orange to light red pigments for plastics. Their advantages are high opacity, bright color, light-fastness, good heat stability and freedom from bleeding.

Molybdenum Disulfide. MoS_2 (molybdic sulfide, molybdenum sulfide) A white, shiny crystalline material used as a lubricant in nylons, fluorocarbons and polystyrene to reduce wear and friction. It also acts as a strength-improving filler.

Mono-. A prefix denoting a single radical.

Monobasic. Pertaining to acids or salts which have one displaceable hydrogen atom per molecule. Such substances having two displaceable hydrogen atoms are called *dibasic,* and those with three displaceable hydrogen atoms are called *tribasic.*

Monocarboxylic Acids. A family of organic acids whose molecules contain a single carboxylic (-COOH) group, many of which are derived from natural fats and oils. They are used in the production of alkyd resins and polyesters. Esters of monocarboxylic acid, especially of oleic, stearic, pelargonic and ricinoleic acids, are widely used as plasticizers.

Monochloroethylene. See VINYL CHLORIDE.

Monodispersity. The state of uniformity in molecular weight of all molecules of a substance.

Monofilament. (Monofil) A single filament of indefinite length. Monofilaments are generally produced by extrusion. Their outstanding uses are in the fabrication of bristles, surgical sutures, fishing leaders, tennis-racquet strings, screen materials, ropes and nets; the finer monofilaments are woven and knitted on textile machinery.

Monomer. A relatively simple compound, usually containing carbon and of low molecular weight, which can react to form a polymer by combination with itself or with other similar molecules or compounds.

Monomeric. Pertaining to a MONOMER, which see.

Monomeric Cement. See ADHESIVES.

Monotropic. Descriptive of crystal forms one of which is always metastable with the other.

Montan Wax. (lignite wax) A hard, white wax derived from lignite, used as a mold lubricant.

Morphology. The study of the physical form and structure of a material. This includes a wide range of characteristics, extending from the external size and shape of large articles to dimensions of a crystal lattice.

Motionless Mixers. Tubular devices with internal elements that achieve continuous multiple splitting of streams of material passing through, propelled by external pumps, screws or other pressure-applying means. Used in extrusion, the unit yields a more uniform melt that does not depend on dwell time. The term "*Inline* Static Mixer" is sometimes used for this type of mixer.

Mottle. (n) An irregular distribution or mixture of colorants or colored materials giving a more or less distinct appearance of specks, spots, or streaks of color. Note: Mottling is often purposely achieved although it may occur accidentally due to improper mixing. (ASTM D 883-65T).

Mounting Plate. In blow molding, the plate to which the mold is attached. See also CLAMPING PLATE.

Movable Platen. The large back platen of an injection molding machine to which the back half of the mold is secured during operation. This platen is moved either by a hydraulic ram or a toggle mechanism.

Multifilaments. Manufactured fiber yarns composed of many fine continuous filaments or strands.

MVT. Abbreviation for MOISTURE VAPOR TRANSMISSION, which see.

Mw. (Weight-average molecular weight) The sum of the total weights of molecules of each size multiplied by their respective weights divided by the total weight of all molecules.

Mylar. Du Pont's tradename for film composed of polyethylene glycol terephthalate.

Myristoyl Peroxide. $(C_{13}H_{27}CO)_2O_2$ Soft granular powder, used as a polymerization catalyst for vinyl monomers.

n-. Abbreviation for normal, designating those hydrocarbons or hydrocarbon radicals whose molecules contain a single unbranched chain of carbon atoms.

Nacreous. Pertaining to, or having the appearance of, mother-of-pearl. See PEARLESCENT PIGMENTS.

Nacreous Pigments. See PEARLESCENT PIGMENTS.

nano-. (n) The SI-approved prefix for a multiplication factor to replace 10^{-9}.

Nanometer. (nm) A term sometimes used in place of millimicron (Mμ) in expressing the wavelength of light.

Naphthalene. $C_{10}H_8$ (naphthalin, tar camphor) An aromatic hydrocarbon with two ortho-condensed benzene rings, derived from coal tar oils or petroleum fractions. Its familiar application as an insectide, e.g. in the form of "moth balls," is now secondary in importance to its use as an intermediate in the production of phthalic anhydride which in turn is used for making plasticizers, alkyd resins and polyester resins.

1,5-Naphthalene Diisocyanate. (NDI) An isocyanate used in the production of urethane elastomers and foams.

Natta Catalysts. Catalysts used in STEREOSPECIFIC (which see) polymerization reactions, particularly catalysts made from titanium chloride and aluminum alkyl or similar materials by a special process including grinding the materials together to produce an active catalyst surface.

NBR. Abbreviation for ACRYLONITRILE-BUTADIENE COPOLYMERS, which see.

NCNS Resins. See TRIAZINE RESINS.

NDOP. Abbreviation for n-decyl, n-octyl phthalate. See n-OCTYL, n-DECYL PHTHALATE.

Neck-In. (n) In extrusion coating, the difference between the width of the extruded web as it leaves the die and the width of the coating on the substrate.

Necking. (n) The localized reduction in cross-section which may occur in a material under stress. (ASTM D 883-75a). This phenomenon can occur in extrusion under certain conditions as the extrudate leaves the die, but the term most often refers to the cold drawing of fibers at temperatures below their melting points. Fibers of crystalline and some non-crystalline thermoplastics, e.g. polyethylene, exhibit necking at a critical stress near the yield point.

Neck Insert. (finish insert) In blow molding of bottles, a removable part of the mold that forms a specific neck finish of the bottle.

Needle Blow. A specific blow molding technique where the blowing air is injected into the hollow article through a sharpened hollow needle which pierces the parison.

Negative Catalyst. (inhibitor, retarder) An agent which reduces the speed of a reaction.

Neo-. (1) Prefix meaning *new* and denoting a compound related to an older one. (2) A prefix denoting a hydrocarbon in which at least one carbon atom is connected directly to four other carbon atoms.

Neopentyl Glycol. $HOCH_2C(CH_3)_2CH_2OH$ (NPG, 2,2′-dimethyl-1,3-propanediol) An important intermediate used in the production of alkyd and polyester resins, urethane foams and elastomers, and polyester plasticizers. Gel coats based on NPG for reinforced polyesters have improved flexibility, hardness, abrasion resistance and resistance to weathering.

Neopentyl Glycol Diacrylate. (NPGDA) A highly reactive crosslinking monomer used in photocurable coatings. It provides solvent and strain resistance as well as improved response to light.

Neopentyl Glycol Dibenzoate. A solid plasticizer for rigid PVC.

Neoprene. The generic name of polymers of chloroprene ($CH_2 = CHCCl = CH_2$; 2-chloro-1,3-butadiene). They are available as dry solids and latices, and are vulcanizable to tough products with good resistance to oil, solvents, heat and weathering. The original neoprene, produced by Du Pont under the tradename "Duprene," was America's first successful synthetic rubber.

Nepheline Syenite. A naturally occurring mineral composed mainly of feldspar and nephelite. As a filler in PVC compounds, it has the unique property of contributing almost no opacity, so that it can be used in nearly transparent compounds. It is also used as a filler in epoxy and polyester resins.

Nesting. In reinforced plastics, the placing of plies of fabric so that the yarns of one ply lie in the valleys between the yarns of the adjacent ply.

Nest Plate. A retainer plate with a depressed area for cavity blocks used in injection molding.

Netting Analysis. The stress analysis of filament-wound structures that neglects the strength of the resin and assumes that the filaments carry only axial tensile loads and possess no bending or shearing stiffness. (ASTM D 883-75a).

Network Polymers. Polymers obtained by the polymerization of monomers having two or more functional groups which become interconnected with sufficient interchain bonds to form a large three-dimensional network. The networks can be formed during polymerization, or may be made by cross-linking the polymers after they have been formed. The vulcanization of rubber is an example of the formation of a network polymer from a preformed polymer. Copolymers of ethylene and propylene can be made into network polymers by crosslinking with ionizing radiation after reactive sites have been prepared by treating with heat or peroxides.

Network Structure. An atomic or molecular arrangement in which primary bonds form a three-dimensional network.

Neutralization. (chemical) A reaction in which the hydrogen ion of an acid and the hydroxyl ion of a base unite to form water, the other product being a salt.

Newton. The unit of force which, when applied to a body having a mass of one kilogram gives it an acceleration of one meter per second squared.

Newtonian Flow. A flow characteristic evidenced by viscosity that is independent of shear rate, that is the rate of shear is directly proportional to the shearing force. Water and thin mineral oils are examples of fluids that possess Newtonian flow.

Newtonian Fluid. A liquid material that has Newtonian flow characteristics. Such fluids have no yield value.

NHDP. Abbreviation for n-hexyl, n-decyl phthalate. See n-OCTYL, n-DECYL PHTHALATE.

Nickel-Azo Yellow. An azo pigment based on a nickel complex of an azo dyestuff.

NIOSH. Abbreviation for National Institute for Occupational Safety and Health, a division of the Center of Disease Control, Public Health Service, Department of Health, Education and Welfare. This agency conducts investigations and research projects on industrial safety and makes recommendations for the guidance of OSHA. However, it does not enforce its own findings or OSHA regulations.

Nip. The V-shaped gap between a pair of calender rolls where incoming material is "nipped" and drawn between the rolls.

Nip Rolls. In film blowing, a pair of rolls situated at the top of the tower which close the blown film envelope, seal air inside of it, and regulate the rate at which the film is pulled away from the extrusion die. One roll is usually covered with a resilient material, the other being bare metal with internal cooling means.

Nitriding. A process used on ferrous alloy extruder or injection molding screws to improve their wear resistance. It can also be used on cylinders or barrels but is usually used on the screws only. In the method termed gas nitriding, the parts to be nitrided are placed in a high-temperature atmosphere containing ammonia gas. The steel reacts with the ammonia to form nitrides that increase the surface hardness to about 65 to 70 Rockwell C. The total depth of the wear-resistant layer is about 20 to 25 mils, but the maximum surface hardness drops off sharply after 5 to 10 mils have been removed by surfacing treatment and wear. Other methods of nitriding employ immersion in liquids and powders, and during recent years a process called ion nitriding has been developed. In this process nitrogen is ionized with the help of an electrical field in a vacuum. The positive nitrogen ions are attracted by the cathodic workpieces and are impinged into the surface with high kinetic energy.

Nitrile Barrier Resins. (High Nitrile Polymers) A family of polymers generally containing greater than 60% acrylonitrile, along with co-monomers such as acrylates, methacrylates, butadiene and styrene. Both straight copolymers and copolymers grafted onto elastomeric backbones are available. The unique property of these polymers is outstanding resistance to passage of gases and water vapor, making them useful in packaging applications.

Nitrile Rubbers. See ACRYLONITRILE-BUTADIENE COPOLYMERS.

Nitrocellulose. (guncotton, cellulose nitrate) This term is widely used in the plastics industry for CELLULOSE NITRATE, which see, and plastics based theron.

Nitroethane. $CH_3CH_2NO_2$ A colorless liquid used as a solvent for cellulosic, vinyl, alkyd and other resins.

Nitrogen. Nitrogen gas is the most widely used blowing agent for injection molded structural foams. It is less expensive than most chemical blowing agents, leaves no residue, and is easy to handle. Nitrogen is added to the polymer melt by pumping it directly into the barrel of the injection machine.

Nitromethane. CH_3NO_2 A colorless liquid made by reacting methane with oxides of nitrogen or nitric acid under pressure, and used as a solvent for cellulosic, vinyl, alkyd and other resins.

Nitroso Rubber. A copolymer of tetrafluoroethylene and trifluoronitrosomethane.

NODA. Abbreviation for n-octyl n-decyl adipate. See OCTYL DECYL ADIPATE.

NODP. Abbreviation for n-OCTYL, n-DECYL PHTHALATE, which see.

NODTM. Abbreviation for TRI(n-OCTYL n-DECYL) TRIMELLITATE, which see.

Nomogram. (nomograph) A graph containing several (usually three) parallel scales graduated for different variables so that when a straight line connects values of any two, the related value may be read directly from the third at a point intersected by the line. Nomograms are presented frequently in plastics literature to assist in estimating data that normally would require intricate calculations.

Nonionic. Pertaining to a material, atom, radical or molecule which is not capable of being electrically charged. See also IONIC.

Nonpolar. Having no concentrations of electrical charge on a molecular scale, thus incapable of significant dielectric loss. Examples of nonpolar resins are polystyrene and polyethylene.

Nonrigid Plastic. For purposes of general classification, a plastic that has a modulus of elasticity either in flexure or in tension of not over 700 kg per sq. cm. (10,000 psi) at 23 °C and 50 per cent relative humidity when tested in accordance with ASTM D 790, D 747, D 638 or D 882. (ASTM D 883-75a).

Non-Toxic Materials. The toxicity status of resins and additives used for food contact and packaging changes frequently, and in many cases percentages and conditions of end use are stipulated. Thus, no such materials are listed under this heading. The current toxicity status of resins, plasticizers, stabilizers and other additives with respect to their permissible use in contact with foods may be found in the Code of Federal Regulations, Title 21, Chapter 1, Part 121, available from the Superintendent of Documents, Government Printing Office, Washington, D.C. 20402. Cost of entire volume containing Parts 10 through 128 was $5.20 for the 1976 edition. The publication is revised each year.

Nonwoven Fabrics. Staple lengths of natural or synthetic fibers mechanically positioned into a random pattern, then bonded with suitable resins to form sheets.

Non-Woven Mat. A mat of glass fibers in random arrangement, used in reinforced plastics.

Nor-. A prefix used to indicate the parent compound from which a substance may be derived, usually by removal of one or more carbon atoms with attached hydrogen.

Normal Salt. An ionic compound containing neither replaceable hydrogen nor hydroxyl ions.

Normal Solution. A solution containing one gram molecular weight of the dissolved substance divided by the hydrogen equivalent of the substance (that is, one gram equivalent) per liter of the solution.

Noryl. Registered trademark of the General Electric Company for a modified type of POLYPHENYLENE OXIDE, which see.

Notch Sensitivity. The extent to which the sensitivity of a material to fracture is increased by the presence of a surface inhomogeneity such as a notch, a sudden change in section, a crack, or a scratch. Low notch sensitivity is usually associated with ductile materials, and high notch sensitivity with brittle materials.

Novaculite. A very fine-grained type of quartz found in Arkansas, Georgia, Massachusetts, North Carolina, Oklahoma and Tennessee. A variety known as *altered novaculite,* typically about 99.5% quartz, is a solid crystalline substance with the basic hardness of quartz but is more easily reduced to very fine particles. These particles have a great many broken Si-O bonds at their interfaces, which will combine with water to create surface oxides called *silanols.* Such novaculites are useful as semi-reinforcing fillers in silicone rubber, epoxy resins, urethane foams and PVC.

Novolaks. (novolacs) According to ASTM, a novolak is a phenolic-aldehyde resin which, unless a source of methylene groups is added, remains permanently thermoplastic. (See also Resinoid and Thermoplastic) (ASTM D 883-75a) For a preferred definition, see PHENOLIC NOVOLAKS. However, the term is also used in connection with epoxy resins. See EPOXY-NOVOLAK RESINS.

Nozzle. In injection or transfer molding, the orifice-containing plug at the end of the injection cylinder or transfer chamber which contacts the mold sprue bushing and conducts the molten resin into the mold. The nozzle is shaped to form a seal under pressure against the sprue bushing. its orifice is shaped by tapering to maintain the desired flow of resin, and sometimes contains a check valve to prevent back-flow, or an on-off valve to interrupt the flow at any desired point in the molding cycle.

Nozzle Manifold. A series of injection nozzles mounted on a common manifold, each nozzle positioned so as to feed a single cavity in the mold. Such manifolds are used to eliminate runners in molds such as for cup-shaped articles, when it is desired to gate the cavities at the centers of the bottoms.

Nozzles, Mold Gating. In injection molding, nozzles whose tips are parts of mold cavities, thus feeding material directly into cavities, eliminating the need for sprue and runner systems.

NR. Abbreviation for natural rubber. (British Standards Institution). See also POLYISOPRENE.

Nuclear Magnetic Resonance. (NMR) The spinning motion of atomic nuclei imparted by a magnetic field.

Nuclear Magnetic Resonance Spectroscopy. (NMR studies) Determinations of the number of hydrogen atoms in a complex molecule and the characteristic grouping in which they occur, conducted by placing the specimen in a strong constant magnetic field, then applying a perpendicular r.f. alternating magnetic field. At certain frequencies of the latter field a hydrogen atom nucleus will absorb and emit energy, the frequency and amount of which are indicative of the characteristic grouping in which the atom is located — e.g. a CH_3, CH_2 or an -OH group.

Nucleating Agents. Chemical substances which when incorporated in plastics form nuclei for the growth of crystals in the polymer melt. In polypropylene, for example, a higher degree of crystallinity and more uniform crystalline structure is obtained by adding a nucleating agent such as adipic and benzoic acid or certain of their metal salts. Colloidal silicas are used as nucleating agents in nylon, seeding the material to produce more uniform growth of spherulites.

Number-Average Molecular Weight. (M_n) The average molecular weight of a high polymer expressed as the first moment of a plot of the number of molecules in each molecular weight range against the molecular weight. In effect, this is the total molecular weight of all molecules divided by the number of molecules. See also MOLECULAR WEIGHT, MOLECULAR WEIGHT DISTRIBUTION, WEIGHT-AVERAGE MOLECULAR WEIGHT.

Nylon. Generic name for all long-chain polyamides which have recurring amide groups (-CO-NH-) as an integral part of the main polymer chain. Nylons are synthesized from intermediates such as dicarboxylic acids, diamines, amino acids and lactams, and are identified by numbers denoting the number of carbon atoms in the polymer chain derived from specific constituents, those from the diamine being given first. The second number, if used, denotes the number of carbon atoms derived from a diacid. For example, in Nylon 6/6 the two numbers refer to the number of carbon atoms in hexamethylenediamine and adipic acid, respectively. However, in the literature these numbers may otherwise appear as 66, 6.6, 6,6 or 6-6. It has been proposed that the system should be standardized by use of the slash mark between the two numbers. Nylon molding powders can be processed by injection molding, extrusion, and blow molding. Finely powdered forms of nylon are available for fluidized bed coating, rotational molding and other powder processes. A recently developed casting process employs molten caprolactam monomer to which catalysts are added, polymerization occurring in the mold after pouring without additional heat or pressure. Large solid castings and rotationally cast parts have been made by this method. Various types of nylons are described in the subsequent listings. See also POLYCYCLAMIDE, INTERFACIAL POLYMERIZATION.

Nylon 3. (polypropiolactam) A type of nylon that has been prepared and explored experimentally, but has not yet been used commercially.

Nylon 4. (Polypyrrolidone) A polymer of 2-pyrrolidone. Early attempts to commercialize Nylon 4 failed because much of the material was of low molecular weight and decomposed at a relatively low temperature, making it unsuitable for melt spinning. Improved catalyst systems resulted in a polymer with a molecular weight of about 400,000 and a melting point of 265 °C. The newer Nylon 4's have better heat stability than other nylons. Its moisture absorption is higher than that of nylon 6 and -6/6. Nylon 4 can be molded and extruded. Artificial leathers and papers have been produced from slurries of Nylon 4 fibers.

Nylon 6. (polycaprolactam) A type of nylon made by the polycondensation of CAPROLACTAM, which see. It is used for fibers, including those used in tires, and as a thermoplastic molding powder. Nylon 6 is as structurally sound as type 6/6 at room temperature, but it picks up moisture more rapidly and loses strength more rapidly as moisture and temperature increase. Nylon 6 is available in many grades in a broad range of molecular weights, suitable for injection molding, extrusion, blow molding and rotational molding. Parts can be machined, welded, and bonded with adhesives. It is the second most widely used polyamide resin in the U.S.

Nylon 6/6. (polyhexamethyleneadipamide) A type of nylon made by condensing hexamethylenediamine with adipic acid, first prepared by Carothers in 1936. It is the most important of the polyamides commercially, being used extensively for fibers and the most widely used type in other applications. The bulk polymer is a tough, white, translucent crystalline material which melts at 265 °C. Nylon 6/6 is the strongest of the nylons over the widest range of temperature and moisture, and exhibits no appreciable flow below the melting point.

Nylon 6/10. A type of nylon made by condensing hexamethylenediamine with sebacic acid, used for brush bristles and monofilaments. It has a lower water absorption and lower melting point than nylon 6 or nylon 6/6. When a small amount of an alkyl-substituted hexamethylenediamine is added to the condensation mixture, a more elastic polymer known as *elastic nylon* is obtained.

Nylon 6/12. A new type of nylon introduced by Du Pont in 1970, made from hexamethylenediamine and a 12-carbon dibasic acid. Nylon 6/12 is characterized by retention of physical and electrical properties over a wide humidity range, good dimensional stability and low moisture absorption. It is expected to replace nylon 6/10 in most applications, and to compete with phenolics, polycarbonates and DAP in electrical, military and other engineering-type applications.

Nylon 6/T. (polyhexamethylene terephthalamide) The major member of an aliphatic-aromatic family of polyamides, none of which have gained commercial importance because they are difficult to prepare and process.

Nylon 7. (polyenanthamide) A type of nylon known commercially in the U.S.S.R. as Enant, not yet produced in commercial quantities in the U.S. It can be produced by the polymerization of enantholactam, or by the melt polycondensation of 7-aminoheptanoic acid. Its properties are similar to those of nylon 6.

Nylon 8. (polycapryllactam) A type of nylon made by polymerization of capryllactam. Its low melting temperature (200 °C) and high cost of raw materials have limited the usefulness of this polymer. Not to be confused with a type of nylon marketed as "Type 8," which is actually a chemically-modified nylon 6/6.

Nylon 9. (polypelargonamide) A type of nylon made by melt condensation of aminopelargonic acid, or by more recent techniques derived from 9-aminononanoic acid. Nylon 9 has attractive properties, offering tensile yield and flexural strengths approaching those of Nylon 6, while having low water absorption closely approaching that of Nylon 11 and 12. It is, however, in limited use.

Nylon 10. An experimental polymer prepared from aminodecanoic acid. Research on the polymer has been limited due to difficulties in preparing the monomer.

Nylon 11. A type of nylon produced by polycondensation of the monomer 11-amino-undecanoic acid, a derivative of castor oil. It is available in the form of fine powders for rotational molding and other powder processes; and in the form of pellets for extrusion or molding. Like nylon 12, nylon 11 has properties intermediate between those of nylon 6 and polyethylene: good impact strength, hardness and abrasion resistance, but other mechanical properties lower than most other nylons. However, due to its exceptionally low water absorption the dimensional stability of Nylon 11 is high. A modified type of nylon 11 trademarked Rilsan N is said to be flexible, transparent and self-extinguishing.

Nylon 12. A type of nylon made by the polymerization of lauric lactam (dodecanoic lactam) or cyclododecalactam, with 11 methylene units between the linking -NH-CO- groups in the polymer chain. Its mechanical properties are intermediate between those of conventional nylons and polyethylene, and it is the lowest in water absorption (1.5%) and specific gravity (1.01) of all the nylons.

Nylon 55. A blend of aliphatic-, cycloaliphatic-, and aromatic-based polyamides. The material is clear in thick cross sections, has low water absorption, good dimensional stability, solvent resistance, and can be processed economically by injection molding or extrusion.

Nylon Fiber. Generic name for a manufactured fiber in which the fiber-forming substance is any long-chain synthetic polyamide having recurring amide groups (-CONH-) as an integral part of the polymer chain. (Federal Trade Commission.) Nylon was the first fiber of major commercial importance to be made of wholly synthetic material. Carothers' pioneering research in 1929 culminated in Du Pont's introductory marketing of nylon stockings in 1940.

Nylon Monofilaments. Relatively coarse strands of nylon used for fishing lines, brush bristles, racket strings, surgical sutures, etc.

Nylon MXD6. (poly meta-xylylene adipamide) A type of nylon with lower elongation at break than nylon 6 or 6/6, but capable of attaining these properties by reinforcement with glass fibers. The resin has low melt viscosity, good flexural strength and modulus, and resists alkalis and hydrolytic degradation.

Nylon, Nucleated. A nylon polymerized in the presence of a nucleating agent, such as about 0.1% of finely dispersed silica, which promotes the rate of growth of spherulites and controls their number, type and size. Nucleated nylons have increased tensile strength, flexural modulus, abrasion resistance and hardness; but lower impact strength and elongation.

Nylons, Transparent. Departing from the traditional nylons based on aliphatic base units (e.g. chains), such as the amide group (-CO-NH-) and bearing numbers indicating the number of carbon atoms, Dynamit Noble broke the tradition in the early 1970's by introducing a transparent polyamide based on aromatic (ring) base units. The first such nylon, trademarked Trogamid T, is made by polycondensation of terephthalic acid with a 1:1 mixture of 2,2,4-trimethyl and 2,2,4-trimethylhexamethylene-1,6-diamine. This crystal-clear, amorphous polyamide has excellent resistance to stress cracking and a glass transition temperature comparable with that of polycarbonates. Subsequently, Hoechst entered the field with a new, crystal-clear polyamide reportedly based on terephthalic acid with the isomeric bis-aminomethylboranes and up to 70% epsilon-caprolactam. The Phillips Petroleum version, a copolyamide named PACP 9/6, is made from a 64:40 mixture of azelaic acid and adipic acid with the diamine. ATO Chimie has developed another amorphous polyamide trademarked Rilsan N, believed to be a modification of NYLON 11 (which see).

Nylon "Zytel ST." A nylon introduced by Du Pont, identified by the initials for "Super Tough" rather than the conventional numbering system. Claimed to be the most rugged engineering resin ever developed, the new Zytel ST is said to be superior to polycarbonate in impact strength.

o-. Abbreviation for ORTHO-, which see.

OBP. Abbreviation for OCTYL BENZYL PHTHALATE, which see.

OBSH. Abbreviation for 4,4′-OXYBIS(BENZENESULFONYL HYDRAZIDE), which see.

Octahedrite. A term sometimes used for the anatase form of TITANIUM DIOXIDE, which see.

Octakis (2-Hydroxypropyl) Sucrose. A viscous, straw-colored liquid used as a crosslinking agent for urethane foams and a plasticizer for cellulosics.

Octyl. The general term for all 8-carbon radicals having the formula C_8H_{17}-, often used interchangeably with 2-ETHYLHEXYL, which see.

Octyl Benzyl Phthalate. (OBP) A plasticizer for PVC, cellulosics, polystyrene and PVB. It is similar to BBP but has lower volatility. It resists oil extraction.

Octyl Decyl Adipate. $C_8H_{17}OCO(CH_2)_4COOC_{10}H_{21}$ (NODA, n-octyl, n-decyl adipate; iso-octyl iso-decyl adipate) A plasticizer for cellulosics, synthetic rubbers and vinyl resins. It imparts good low temperature flexibility, resistance to extraction by water, and permanence. NODA is also useful at low concentrations in polypropylene as a processing aid.

n-Octyl, n-Decyl Phthalate. (octyl-decyl phthalate, ethylhexyl-decyl phthalate, NODP) One of a family of phthalate plasticizers derived from normal C_6 to C_{10} alcohols. The other members are n-hexyl, n-octyl, n-decyl phthalate (NHDP) and n-decyl, n-octyl phthalate (NDOP). These plasticizers may be used interchangeably in PVC compositions, to which they impart somewhat better drape, flexibility and low temperature resistance than does DOP. They are also compatible with vinyl chloride-acetate copolymers, cellulose nitrate, ethyl cellulose, CAB, polystyrene, acrylic resins, butadiene-acrylonitrile resins, neoprene and chlorinated rubber.

n-Octyl-n-Decyl Trimellitate. An ester of trimelletic acid, used as a plasticizer for vinyl chloride polymers and copolymers. The trimellitate plasticizers are used especially for non-fogging applications, and in adhesives or laminates where low migration properties are important.

Octyl Diphenyl Phosphate. An alkyl aryl phosphate plasticizer for vinyl and other resins, with good permanence and low temperature properties. It imparts flame resistance and is the only phosphate approved by FDA for use in food packaging.

Octyl Epoxy Tallate. A monomeric plasticizer and heat- and light-stabilizer for vinyls and cellulosics. It imparts good low-temperature flexibility, has low volatility, and is used primarily in coated fabrics, garden hose, film and sheeting, and slush-molded parts.

Octyl-Isodecyl Phthalate. See ETHYLHEXYL-ISODECYL PHTHALATE.

n-Octyl Methacrylate. $H_2C:C(CH_3)COOC_8H_{17}$ A monomer for acrylic resins.

para-Octylphenyl Salicylate. $C_6H_4OHCOOC_6H_4C_8H_{17}$ A white crystalline solid material, used as an ultra-violet absorber in polyolefins and cellulosics. It is reported to have increased the outdoor weathering resistance of polyethylene by 400%.

Octyl Phosphate. See Trioctyl Phosphate.

Offset Molding. See JET MOLDING.

Offset Printing. A printing process in which the image to be printed is first applied to an intermediate carrier such as a roll or plate, then is transferred to a plastic film or molded article.

Offset Yield Strength. The stress at which the strain exceeds by a specified amount (the offset) an extension of the initial proportional portion of the stress-strain curve. It is expressed in force per unit area, usually pounds per square inch. (ASTM D 638-60T).

OHM. (Ω) The unit of electric resistance, equal to one volt divided by one ampere. The new SI definition for one ohm is the electric resistance between two points of a conductor when a constant difference of potential of one volt, applied between these two points, produces in this conductor a current of one ampere, this conductor not being the source of any electromotive force.

OIDP. Abbreviation for octyl-isodecyl phthalate. See ETHYLHEXYL-ISODECYL PHTHALATE.

Oil Absorption. The percentage increase in weight of a specimen after immersion in oil for a specified time. This factor is of significance with regard to fillers used in plasticized thermoplastics, which can absorb plasticizer from the compound.

Oil-Soluble Resin. A resin which is capable of dissolving in, dispersing in or reacting with drying oils at moderate temperatures. Such oil-modified resins are used for producing homogeneous films with modified characteristics.

Oleamide. cis-$CH_3(CH_2)_7CH:CH(CH_2)_7CONH$ An ivory-colored powder used as a lubricant in extruding polyethylene.

Olefiant Gas. See ETHYLENE.

Olefin Fiber. Generic name for a manufactured fiber in which the fiber-forming substance is any long chain synthetic polymer composed of at least 85% by weight of ethylene, propylene or other olefin units. (Federal Trade Commission)

Olefin Plastics. See POLYOLEFINS.

Olefins. The group of unsaturated hydrocarbons of the general formula C_nH_{2n}, and named after the corresponding paraffins by the addition of "ene" or "ylene" to the stem. Examples are ethylene, propylene and butenes. Polymers of olefins are sometimes called *olefin plastics* or *polyolefins*.

Oleoresins. Mixtures of a natural resin obtained from a vegetable plant with an essential oil obtained from the same plant. They are usually semi-solids, and are sometimes called *balsams*.

Oligomer. (n) A polymer consisting of only a few monomer units such as a dimer, trimer, tetramer, etc., or their mixtures. (ASTM D 883-75a). Other definitions in the literature place the upper limit of repeating units in an oligomer at about ten. The term *telomer* is sometimes used synonymously with oligomer.

One-Shot Molding. A urethane foam molding process in which the reactants, usually an isocyanate, a polyol and catalyst, are fed in separate streams to a mixing head from which the mixed reactants are discharged into a mold. The polyol and catalyst are sometimes combined along with other additives, but the isocyanate is always fed separately to the mixing head.

Opalescence. The limited clarity of vision through a sheet of transparent plastic at any angle, because of diffusion within or on the surface of the plastic.

Open Cell Foamed Plastic. A cellular plastic in which most of the cells are interconnected in a manner such that gases can travel freely from one cell to another. See also CELLULAR PLASTIC.

Optical Brighteners. See BRIGHTENING AGENTS.

Optical Distortion. Any apparent alteration of the geometric pattern of an object when seen either through a plastic or as a reflection from a plastic surface. (ASTM D 883-65T).

Orange Peel. An uneven surface texture of a plastic article or its finish coating, somewhat resembling the surface of an orange. The condition is often caused by uneven wear of the mold surface due to overpolishing, overheating or overcarburizing of the mold.

Organic Peroxides. See PEROXIDES.

Organic Pigments. Pigments derived from naturally-occurring or synthetic organic substances, characterized by good brightness and brilliance and (usually) transparency. They are generally more resistant to chemicals than inorganic pigments, but are less resistant to heat, light and solvents.

Organometallic Compounds. Compounds containing carbon and a metal, excluding ordinary metallic carbonates such as calcium carbonate and also excluding metallic salts of common organic acids. Many organometallic compounds are used as polymerization catalysts and stabilizers, most notably the ORGANOTIN STABILIZERS, which see.

Organopolysiloxanes. See SILICONES.

Organosol. A suspension of a finely divided resin in a plasticizer together with a volatile organic liquid, the volatile liquid comprising at least 5% of the total weight. The resin used is most frequently PVC, but the term applies to such suspensions of any resin. An organosol can be prepared from a PLASTISOL, which see, merely by adding a volatile diluent or a solvent which serves to lower the viscosity and evaporates when the compound is heated. The diluents most often used in organosols are primary alcohols (methanol, propanol, butanol, pentanol, octanol); commercial mixed diluents such as Apco thinner, VM & P naptha, Stoddard solvent, Varsol #2, Solvesso #100 and #150; and isoparaffins (Isopars).

Organotin Stabilizers. An important class of stabilizers for PVC, notable for their high efficiency, compatibility and impartation of clarity. The family includes sulphides and oxides of tin-alkyl or aryl, organotin salts of carboxylic acid, organotin mercaptides, and trialkyl or triaryl tin alcoholates. Certain dioctyl-tin mercaptides and maleate compounds have been approved for food contact use. See also DI-n-OCTYLTIN MALEATE POLYMER and DI-n-OCTYLTIN, S,S'-BIS(ISOOCTYL MERCAPTOACETATE).

Orientation. The process of stretching a hot plastic article to realign the molecular configuration, thus improving mechanical properties. When the stretching force is applied in one direction, the process is called *uniaxial orientation*. When stretching is in two directions, the term *biaxial orientation* is used. Upon reheating, an oriented film will shrink in the direction(s) of orientation. This property is useful in applications such as shrink packaging, and for improving the strength of molded or extruded articles such as pipe and fibers.

Orientation Release Stress. The internal stress remaining in a plastic sheet after orientation, which can be relieved by reheating the sheet to a temperature above that at which it was oriented. The orientation release stress is measured by heating the sheet and determining the force per unit area exerted by the sheet in returning to its preorientation dimensions.

Orifice. In extrusion, the opening in the die formed by the orifice bushing (ring) and mandrel.

Orr's White. See LITHOPONE.

Ortho-. A prefix denoting a specific position of a substituting radical or group on a benzene ring.

Orthophthalate Plasticizers. The family of plasticizers derived by reacting phthalic anhydride with an alcohol. They include the widely-used plasticizers DOP, DIDP and DTDP.

Orthotropic. Having three mutually perpendicular planes of elastic symmetry.

OSHA. Abbreviation for Occupational Safety and Health Administration, the Federal Agency established by the Department of Labor, Bureau of Labor Standards, to enforce occupational safety and health standards. The standards are known as Part 1910 of amended Chapter XVII of Title 29 of the Code of Federal Regulations established in April 13, 1971 (36 F.R. 7006) and amended thereafter.

Osmometer. An instrument for measuring osmotic pressure. The essential elements are a membrane which is permeable to solvents but impermeable to polymer molecules of a specific size range, reservoirs on each side of the membrane containing respectively the polymer solution and a solvent, means for holding the temperature of the instrument at a constant point and means for measuring the differential osmotic pressure between sides of the membrane. Osmometers are used for measuring the number-average molecular weights of polymers in the range of 20,000 to one or two million. Membrane materials include cellophane, cellulose acetate, polyvinyl alcohol and PCTFE.

Osmosis. The passage of solvent from a mass of pure solvent into a solution, or from a less to a more concentrated solution, through a membrane which is permeable to the solvent but not to the solute.

Osmotic Pressure. The hydrostatic pressure at which the flow of solvent through the membrane of an osmometer just stops. This pressure is related to the number of polymer molecules in dilute solution, and can be used in calculations of molecular weight.

Outgassing. A term used in the vacuum metallizing industry for the evaporation under vacuum of a volatile substance such as moisture, solvent or plasticizer, which causes loss of vacuum, and darkening and poor adhesion of the metal coating to the plastic substrate.

Outsert Molding. A term coined to distinguish the process of molding small plastic parts in a large metal plate from the "insert molding" process in which small metallic parts are incorporated in a plastic molding. The plate is indexed in the injection mold by a peg-and-hole system, with the plastic parts injected through blanks prepunched in the metal plate. The process makes it feasible to use plastics-metal combinations where economics formerly dictated all-metal components.

Ovaloid. A surface of revolution symmetrical about the polar axis which forms the end closure for a filament-wound cylinder.

Overcoating. In extrusion coating, the practice of extruding a web beyond the edge of the substrate web.

Overcure. A thermal decomposition in a thermosetting resin or vulcanizable elastomer due to overheating or excessive molding time.

Overflow Groove. A small groove used in molds to allow material to flow freely to prevent weld lines and low density, and to dispose of excess material.

Overlay Sheet. (top sheet, surfacing mat) A nonwoven fibrous mat of glass or synthetic fiber used as the surfacing sheet in decorative laminated plastics. Its function is to provide a smoother finish, hide the fibrous pattern of the laminate, or provide a decorative motif when printed on the underside.

Overspray. The roughness of a film of paint or lacquer due to dry particles deposited and left undissolved on a previously sprayed semi-dried film. Overspray is encountered particularly with surfaces in more than one plane, such as auto bodies, television cabinets, etc. The remedy for overspray is a well-balanced solvent system containing enough high boiler to minimize the dry particles overshooting a surface.

Oxamide. $NH_2COCONH_2$ A white powder used as a stabilizer for cellulose nitrate.

Oxetane Resins. See CHLORINATED POLYETHER.

Oxidation. Any process which increases the proportion of oxygen or acid-forming element or radical in a compound.

Oxidative Coupling. A process defined as a reaction of oxygen with active hydrogen atoms from different molecules producing water and a dimerized molecule. If the hydrogen-yielding substance has two active hydrogen atoms, polymerization results. This process is used in the polymerization of phenols, particularly POLYPHENYLENE OXIDE, which see.

Oxidative Dehydrogenation. A chemical process used in making monomers such as styrene, butadiene and vinyl chloride. "Oxydehydro" processes involve (1) removal of hydrogen from a hydrocarbon by oxygen, forming water; or (2) removal of hydrogen from a hydrocarbon by a halide to form the hydrogen halide, then regeneration of the halide with oxygen.

Oxirane. A synonym for ethylene oxide, H_2COCH_2.

Oxirane Group. The $=COC=$ group, also called an *epoxy group*.

Oxirane Value. (oxirane oxygen) The percent of oxygen absorbed by an unsaturated raw material during epoxidation; a measure of the amount of epoxidized double bond. The oxirane value and the IODINE VALUE, which see, are used in evaluating epoxy plasticizers. A high oxirane value and low iodine value are considered to be essential for good performance, but these are not the only criteria.

Oxo Process. A chemical process utilizing a reaction known as *oxonation* or *hydroformylation,* in which hydrogen and carbon monoxide are added across an olefinic bond to produce aldehydes containing one more carbon atom than the olefin. The aldehydes produced by this process can be reduced to alcohols used for making many ester-type plasticizers.

Oxy-. A prefix denoting the -O- radical or (primarily in Europe) the -OH radical.

para-Oxybenzoyl Copolyesters. Subsequent to the development of the homopolyester poly(p-oxybenzoyl) trademarked Ekonol® , the Carborundum Co. developed a family of aromatic p-oxybenzoyl copolyesters that are readily moldable. These copolymer systems trademarked Ekkcel® consist of mixtures of p-oxybenzoyl with units from aromatic dicarboxylic acids and aromatic diphenols.

para-Oxybenzoyl Polymer. A new polymer based on p-hydroxybenzoic acid (derived from phenol and carbon dioxide), announced early in 1970 under the trademark "Ekonol" derived from the name of its developer, Dr. James Economy of the Carborundum Company. Ekonol is technically a thermoplastic in the absence of crosslinking, but has unique thermal properties. It retains high stiffness at temperatures above 600 °F, and at temperatures around 800 °F undergoes a second order transition and becomes malleable so that it can be forged like metals. Decomposition occurs at 932 °F. Other properties of the polymer are high dielectric strength, elastic modulus, thermal conductivity, resistance to wear and solvents, self-lubricity, and good machinability. The inherent lubricity can be varied by blending with graphite, boron nitride, molybdenum disulfide and Teflon. Ekonol is available in stock shapes which can be machined, and in powder form processable by plasma spray or powder metallurgy. It can also be blended with metals to form plastic-metal alloys. Initial applications are expected to be in the manufacture of bearings, high-temperature electrical and electronic parts, chemical handling systems, and many critical machine parts. See also para-OXYBENZOYL COPOLYESTERS.

4,4′-Oxybis-(Benzene Sulfonyl Hydrazide). (OBSH) The most important of the sulfonyl hydrazide family of blowing agents, a white crystalline solid melting at 164 °C yielding nitrogen upon decomposition. It is used most widely in rubber-resin blends, due to its ability to simultaneously expand and act as a cross-linking curing agent, but is also used in polyethylene, PVC, phenolics and epoxy resins. Favorable characteristics are low odor, non-toxicity, and freedom from discoloration.

Oxygen Index Flammability Test. A relatively new flammability test developed by C. P. Fenimore and F. J. Martin, based on the principle that a certain volumetric concentration of oxygen is necessary to maintain combustion of a specimen after it has been ignited. The plastic specimen is clamped in a tube into the bottom of which is introduced a mixture of oxygen and nitrogen, the relative concentrations of which can be gradually varied at measured rates. A hydrogen flame is applied to the top of the sample until it ignites, then is withdrawn. If the flame extinguishes, the concentration of oxygen is increased and the sample is re-ignited until it finally continues to burn. The concentration of oxygen at this point is the index of flammability, calculated as

$$\text{Oxygen Index} = \frac{O_2}{O_2 + N_2}$$

wherein O_2 is the minimum volumetric concentration of oxygen which will just support combustion, and N_2 is the associated nitrogen concentration. The test is rapidly gaining acceptance due to its accuracy and simplicity.

Oxymethylene. See FORMALDEHYDE.

p-. Abbreviation, in certain cases, for PARA-, which see.

PA. (1) Abbreviation for PHTHALIC ANHYDRIDE, which see. (2) Abbreviation for polyamides. See NYLON.

PAA. Abbreviation for POLYACRYLIC ACID, which see.

Paddle Agitators. One of the more simple types of mixing equipment for plastics in the form of dispersions, pastes and doughs. The most common form comprises a set of rotating blades driven by a vertical shaft intermeshing with a set of fixed blades.

Paint. A dispersion of pigment in a liquid vehicle which may be applied to surfaces to form a thin adherent protective or decorative coating. The liquid vehicle usually consists of a film-forming resin dissolved in a solvent. Resins most frequently used in paint are phenolics, polyesters, ureas, melamines, cellulosics, acrylics, vinyls, alkyds and epoxies. See also LACQUER and ENAMEL.

Painting Of Plastics. Plastic articles are painted not only to enhance their appearance, but also to provide a desired surface property lacking on the article. For example, electrical properties and resistance to water, solvents, chemicals and abrasion may be improved by painting. Adhesion of paints to plastics is achieved by intermolecular attraction, solvent etching, or a combination of both. The methods used for applying paints to plastics are spraying, with or without masks; screen printing; spray-and-wipe (fill-in' marking); dip coating; roller coating and flow-coating.

PAK. Abbreviation for polyester alkyd resins. See ALKYD RESINS.

PAN. Abbreviation for polyacrylonitrile, a synthetic elastomer. See ACRYLONITRILE.

Paneling. Distortion of a plastic container occurring during aging or storage, caused by the development of a reduced pressure inside the container.

PAPA. Abbreviation for POLYAZELAIC POLYANHYDRIDE, which see.

Paper Chromatography. See CHROMATOGRAPHY.

PAPI. Abbreviation for polymethylene polyphenyl isocyanate. See DI-ISOCYANATES.

Para-. A chemical prefix from the Greek word meaning *beside* or *beyond,* denoting a relation of some kind to another compound such as (1) a higher hydrated form of an acid, (2) a polymeric form, as in *para*ldehyde, (3) the specific position of a substituting radical or group on the benzene ring, or (4) the relation of the 1 and 4 positions in benzene. The abbreviation p- is used in senses (3) and (4).

Paraffin, Chlorinated. See CHLORINATED PARAFFINS.

Paraformaldehyde. A very low molecular weight polymer of formaldehyde, or a mixture of polyoxymethylene glycols, in the form of a white solid which is easily depolymerized by heating to yield formaldehyde gas and water vapor. It is thus a convenient form in which to handle and ship formaldehyde for industrial processes such as the manufacture of ACETAL RESINS, which see. See also s-TRIOXANE.

Parallel-Laminated. Pertaining to a laminate in which all layers of material are oriented approximately parallel with respect to the grain or the strongest dimension in tension.

Parallels. (1) Spacers placed between the steam plate and press platen to prevent the middle section of the mold from bending under pressure. (2) Pressure pads or spacers between the steam plates of a mold to control height when closed to prevent crushing the parts of the mold when the land area is inadequate.

Parameter. In mathematics, a parameter is defined as a variable entering into the mathematical form of any distribution such that the possible values of the variable correspond to different distributions. In the plastics literature, the term is used loosely to denote a specified range of variables, characteristics or properties relating to the subject being discussed; or an arbitrary constant.

Paraphthalate Plasticizers. A family of plasticizers developed in the mid-1970's, derived by reacting terephthalic acid with an alcohol. They are similar in plasticizing capacity to the *orthophthalates,* which see, while offering improved performance in areas such as volatility, low-temperature flexibility, electrical and lacquer-marring. With the exception of dioctyl terephthalate (DOTP), a liquid plasticizer suitable for plastisols, most paraphthalates are solids when prepared with alcohols having an average chain length over six carbon atoms.

Parison. The hollow tube or other preformed shape of thermoplastic compound which is inflated inside the mold in the process of blow molding. Most commonly, the parison is extruded immediately before blowing. In a recently developed modification, parisons are injection molded and stored, to be reheated before blowing. In the earliest application of blow molding, a pair of calendered sheets joined at the horizontal edges was used as the parison.

Parison Programmer. A device which varies the thickness of a parison in pre-determined local areas while it is being extruded, in order to compensate for variations in thickness of certain areas of the finished product which would result from the use of a uniform parison. For example, in a blow molded bottle with a neck portion much smaller in diameter than that of the main body, the neck wall would be much thicker than the body wall if blown from a uniform parison. The parison programmer causes a reduction in thickness of the portion of the parison that will be blown into such a smaller diameter. Programming is accomplished by varying the die gap, by moving either the die or the mandrel, the motions being controlled by a timer or an electromechanical synchronizer to assure that the programmed segments of the parison always coincide with the mold.

Parison Swell. In blow molding, the tendency of the parison to enlarge as it emerges from the die. It is expressed as the ratio of the cross-sectional area of the parison to the cross-sectional area of the die opening.

Parkesine. The name given to the first, however commercially unsuccessful, product resulting from the plasticization of cellulose nitrate by dissolving the polymer in a solvent, mixing in castor oil, and driving off the solvent. The process was developed by Alexander Parkes, and was the forerunner of Celluloid, which was perfected by John Wesley Hyatt in 1870 by using camphor as the plasticizer.

Parting Agent. (release agent, mold lubricant, mold release) A lubricant, often wax, silicone or fluorocarbon fluid, used to coat a mold cavity to prevent the molded piece from sticking to it, and thus to facilitate its removal from the mold. Parting agents are often packaged in aerosol cans for convenience in application.

Parting Line. (flash or flash line) (1) The mark on a molded or cast article caused by flow of material into the crevices between mold parts. In this sense, the term is equivalent to *flash*. (2) A line established on a three-dimensional model form which a mold is to be prepared, to indicate where the mold is to be split into two halves or several components.

Partitioned Mold Cooling. A large diameter hole drilled into the mold (usually the core) and partitioned by a metal plate extending to near the bottom end of the channel. Water is introduced near the top of one side of the partition and removed on the other side.

Parylenes. A new group of thermoplastics introduced in 1965 by Union Carbide. The basic member of the group, Parylene N, is a polymer of para-xylene. It is a highly crystalline, linear material recommended for electrical applications. The second member of the group, Parylene C, is produced from the same monomer modified by substitution of a chlorine atom for one of the hydrogen atoms on the benzene ring (mono-chloro-para-xylene). It was formulated primarily for moisture resistance. The most recent member of the group, Parylene D, has two chlorines substituted for a hydrogen atom on each aromatic ring on the polymer chain. It was developed for high temperature applications. Polymerization is effected by the vapor phase polymerization of the basic monomer in an aqueous system, yielding particulate polymers; or onto a surface from which it may or may not be removed. The polymers exhibit remarkable thermal stability at high temperatures in inert atmospheres, flexibility at cryogenic temperatures down to −320°F, chemical and solvent resistance, and excellent dielectric properties. However, the high price limits the use of parylenes to extremely thin films or depositions, such as conformal coatings on electrical components. Coatings and membranes as thin as 250 Angstroms (1 micro-inch) can be formed by the vapor phase deposition process.

Pascal. The International System of Units term for the pressure or stress of one newton per square meter. (Newton is that force which, when applied to a body having a mass of one kilogram gives it an acceleration of one meter per second squared.) The standard abbreviation of Pascal in the metric system is Pa. It is to replace all units of force per unit area such as psi ($= 6.894,757 \times 10^3$ Pa), atmosphere ($= 1.013,251 \times 10^5$ Pa), and lb f/ft^2 ($= 4.788026 \times 10^3$ Pa).

Pascal's Law. Pressure exerted at any point upon a confined liquid is transmitted undiminished in all directions.

Paste, PVC. A term sometimes used for PLASTISOL, which see.

Paste Resin. A term sometimes used for PVC resins used in making vinyl dispersions such as plastisols. See DISPERSION RESINS.

PAT. Abbreviation for POLYAMINOTRIAZOLES, which see.

PBI. Abbreviation for POLYBENZIMIDAZOLES, which see.

PBMA. Abbreviation for poly-n-butyl methacrylate. See ACRYLIC RESINS.

PB. Abbreviation for polybutene-1. See POLYBUTYLENE RESINS.

PBAN. Abbreviation for polybutadiene-acrylonitrile copolymers.

PBS. Abbreviation for polybutadiene-styrene copolymers.

PBT. See POLYBENZOTHIAZOLE, POLYBUTYLENE TEREPHTHALATE.

PBTP. Abbreviation for polybutylene terephthalate.

PC. Abbreviation for polycarbonate. See POLYCARBONATE RESINS.

PCTFE. Abbreviation for POLYCHLOROTRIFLUOROETHYLENE, which see.

PDAP. Abbreviation appearing in ASTM D 1600 for DIALLYL PHTHALATE, which see. The abbreviation DAP is widely used for both the monomeric and polymeric forms of diallyl phthalate.

PE. Abbreviation for POLYETHYLENE, which see. Also used for *pentaerythritol.*

Peanut Hull Flour. Ground peanut hulls, treated by drying to remove water and sometimes by extraction of volatiles with caustic or toluene, have been used as low-cost fillers in polyethylene. Physical properties are not greatly changed from polyethylenes filled with wood flour.

Pearlescent Pigments. (pearl essence pigments, nacreous pigments) Pigments with crystalline, transparent particles in the form of parallel platelets which impart a mother-of-pearl appearance to plastics. The thin platelets have a high refractive index. Each crystal reflects only a portion of incident light reaching it, transmitting the remaining light to the crystal below. The simultaneous reflection of light from many parallel layers produces the characteristic pearly luster, the brilliance of which depends on the uniformity and parallel alignment of the crystals. Natural pearlescent pigments are composed primarily of guanine crystals derived from fish scales. They are expensive, but non-toxic. The synthetic pearlescents are based on crystallized lead or bismuth compounds or platelets of mica coated with a dye or pigment.

Pearl Polymerization. See SUSPENSION POLYMERIZATION.

Pear Oil. See AMYL ACETATE.

Pebble Mill. See BALL MILL.

Peelers. Machines for slitting large rolls or blocks of foamed plastics into thin sheets, by rotating the blocks into a horizontally mounted band saw blade. Sheets as thin as $1/16''$ may be produced by this method.

Peel Ply. The outside layer of a laminate which is removed or sacrificed to achieve improved bonding of additional layers.

Peel Strength. The force required to peel apart two sheets of material that have been joined with an adhesive. There is no set formula for obtaining maximum peel strengths based on properties and thicknesses of various materials and adhesives.

Peel Test. See "SCOTCH TAPE" TEST.

PEG. Abbreviation for POLYETHYLENE GLYCOL, which see.

Pellets. (molding powders) Tablets or granules of uniform size, consisting of resins or mixtures of resins with compounding additives which have been prepared for molding operations by shaping in a pelletizing machine or by extrusion and chopping into short segments.

Peltier Effect. The phenomenon opposite to that which produces an electric current when a thermocouple junction is heated. When a current flows across the junction of two unlike metals it causes an absorption or liberation of heat. If the current flows in the same direction as the current at the hot junction in a thermoelectric circuit, heat is absorbed. Contrarywise, heat is liberated.

Pendulum Impact Strength. See IMPACT RESISTANCE.

Pentacites. Alkyd resins formed by use of pentaerythritol as the polyhydric alcohol.

Pentaerythritol. $C(CH_2OH)_4$ A white crystalline powder derived by reacting acetaldehyde with an excess of formaldehyde in an alkaline medium. It is used in the production of alkyd resins and chlorinated polyethers.

1,5-Pentanediol. $HOCH_2(CH_2)_3CH_2OH$ (pentamethylene glycol) A colorless liquid used in the production of polyester and urethane resins.

PEO. Abbreviation for POLYETHYLENE OXIDE, which see.

Percentage Elongation. See ELONGATION.

Perchloropentacyclodecane. $C_{10}Cl_{12}$ A solid material used as a flame retardant in epoxy resins, often in conjunction with antimony trioxide.

Perfluoroalkoxy Resins. A class of melt-processible fluoroplastics in which perfluoroalkyl side chains are connected to the carbon-fluorine backbone of the polymer through flexible oxygen linkages. Introduced by Du Pont in 1972, PFA resins have the desirable properties typical of fluoroplastics plus a superior creep resistance, and are more easily processed by extrusion and injection molding.

Perfluoroelastomers. Introduced by Du Pont in early 1977 under the trademark Kalrez, this new elastomer is said to combine the properties of a conventional fluoroelastomer such as Du Pont's Viton with those of fluorocarbon plastics like TFE. Its early applications are for O-rings and gaskets that must withstand temperatures up to 550°F and high resistance to chemicals and solvents.

Perfluoroethylene. See TETRAFLUOROETHYLENE.

Perforating. Processes by which plastic film or sheeting is provided with holes ranging from relatively large diameters for decorative effects (by means of punching or clicking) to very small, even invisible, sizes. The latter are attained by passing the material between rollers or plates, one of which is equipped with closely-spaced fine needles; or by spark erosion.

Permanence. The property of a plastic which describes its resistance to change in characteristics with time and environment. (ASTM D 883-65T).

Permanent Set. The increase in length, expressed as a percentage of the original length, by which an elastic material fails to return to original length after being stressed for a standard period of time.

Permanent White. See BARIUM SULPHATE.

Permeability. (1) The passage or diffusion of a gas, vapor, liquid, or solid through a barrier without physically or chemically affecting it. (2) The rate of such passage. See also GAS TRANSMISSION RATE.

Permittivity. (specific inductive capacity, dielectric constant) The ratio of the capacitance of a given configuration of electrodes with a given material as the dielectric to the capacitance of the same electrode configuration with air as the dielectric.

Permselective Membrane. A thin film which will preferentially permit gases of different kinds to pass through the film at different rates. For common gases such as hydrogen, oxygen, nitrogen and carbon dioxide, silicone rubber is the most permeable polymer known. Under development by General Electric in the late 1970's is a system which might be of great importance in solving the energy problem. Silicone rubber membranes as thin as 500 Angstrom units are 2.2 times as permeable to oxygen as to nitrogen. Oxygen-enriched air produced by pumping air under low pressure through the membrane can be used to reduce fuel consumption up to

50%. Although not yet commercially developed for combustion uses, the principle has been used for oxygen therapy in medical treatment.

Peroxides. (peroxy compounds) A peroxide is a compound which contains at least one pair of oxygen atoms bonded by a single covalent bond. The peroxides most widely used in the plastics industry are organic peroxides, thermally decomposable compounds analogous to hydrogen peroxide, H_2O_2, in which one or both of the hydrogen atoms are replaced by an organic radical. As they decompose, they form free radicals which can initiate polymerization reactions and effect cross linking. The rate of decomposition can be controlled by means of promoters or accelerators added to the system to increase the decomposition rate; or by inhibitors when it is desired to retard the decomposition. Peroxides are used in curing systems for thermosetting resins, and in polymerization reaction mixtures for many thermoplastics. Organic peroxides are often incorrectly described as catalysts.

Peroxyesters. (peresters, t-alkyl peroxyesters) A family of liquid catalysts used for crosslinking of polyethylene, polymerization of vinyls, high-temperture crosslinking of DAP-modified polyesters, curing of styrene-modified polyesters, and styrenation of alkyd paints. They are aliphatic, not prone to yellowing and bleaching, and have good solubility and compatibility characteristics. One example of the numerous compounds in the family is t-Butyl peroxypentanoate.

Persorption. The adsorption of a substance in pores only slightly wider than the diameter of absorbed molecules of the substance.

Perylene Pigments. Diimides of perylene-3,4,9,10-tetra-carboxylic acid. Scarlet and vermillion varieties, resistant to bleeding, light, heat and chemicals, are used in plastics.

PET. Abbreviation used by some authors for polyethylene terephthalate. See PETP.

Peta-. (P) The SI-approved prefix for a multiplication factor to replace 10^{15}.

PETP. Abbreviation approved by ISO/TC 61 and British Standards Institute for POLYETHYLENE TEREPHTHALATE, which see.

Petrochemicals. Chemicals derived directly or indirectly from petroleum or natural gas.

Petroleum Resins. See HYDROCARBON PLASTICS.

PF. Abbreviation for PHENOL-FORMALDEHYDE RESINS, which see.

PFA. See PERFLUOROALKOXY RESINS.

pH. A number which expresses the degree of acidity or alkalinity of a solution. The original definition, not considered entirely correct, was the logarithm of the reciprocal of the hydrogen ion concentration in gram equivalents per liter of solution. 7 on the pH scale is neutral, acid solutions are less than 7, and basic solutions are more than 7.

Phase Angle. See DIELECTRIC PHASE ANGLE.

Phenol. (1) The generic name for the class of organic compounds in which one or more hydroxyl (OH) groups are attached directly to an aromatic ring. (2) The specific name for C_6H_5OH (carbolic acid, phenylic acid, benzophenol, monohydroxybenzene). The specific substance, phenol, was first derived from coal tar, but today is most commonly synthesized from benzene or toluene. About 60% of all phenol manufactured in the United States is used for the production of phenolic resins. Other applications in the plastics industry are for the production of bisphenol A and caprolactam.

Phenol-Aralkyl Resins. A family of thermosetting resins produced by the condensation of aralkyl ethers and phenols. They are cured to hard, intractible resins by one of two methods: (a) heating with hexamethylene tetramine or (b) heating with selected epoxy compounds and an accelerator. Resins of this family marketed by Albright & Wilson Ltd. under the trademark Xylok are available as solutions in ketonic solvents for manufacturing reinforced composites, and as powders for the formulation of molding compounds. Having a useful life at 482 °F approximately eight times that of phenolics, the phenol-aralkyl resins are used in high-performance, high temperature electrical applications.

Phenol-Formaldehyde Resins. (PF resins) The most important of the PHENOLIC RESINS, which see. Made by condensing phenol with formaldehyde, these were the first synthetic thermosetting resins to be developed, and were marketed under the "Bakelite" trade name.

Phenolic Foams. There are two basic types of phenolic foams — the *syntactic* type, and the *reaction* type. The syntactic foams comprise hollow microspheres of phenolic resin mixed with a polyester or an epoxy resin to the consistency of putty, which can be applied to surfaces by troweling, molded, or pressed into sandwich core structures. The reaction-type foams are generated by heating a water-containing liquid phenolic resin, a blowing agent, an acid catalyst and a surfactant. Heat liberated by the reaction of the acid with the phenolic resin vaporizes the blowing agent and water so that the foam structure is produced during curing of the resin. The rising gas particles cause elongated cells, so that the final foam has a grain structure such as that of wood, with strength depending on the direction of application of the stress.

Phenolic Novolaks. (novolaks, novolacs) Thermoplastic, soluble resins obtained by reacting a phenol with an aldehyde, usually formaldehyde, in the proportion of less than one mol of the phenol with one mol of the aldehyde, in the presence of an acid catalyst. When a source of methylene groups is added, linkage between the methylene bonds and the benzolic rings occurs, and the resins can react with diamines or diacids (e.g. hexamethylenetetramine) to form thermosetting insoluble resins. Without such source of methylene groups, the resins remain permanently thermoplastic. See also RESINOID, EPOXY-NOVOLAK RESINS.

Phenolic Resins. A family of thermosetting resins made by reacting a phenol with an aldehyde. The phenols used commercially are phenol, cresols, xylenols, p-t-butylphenol, p-phenylphenol, bisphenols, and resorcinol. The aldehydes most commonly used are formaldehyde and furfural. In the uncured and semi-cured condition, phenolic resins are used as adhesives, casting resins, potting compounds and laminating resins. Phenolic molding powders, compounded with reinforcements or fillers and curing agents, e.g. hexamethylene tetramine, are formed into a large variety of low-cost products by compression molding, transfer molding and, to a limited extent, continuous extrusion. A modified type of injection molding, using a reciprocating screw rather than a plunger, is also used for molding phenolic products. Phenolic moldings have good mechanical properties, electrical characteristics, and resistance to high temperatures.

Phenoxy Resins. (polyhydroxyethers) Linear thermoplastic resins made by reacting an exact stoichiometric equivalent of epichlorohydrin with bisphenol A and sodium hydroxide in dimethyl sulfoxide. The phenoxy resins are chemically similar to epoxy resins, but contain no epoxy groups, have higher molecular weights, and are true thermoplastics. However, the presence of many free hydroxyl groups permits cross-linking with isocyanates, anhydrides, triazines and melamine. Their principal advantages are good processability, low mold shrinkage, excellent dimensional stability and creep resistance. Crystal clear, water white grades are available for extrusion, blow molding and injection molding, which grades have been FDA approved for food contact use.

Phenylene Oxide Resins. See POLYPHENYLENE OXIDE.

Phenylethylene. See STYRENE.

Phenylformic Acid. See BENZOIC ACID.

Phenyl Group. The group C_6H_5-, existing only in combination.

Phenylsilane Resins. Thermosetting copolymers of silicone and phenolic resins, available in solution form.

Phosphate Plasticizers. A group of plasticizers derived from phosphoric acid, used mainly in conjunction with other plasticizers to impart flame resistance. The *aromatic phosphates* (see TRICRESYL-, and CRESYL DIPHENYL PHOSPHATE) also impart good permanence and resistance to greases and oils, but have poor low temperature properties. The *aliphatic phosphates* (see TRIOCTYL-, TRI-BUTOXYETHEL- and OCTYL DIPHENYL PHOSPHATE) impart good low temperature flexibility.

Phosphazene Polymers. A family of experimental resins built on long chains of alternating phosphorus and nitrogen atoms. Early applications have been in fuel hoses that remain flexible in sub-zero arctic weather, prosthetics for reconstruction surgery, and fabric waterproofings. More recent research has shown that the material can be reacted in solution with various nucleophilic reagents to form a host of thermoplastic materials and elastomers with a rich variety of useful purposes. It has been predicted that perhaps the biggest application of phosphazene polymers may be in the textile field as flame retardants. Foams can also be made from the material with good flame retardance.

Phosphorescent Pigments. A family of pigments, generally inorganic sulphide crystals of fairly large and controlled size, which absorb the energy of incident light and then gradually re-emit it as radiation of a color specific to each pigment. The phosphorescence gradually decreases in the absence of incident light.

Photodegradation. Degradation of plastics due to the action of light. Most plastics tend to absorb high-energy radiation in the ultraviolet portion of the spectrum, which activates their electrons to higher reactivity and causes oxidation, cleavage and other degradative reactions. Prior to the growth of widespread concern over environmental aspects of discarded plastics in the 1970's, photodegradation was an undesirable mechanism to be overcome by the use of protective additives. It then became desirable to induce delayed photodegradation of packaging materials and other articles that are discarded after use. The two most common methods of promoting degradation by UV light are incorporation of additives acting as photoinitiators or photosensitizers, and copolymerization that forms weak links at designated sites along the polymer backbone. See also ULTRAVIOLET STABILIZERS, STABILIZER.

Photoelasticity. Changes in the optical properties of isotropic, transparent dielectrics when subjected to stress.

Photopolymer. A plastic compound containing an agent that undergoes some kind of change upon exposure to light, so that images can be formed on its surface by a photographic process.

Photopolymerization. A polymerization rection brought about by exposure of the monomer or mixture of monomers to natural or artificial light, with or without a catalyst. Styrene, vinyl chloride and methyl methacrylate are examples of monomers that can be photopolymerized.

PHR. Abbreviation for *parts* per *hundred* parts of *resin*. For example, as used in plastics formulations, 5 PHR means that 5 pounds of an ingredient would be added to 100 pounds of resin.

Phthalate Esters. The most important group of plasticizers, produced by the direct action of alcohol on phthalic anhydride. The phthalates are the most widely used of all plasticizers, and are generally characterized by moderate cost, good stability, and good all-around properties.

Phthalic Anhydride. $C_6H_4(CO)_2O$ (acid phthalic anhydride) A white crystalline substance derived by oxidation of naphthalene or ortho-xylene, shipped in flake or molten form. Its major use is in the production of phthalate esters for plasticizing vinyl and cellulosic resins. It is also an important intermediate in the manufacture of alkyd and unsaturated polyester resins, and a curing agent for epoxy resins.

Phthalocyanine Pigments. A group of pigments based on phthalocyanine, $(C_6H_4C_2N)_4H_4$, and modifications thereof. The metal-free parent material is blue-green in color. Copper phthalocyanine, $(C_6H_4C_2N)_4Cu$, and modifications thereof containing small amounts of chlorine, produce red shades of blue. Chlorinated copper phthalocyanines in which 14 to 16 hydrogen atoms are replaced by chlorine yield green shades. Certain insoluble forms of phthalocyanine blue and green pigments have been approved for use in food-contacting plastics. They are extremely lightfast, non-bleeding in most vehicles, high in tinctorial strength, and resistant to heat, acids and alkalis.

Physical Catalyst. Radiant energy capable of promoting or modifying a chemical reaction.

PI. Abbreviation for the trans-1,4- type of POLYISOPRENE, which see. (British Standards Institution).

PIA. Abbreviation for Plastics Institute of America.

PIB. Abbreviation for polyisobutylene. (British Standards Institution). See POLYBUTENES.

Pickup Groove. See HOLD-DOWN GROOVE.

pico-. (p) The SI-approved prefix for a multiplication factor to replace 10^{-12}.

Pigments. General term for all colorants, organic and inorganic, natural and synthetic, which are insoluble in the medium in which they are used. Organic pigments are those which contain carbon as the basic part of the molecule. The inorganic pigments, most of which are derived from natural minerals, contain a metal as the basic part of the molecule. See also COLORANTS.

Pill. A term sometimes used for PREFORM, which see.

Pimelic Ketone. See CYCLOHEXANONE.

Pin. See MANDREL.

Pinch-Off. In blow molding, a raised edge around the cavity in the mold which seals off the part and separates the excess material as the mold closes around the parison.

Pinch-Off Blades. The part of the mold which compresses the parison to effect sealing of the parison prior to blowing and to permit easy removal and cooling of the flash.

Pinch-Off Land. The width of the pinch-off blade which effects sealing of the parison.

Pinch-Off Tail. The bottom of the parison that is pinched off when the mold closes.

Pin Gate. (pinpoint gate) In injection molding, an opening or orifice generally 0.030 inches or less in diameter or maximum dimension, through which material flows from the sprue into the mold cavity. Such a gate leaves a small, easily-removed mark on the part, but due to tendency of material to freeze in the restricted orifice its use is limited to small parts and to resins with good flow properties. See also GATE.

Pink Staining. A pink colored stain that sometimes appears on vinyl coated fabrics of white and pastel colors when exposed to ground for prolonged periods. It has been attributed to growth of fungi of the species *Penicillium,* and can be prevented by treatment of the fabric with anti-fungus agents, e.g. n-(trichloromethylthio) phthalimide or an arsenic compound. More recently, pink staining has been found to be caused by the bacterium *Streptomyces rubrireticuli.* See also BIOGRADATION.

Pipe Train. A term used in extrusion of pipe which denotes the entire equipment assembly, e.g. extruder, die, cooling bath, haul-off and cutter.

PIS. Abbreviation for polyisobutylene. See POLYBUTENES.

Piston. See FORCE PLUG.

Pit. An imperfection, a small crater in the surface of the plastic, with its width of approximately the same order of magnitude as its depth. (ASTM D 883-75a).

Pitch. With respect to extruder screws, the distance from any point on the flight of a screw line to the corresponding point on an adjacent flight, measured parallel to the axis of the screw line or threading.

Planar Helix Winding. A winding in which the filament path on each dome lies on a plane which intersects the dome, while a helical path over the cylinder section is connected to the dome paths.

Planar Winding. A winding in which the filament path lies on a plane which intersects the winding surface.

Planetary Screw Extruder. See EXTRUDER, PLANETARY SCREW.

Planishing. See PRESS POLISHING.

Plasma Etching. Many plastics require pretreating prior to electroplating, such as chromic-sulfuric acid baths which are hazardous and contribute to pollution. A new process for etching such plastics exposes the plastic substrate to a gas plasma in a vacuum, producing chemical and physical changes that yield bondability and wetability properties equivalent to those produced by chemical pretreatments. Although many gases may be used, bottled oxygen has been found to be best. An r-f source inside the high-vacuum chamber generates the plasma (an ionized gas consisting of an equal number of positive ions and electrons). The process is effective on nylons, ABS, and phenylene oxide-based plastics.

Plasma Spray Coating. A spray coating process developed to apply sinterable plastics such as PTFE to metals and ceramics. A special spray gun produces a rotating jet of hot, ionized gas particles (plasma) with laminar flow characteristics. Plastic powder supplied to the gun is channeled within the gun so that it emerges as a layer on the periphery of the plasma jet where temperatures are lower than those in the center of the jet. The process is capable of producing coatings as thin as 0.0001 inch on unprimed but clean substrates, without after-baking. Substrates must be capable of withstanding the sintering temperatures of the polymer.

Plastic. (adj.) An adjective indicating that the noun modified is made of or pertains to a plastic or plastics. The singular form is customarily used when the noun obviously refers to a single type of plastic (a *plastic* hose), and the plural form is often used when the noun could refer to several types of plastics (the *plastics* industry). However, there has been a trend in Europe to use the plural form exclusively even when it results in ungrammatical phrases such as "a plastics hose." The intent of this ungrammatical pluralization is to distinguish between the synthetic polymers used in the plastics industry and other materials sometimes referred to as "plastic," such as molten glass, modeling wax and clay in the wet, unfired state. The preference of most authors is to use the singular form when it is evident in context that the noun refers to a single material.

Plastic. (n) The basic ASTM definition of a plastic is a material that contains as an essential ingredient one or more organic polymeric substances of large molecular weight, is solid in its finished state, and, at some stage in its manufacture or processing into finished articles, can be shaped by flow. (ASTM D 883-75a). However, this definition is supplemented by a series of notes explaining that materials such as rubber, textiles, adhesives, and paint, which may in some cases meet this definition, are not considered plastics. The terms *plastic, resin* and *polymer* are somewhat synonymous, but *resins* and *polymers* most often denote the basic materials as polymerized, while the term *plastic* or *plastics* encompasses compounds containing plasticizers, stabilizers, fillers and other additives. See also ELASTOMER.

Plasticate. (v) To render a thermoplastic more flexible by means of heat and mechanical working. Sometimes used imprecisely for PLASTIFY (which see), and incorrectly for PLASTICIZE (which see).

Plastic Deformation. A change in dimensions of an object under load that is not recovered when the load is removed. For example, squeezing a piece of putty results in plastic deformation. The opposite of plastic deformation is *elastic deformation,* in which the dimensions return to their original size when the load is removed, e.g. as exemplified by stretching and releasing a rubber band.

Plastic Flow. A fluid movement that is proportional to the pressure in excess of a certain minimum pressure (yield value) necessary to begin the flow.

Plastic Foam. See CELLULAR PLASTIC.

Plasticity. (n) The ability of a material to withstand continuous and permanent deformation by stresses exceeding the yield value of the material without rupture. The opposite of *elasticity.*

Plasticize. (v) To render a material softer, more flexible and/or more moldable by the addition of a plasticizer. Should not be confused with PLASTIFY and PLASTICATE, which see.

Plasticizer. A substance or material incorporated in a material (usually a plastic or an elastomer) to increase its flexibility, workability or distensibility. A plasticizer may reduce the melt viscosity, lower the temperature of a second order transition or lower the elastic modulus of the product. (IUPAC) Most plasticizers are nonvolatile organic liquids or low-melting point solids, which function by reducing the normal intermolecular forces in a resin thus permitting the macromolecules to slide over one another more freely. Plasticizers are classified in several ways, according to their compatibility (see PRIMARY- and SECONDARY PLASTICIZERS); their general structure (monomeric or polymeric); their function (flame-retardant, high-temperature, low temperature, non-toxic (see NON-TOXIC MATERIALS), stabilizing, cross-linking, etc.; and chemical nature (see ADIPATE PLASTICIZERS, CHLORINATED PARAFFIN, EPOXY PLASTICIZERS, PHOSPHATE PLASTICIZERS and PHTHALATE ESTERS). Many thousands of compounds have been developed as plasticizers, of which perhaps less than 100 are in widespread commercial use today. About 70% (in 1965) of all plasticizers produced are used in vinyls, in which field the three "workhorse" plasticizers are DOP, DIOP and DIDP.

Plasticizer-Adhesives. Additives, partially replacing plasticizers, which improve the adhesion of plastics coatings to substrates. For example, polymerizable monomers such as DAP or triallyl cyanurate are added to PVC plastisols to improve their adhesion to metals.

Plasticizers, Polymerizable. A special class of plasticizers, unique in that they function as plasticizers only during and before the processing step, is constituted by the monomers added to plastisols to increase their fluidity, which cure in the presence of catalysts to become rigid in the fused plastisol article. Among such monomers are polyglycol dimethylacrylates (Union Carbide's MG-1), dimethyacrylates of 1,3-butylene glycol and trimethylolpropane, and several monomers with proprietary names such as X-970, X-980 (Rohm & Haas Co.), and "Santoset" (Monsanto Chemical Co.). These polymerizable plasticizers, sometimes called *reactive plasticizers,* enable very rigid articles to be made by casting fluids which otherwise would have to contain so little conventional plasticizer as to be too viscous to flow and process.

Plasticizers, Solid. Plasticizers which are solid at room temperature but melt at processing temperatures to improve processibility of the polymer in which they are incorporated. Upon cooling they re-solidify and thus do not soften the finished article. Solid plasticizers are used in rigid PVC, one of the most commonly used being diphenyl phthalate.

Plastic Memory. See MEMORY.

Plasticorder. (plastigraph) See BRABENDER PLASTOGRAPH.

Plastic Paper. (synthetic paper) Paper-like products in which the skeletal structure is composed of synthetic resin. Three main types are SPUNBONDED SHEET (which see), *film papers* and *synthetic pulps* (synpulps). Film-papers are similar to thin films of oriented polystyrene or polyolefins but they are treated to obtain opacity and ink receptivity. Synpulps are papermaking pulps made usually from polyolefins by new processes which produce fibrous pulps without the use of conventional spinning methods. They have not been developed to the state of commercial acceptance as of 1976.

Plastic, Rigid. See RIGID PLASTIC.

Plastic, Semi-Rigid. See SEMI-RIGID PLASTIC.

Plastic Tooling. A term employed for structures composed of plastics, usually reinforced thermosets, which are used as tools in the fabrication of metals or other materials including plastics. Common applications of plastics tooling are sheet metal forming dies, models for Keller Duplicators, drill fixtures, spotting racks, molds for thermoforming thermoplastic sheets, and molds for lay-ups or contact pressure molding of reinforced plastics. The tools are formed by the ordinary processes used for thermosetting resins, such as laminating, casting and screening.

Plastic Welding. See WELDING.

Plastify. (v) To soften a thermoplastic resin or compound by means of heat alone. Should not be confused with PLASTICIZE or PLASTICATE, which see.

Plastigel. A vinyl compound similar to a plastisol, but containing sufficient gelling agent and/or filler to produce a putty-like consistency. It may be molded to a shape-retaining form at room temperature, then heated and cooled to impart permanency.

Plastisol. A suspension of a finely divided vinyl chloride polymer or copolymer in a liquid plasticizer which has little or no tendency to dissolve the resin at normal temperatures but becomes a solvent for the resin when heated. At the proper temperature, the resin is completely dissolved in the plasticizer, forming a homogeneous plastic mass which upon cooling is a more or less flexible solid. Additives such as fillers, stabilizers and colorants are also usually present. Plastisols modified with volatile solvents or diluents which evaporate upon heating are known as ORGANOSOLS, which see. When gelling or thickening agents are added to create a putty-like consistency at room temperature, the dispersion is called a *plastigel*. The coined term *rigidsol* is used to denote plastisols modified with polymerizable or crosslinkable monomers so that the fused products are rigid rather than flexible. Plastisols are used in the processes of rotational casting, slush casting, dipping, spraying, film casting and coating. The term "plastisol" was first proposed in 1943 by J. F. Suter of Carbide and Carbon Chemicals Corp., and within a few years was adopted all over the world.

Plastograph. See BRABENDER PLASTOGRAPH.

Plastomers. A term coined by Phillips Petroleum Co. for its line of thermoplastic elastomers trademarked Solprene. They are solution-polymerized block copolymers of butadiene and styrene in different ratios. While they are used directly as compounds for injection molding footwear, extruded or molded products, and in adhesives and sealants, it is said that their greatest promise lies in blends to upgrade the performance and economics of other plastics and elastomers.

Plastometer. See RHEOMETER.

Plate Dispersion Plug. Two perforated plates held together with a connecting rod which are placed in the nozzle of an injection molding machine to aid in dispersing a colorant in a resin as it flows through the orifices in the plates.

Plate-Mark. Any imperfection in a pressed plastic sheet resulting from the surface of the pressing plate. (ASTM D 883-75a).

Platens. The mounting plates of a press to which the entire mold assembly is bolted.

Plate-Out. An objectionable coating gradually formed on metal surfaces of molds, calendering and embossing rolls during processing of plastics due to extraction and deposition of some ingredient such as a pigment, lubricant, stabilizer or plasticizer. In the case of vinyls, which are especially likely to exhibit this condition, plate-out can be reduced by using highly compatible stabilizers such as barium phenolates and cadmium ethylhexoate. Resins can play a role in the plate-out problem, although the degree and mechanisms of resin contribution to plate-out are controversial.

Platform Blowing. A special technique for blow molding large parts. To prevent excessive sag of the heavy parison the machined employs a table which, after rising to meet the parison at the die, descends with the parison but at a slightly lower rate than the parison extrusion speed.

Plating. See ELECTROPLATING ON PLASTICS.

Plug-and-Ring Forming. A method of sheet forming in which a plug, functioning as a male mold, is forced into a heated plastic sheet held in place by a clamping ring. See also THERMOFORMING.

Plug Forming or Plug-Assist Forming. A thermoforming process in which a plug or male mold is used to partially preform the part before forming is completed by means of vacuum or pressure. See also THERMOFORMING.

Plunger. The part of a transfer or injection molding press that applies pressure on the unmelted plastic material to push it into the chamber, which in turn forces plastic melt out of the chamber through the nozzle at the front. See also RAM, FORCE PLUG, POT PLUNGER.

Plunger Molding. A variation of the TRANSFER MOLDING (which see) process, in which an auxiliary hydraulic ram is employed to assist the main ram. The auxiliary ram forces the material through an opening so small that high frictional heat is developed to rapidly advance the degree of cure of the material, so that complete cure is achieved soon after the mold is filled.

PMAC. Abbreviation for POLYMETHOXY ACETAL, which see.

PMAN. Abbreviation for POLYMETHACRYLONITRILE, which see.

PMCA. Abbreviation for poly(methyl-α-chloroacrylate), a member of the acrylic resin family.

PMMA. Abbreviation for POLYMETHYL METHACRYLATE, which see.

PMP. Abbreviation for POLY(4-METHYLPENTENE-1), which see.

Pock Marks. Imperfections on the surface of a blow molded article consisting of irregular indentations caused by insufficient contact of the blown parison with the mold surface. Contributory factors are insufficient blowing pressure, entrapment of air or gases, and condensation of moisture on the mold surface. Note: See also *pit* for an ASTM-approved definition for this term but which does not specify the process for which the term is valid.

Poise. The metric unit of viscosity, named after the French sccientist Poiseuille. One poise is the viscosity of a liquid in which a force of one dyne is necessary to maintain a velocity differential of one centimer per second per centimeter over a surface one centimeter square. The *centipoise* is one hundredth of a poise.

Poiseuille's Equation. An equation for calculating the pressure drop (ΔP) in laminar flow of fluid through a cylindrical tube, frequently used in discussions of polymer melt rheology. The equation is:

$$\Delta P = \frac{8Q\mu L}{\pi R^4}$$

wherein Q = volumetric flow rate, in.^{3}/sec.
   μ = viscosity, lb. sec./in.^{2}
   L = tube length, in.
   R = tube radius, in.

Poisson's Ratio. When a material is stretched, its cross-sectional area changes as well as its length. Poisson's ratio is the constant relating these changes in dimensions, and is obtained by dividing the change in width per unit length by the change in length per unit length.

Polar Winding. A winding in which the filament path passes tangent to the polar opening at one end of a chamber and tangent to the opposite side of the polar opening at the other end.

Polepiece. In reinforced plastics, the supporting part of the mandrel used in filament winding, usually on one of the axes of rotation. (ASTM D 883-75a).

Polishing Roll. A roll, or series of rolls, which have a highly polished chrome plated surface, utilized to produce a smooth surface on sheet as it is extruded or calendered.

Poly-. A prefix denoting many. Thus, the term polymer literally means *many mers,* a mer being the repeating structural unit of any high polymer.

Polyacetals. See ACETAL RESINS.

Polyacetylenes. Dark colored polymers of acetylene, made with Ziegler-Natta catalysts, of limited application. They have been suggested as semi-conductors, high energy binders and plasticizers for solid propellants.

Polyacrylamide. $(CH_2CHCONH_2)_x$ A nonionic, water-soluble polymer prepred by the addition polymerization of acrylamide. The white polymer is readily soluble in cold water, but insoluble in most organic solvents. It is used as a thickening agent, suspending agent, and as an ingredient in adhesives.

Polyacrylate. A thermoplastic resin made by the polymerization of an acrylic compound such as methyl methacrylate. See also ACRYLIC RESINS.

Polyacrylic Acid. (PAA) A polymer of acrylic acid, used as a textile sizing agent.

Polyaddition. See ADDITION POLYMERIZATION.

Polyalcohols. See POLYOL.

Polyalkenamer. A chlorine-containing elastomer developed by Goodyear, but not commercially available as of mid-1976. The elastomer, with properties similar to but somewhat superior to those of neoprene rubber, is a copolymer of (a) the addition reaction product of hexachlorocyclopentadiene and 1,5-cyclooctadiene and (b) a simple hydrocarbon olefin such as cyclopentene.

Polyalkylene Amides. See AMINO RESINS.

Polyalkyleneterephthalates. A family including polyesters which are polycondensates derived from terephthalic acid, whose diol components are variable within a wide range. The principal members of the family are POLYETHYLENE TEREPHTHALATE and POLYBUTYLENETEREPHTHALATE, which see.

Polyallomers. Crystalline thermoplastic copolymers produced from two or more different monomers, usually ethylene and propylene, by alternately polymerizing the monomers employed in the presence of anionic coordination catalysts, resulting in chains containing polymerized segments of both monomers. The polymer chains exhibit degrees of crystallinity normally found only in stereo-regular homopolymers of propylene and ethylene, and the copolymers possess properties different from those of either blends of polyethylene and polypropylene or copolymers prepared by conventional polymerization processes. Among such properties are high impact strength, low density and resistance to fatigue from flexing. The name "polyallomer" is derived from *allomerism,* meaning a similarity of crystalline form with a difference in chemical composition.

Polyallyl Diglycol Carbonate. A high-impact transparent plastic with good abrasion resistance, made from PPG industries' CR-39 monomer. An initial application was for safety screens for hazardous equipment such as punch presses.

Polyamide-Imide Resins. (AI) A family of polymers based on the combination of trimellitic anhydride with aromatic diamines. In the uncured form (ortho-amic acid) the polymers are soluble in polar organic solvents. The imide linkage is formed by heating, producing an infusible resin with thermal stability up to 550 °F. These resins are used for laminating, prepregs and electrical components. Molding resins that behave as thermoplastics can be produced by thermally curing and modifying amide-imide polymers. These molding resins can be processed by compression molding, extrusion and injection molding. However, they exhibit a highly viscoelastic behavior, being extremely viscous at low shear rates and becoming highly fluid at high shear rates.

Polyamide Plastics. (polyamides) See NYLON and other listings commencing with same.

Polyamine-Methylene Resin. A light-amber colored resin derived from diphenylol and formaldehyde, used as an ion-exchange resin.

Polyamine Sulfone. A water soluble copolymer of diallylamine monomer and sulfur dioxide, used as a paint additive, anti-static agent, synthetic fiber modifier and metal plating polishing agent.

Polyaminobismaleimide Resins. (PABM) Thermosetting resins of dark brown color obtained from aromatic diamines and bismaleimides. The prepolymers can be cast and compression or transfer molded. High percentages of fillers can be incorporated. PABMs have flow properties comparable to common thermosetting resins, thermomechanical properties exceeding those of some light alloys, possess excellent dimensional stability, and are flame- and radiation-resistant. They can be adapted to aircraft, electrical/electronic friction, and ablation applications, and to chemical process equipment where resistance to aromatic solvents, refrigerants, and acids is required.

Polyaminotriazoles. (PAT) Fiber-forming polymers made from sebacic acid and hydrazine with small amounts of acetamide.

Polyaryl Ether. A new thermoplastic material introduced in 1969 by Uniroyal under the tradename "Arylon," said to be the first engineering thermoplastic that combines outstanding heat distortion resistance and impact strength with processing ease. The material can be injection molded and extruded under conditions similar to those employed for ABS polymers. Extruded sheet can be thermoformed. Parts can be solvent welded with common commercial solvents, painted without pretreatment, and electroplated. Early applications have been for exterior automobile parts, household appliances, and battery cases for aerospace use.

Polyaryloxysilanes. Newly developed high temperature resistant polymers made up of silicon atoms, oxygen atoms and thermally stable aromatic rings, part organic and part inorganic in nature.

Polyarylsulfone Resin. A thermoplastic resin consisting mainly of phenyl and biphenyl groups linked by thermally stable ether and sulfone groups, available from the 3M Company under the tradename Astrel 360 in pellet or powder form. Its most outstanding property is resistance to high and low temperatures (from —400° to 500 °F). Other properties are good impact resistance, resistance to chemicals, oils and most solvents, and good electrical insulating properties. The resin can be processed by injection molding, extrusion, compression molding, ultrasonic welding and machining.

Polyazelaic Polyanhydride. (PAPA) A carboxyl-terminated polymer of approximately 2300 molecular weight, used as a curing agent for epoxy resins.

Polybenzimidazoles. (PBI) High molecular weight, strong and stable polymers containing recurring aromatic units with alternating double bonds, recently developed by C. S. Marvel. The principal applications to date have been in the aerospace field, as protective coatings on missiles, radar antennas and supersonic jet planes; and in reinforced laminates for critical applications. The first PBI resins were synthesized by condensing 3,3'-diaminobenzidine with diphenylisophthalate or isophthalamide. After further research, using many parent monomers, tetraamines and diacids, PBI's are produced mainly from the condensation of 3.3'.4.4'-tetraaminobiphenyl (diaminobenzidine) and diphenyl isophthalate. The polymers are brown colored amorphous powders, exhibiting a high degree of thermal and chemical stability. They are used for fibers with resistance to high temperatures and flame.

Polybenzothiazole. (PBT) An experimental high-temperature resin obtained by reacting mixed toluides, sulfur and 4-aminophthalimide. Developed by Abex Corporation and designated as AF-R-2506, the resin was available in 1969 in quantities up to one pound. Glass reinforced PBT resin composites have withstood temperatures up to 1000 °F for ten minutes, and up to 650 °F for 250 hours.

Polybiphenylsulfone. A new engineering thermoplastic introduced by Union Carbide in 1976 under the name "Radel." It is said to offer a unique combination of thermal and impact properties — 400 °F heat deflection temperature and notched Izod impact approaching the 15 ft.-lb. range. Radel can be processed on conventional molding and extrusion machinery.

Polyblend. A colloquial term used for physical mixtures of two or more polymers, for example polystyrene and rubber, PVC and nitrile rubber. Such blends usually yield products with favorable properties of both components, sometimes opening markets not available to either of the separate components. The term *alloy* is sometimes used for such blends.

Polybutadiene. A synthetic rubber made from butadiene, $H_2C = CH—CH = CH_2$. The cis-1,4-type has superior abrasion and resilience. The trans-1,4-type is similar to natural rubber.

Polybutadiene-Acrylic Acid Copolymer. A binder used in solid propellants.

Polybutadiene-Type Resins. Unsaturated, thermosetting hydrocarbons cured by a peroxide-catalyzed vinyl type polymerization reaction, or by sodium-catalyzed polymerization of butadiene or blends of butadiene and styrene. Liquid systems, curable in the presence of monomers, are used for casting, encapsulation, and potting of electrical components, and in making laminates. Molding compounds, often containing fillers and modified with other resins or rubbers, may be compression or transfer molded. A new type of butadiene polymer known as syndiotactic 1,2-polybutadiene is introduced in 1974 in Japan under the trademark JSR RB. This resin is described as a thermoplastic having an intermediate property between a crystalline and an amorphous polymer, with good transparency and flexibility without plasticization. In the presence of a photosensitizer such as p,p'-tetramethyl diaminobenzophenone, this polymer may be readily cured by U.V. radiation. Transparent films of the non-toxic polymer are being used for packaging, and cellular forms for shoe soles. Its high gas and water vapor permeability enable contents of packages made from JSR to be sterilized with ethylene oxide inside the package. The polymer is biodegradable.

Polybutenes. The family of polymers of 1-butene, cis-2-butene and trans-2-butene. Depending on molecular weight they range from oils through tacky waxes, crystalline waxes and rubbery solids. See also BUTYL RUBBER and POLYBUTYLENE RESINS.

Polybutylene Adipate Glycol. (PBAG) A polymeric diol used in the production of urethane elastomers, marketed by Witco Chemical Co. under the trademark Formrez F 13-35.

Polybutylene Resins. A group of polymers consisting of isotactic, stereoregular, highly crystalline polymers based on butene-1, first introduced in 1965. Their properties are similar to those of polypropylene and linear polyethylene, with superior toughness, creep resistance and flexibility. Early applications have been in the areas of pipe, wire coating, gaskets and heavy-duty packaging films. Acceptance of polybutylene in the pipe industry has been particularly good because it carries the highest design-stress rating of all flexible, thermoplastic piping material and has proved itself in pressure applications up to 200°F.

Polybutylene Terephthalate. (PBTP, PBT) A member of the polyalkyleneterephthalate family, similar to polyethylene terephthalate in that it is derived from a polycondensate derived from terephthalic acid whose diol component is butanediol rather than glycol as in the case of PET. PBTP can be modified easily to overcome its relatively low temperature limit, making it equivalent to plastics used in construction and appliances. Grades are available for injection and blow molding, extrusion and thermoforming. Properties include high strength, dimensional stability, low moisture absorption, good electrical characteristics and resistance to heat and chemicals when suitably modified.

Poly-n-Butyl Methacrylate. (PBMA) One of the ACRYLIC RESINS, which see.

Polycaprolactam. See NYLON 6.

Polycaprolactone. (PCL) A thermoplastic resin with the linear structure $(CH_2\text{-}CH_2\text{-}CH_2\text{-}CH_2\text{-}CH_2\text{-}COO)_n$, made by polymerizing epsilon-caprolactone. Following its introduction in 1969 by Union Carbide, the major application for PCL was as an additive for other resins to improve their processing characteristics and end use properties. PCL is compatible with most thermosetting and thermoplastic resins and elastomers. It increases impact resistance and aids mold release of thermosets, and acts as a polymeric plasticizer with PVC. It was later found that PCL is uniquely a biodegradable polymer, useful in containers for growing and transplanting trees and other plants. Unmodified PCL is completely consumed by soil microbes. The rate of degradation of a plant container can be pre-determined by incorporating a non-biodegradable polymer to retard attack, or a filler to accelerate attack. Thickness and geometry of the containers are also factors. Thus, the mechanical integrity of the material can be tailored to suit the desired greenhouse growing period. PCL is also useful in the production of polyurethane elastomers and foams, to which it imparts good low temperature properties and high water resistance.

Polycarbonate Resins. (PC) A family of special types of polyesters in which groups of dihydric phenols are linked through carbonate linkages. They can be produced by a variety of methods, of which the most commercially important are (1) phosgenation of dihydric phenols, usually bisphenol A; (2) ester exchange between diaryl carbonates and dihydric phenols, usually between diphenyl carbonate and bisphenol A; and (3) interfacial polycondensation of bisphenol A and phosgene. Bisphenol A polycarbonates within the molecular weight range of 32,000 to 35,000 can be processed by injection molding, extrusion, thermoforming and blow molding. Melt casting and solution casting processes are also employed. Such polycarbonates are characterized by high impact strength, good heat resistance, low water absorption, good electrical properties and non-toxicity. Certain crystal-clear grades have been developed for optical purposes. Other applications include dentures and artificial teeth, non-staining dinnerware, packages for foodstuffs, electrical components, precision parts for instruments and household appliances.

Polycarborane Siloxanes. A material announced in 1971 by Olin Corp. as Dexsil. Stable up to 500°C, the first application was as a partitioning phase in gas separations of high boiling materials.

Polychloroethers. See CHLORINATED POLYETHER.

Polychloroprene. See NEOPRENE.

Polychlorotrifluoroethylene. (PCTFE) A family of polymers made by polymerizing the gas C_2ClF_3 by mass, emulsion or suspension polymerization techniques. The polymers range from oils, greases and waxes of low molecular weight, to the tough, rigid thermoplastics most commonly used in the plastics industry. Unlike PTFE, PCTFE may be processed by conventional techniques such as extrusion, injection molding and compression molding. It is also available as dispersions in xylene or ketones for application by dipping or spraying. The polymers are non-toxic, resistant to heat, chemically inert, and have outstanding electrical properties.

Polycondensation. See CONDENSATION POLYMERIZATION.

Polycyclamide. A suggested generic term for polyamides containing a cycloalkane ring, a series of linear, high-molecular weight polyamides formed by condensing 1,4-cyclohexanebis(methylamine) with one or more dicarboxylic acids. These polymers have a high melting point, but are sufficiently stable to permit melt processing at temperatures above 300°C without thermal decomposition. Their excellent physical and chemical properties indicate their usefulness as fibers, films and moldings. See also NYLON.

Poly(1,4-Cyclohexylene Dimethylene Azelamide). The condensation product of 95 to 99% of trans cyclohexane bis(methylamine) or CBM and azelaic acid, developed and named CBM-9 by The Firestone Tire & Rubber Co. It is claimed to be a non-flatspotting tire reinforcing material competing with nylons 6 and 66.

Poly-1,1-Dihydroperfluorobutyl Acrylate. A white elastomeric gum rubber.

Poly(Dimethylsiloxanes). See DIMETHYL POLYSILOXANES.

Polydispersity. The state of non-uniformity in molecular weight of a molecule of a substance.

Polyester. (alkyd) A general term encompassing all polymers in which the main polymer backbones are formed by the esterification condensation of polyfunctional alcohols and acids.

The term *alkyd* was coined from the *AL* in polyhydric *AL*cohols and the *CID* (modified to KYD) in polybasic a*CID*s. Hence, in a chemical sense the terms alkyd and polyester are synonymous. However, as more commonly used the term alkyd refers to polyesters modified with oils or fatty acids (see ALKYD MOLDING COMPOUNDS). The term polyester is further explained under POLYESTERS, SATURATED and POLYESTERS, UNSATURATED.

Polyester, Aromatic. Two types of materials termed aromatic polyesters are marketed by Carborundum Plastics, Inc. under the trademarks Ekonol and Ekkcel. Ekonol is a homopolyester of repeating p-oxybenzoyl units with a high degree of crystallinity. It does not melt below its decomposition temperature, 550°C, but can be fabricated at or above a reversible crystal-crystal transition temperature range of 300 to 360°C by compression sintering and plasma spray processes. Ekkcel is the term for a series of p-oxybenzoyl copolyesters containing the p-oxybenzoyl unit in combination with moieties derived from aromatic dicarboxylic acids and aromatic bisphenols. They are available in grades for compression molding and injection molding. Applications include electrical connector components, valve seats, high-performance aircraft parts, and automotive parts.

Polyester Fiber. Generic name for a manufactured fiber in which the fiber-forming substance is any long chain synthetic polymer composed of at least 85% by weight of an ester of dihydric alcohol and terephthalic acid. (Federal Trade Commission). The polyester fiber in most common use throughout the world is derived from polyethylene terephthalate. Polyester filaments are produced by forcing the molten polymer at a temperature of about 290°C through spinneret holes about .009″ diameter, followed by air cooling, combining the single filaments into yarns, and drawing. The major end use of polyester fibers is for blending with cotton to enhance crease retention and reduce wrinkling of fabrics. The use of polyester fibers in carpeting and tires is increasing rapidly.

Polyester Plasticizers. A group of plasticizers of the so-called polymeric plasticizer family, characterized by a large number of ester groups in each single molecule. They are synthesized from (1) a dibasic acid, e.g. adipic, azelaic, lauric or sebacic acid; (2) a glycol (dihydric alcohol); and (3) a mono-functional terminator such as a monobasic acid. Polyester plasticizers are noted for their permanence and resistance to extraction.

Polyesters, Saturated. The family of polyesters in which the polyester backbones are saturated and hence unreactive. They comprise low molecular weight liquids used as plasticizers and as reactants in forming urethane polymers; and linear, high molecular weight thermoplastics such as polyethylene terephthalate ("Mylar" and "Dacron"). Usual reactants for the saturated polyesters are (1) a glycol such as ethylene-, propylene-, diethylene-, dipropylene-, or butylene glycol; and (2) an acid or anhydride such as adipic acid, azelaic acid, terephthalic acid or phthalic anhydride. Other types of saturated polyesters are used in high temperature varnishes and adhesives.

Polyesters, Unsaturated. The family of polyesters characterized by vinyl unsaturation in the polyester backbone which enables subsequent hardening or curing by copolymerization with a reactive monomer in which the polyester constituent has been dissolved. The unsaturated polyesters are made by agitating in a heated kettle a mixture of glycols (e.g. propylene- or diethylene glycol); unsaturated dibasic acids or anhydrides (e.g. fumaric acid or maleic anhydride); and, sometimes in order to control the reaction and modify properties, a saturated dibasic acid or anhydride (e.g. isophthalic acid or phthalic anhydride). After removal of water and cooling, the fluid polyester may be dissolved in a reactive monomer in the same kettle, or it may be shipped to users who add the monomer and catalyst in their factories. Styrene is most widely used as the reactive monomer. Others sometimes used are diallyl phthalate, diallyl isophthalate, and triallyl cyanurate. Peroxide catalysts are generally used for the final copolymerization reaction. These unsaturated polyesters are thermosetting and are most widely used in reinforced plastics and in the potting of electrical components. See also WATER-EXTENDED POLYESTERS.

Polyester Terephthalate. See POLYESTERS, SATURATED.

Polyether, Chlorinated. See CHLORINATED POLYETHER.

Polyether Foams. Types of URETHANE FOAMS, which see, that have been made by reacting the isocyanate with a polyether rather than a polyester or other resin component. For rigid foams, polyethers often used are the propylene oxide adducts of materials such as sorbitol, sucrose, aromatics, diamines, pentaerythritol, and methyl glucoside. These range in hydroxyl numbers from 350 to 600. For flexible foams, polyethers with hydroxyl numbers ranging from 40 to 160 are used. Examples are condensates of polyhydric alcohols such as glycerine, sometimes containing small amounts of ethylene oxide to increase reactivity.

Polyethers. Compounds containing many alcoholic hydroxyl groups (see POLYOL), used as reactants in the production of urethane foams. One type of polyether, widely used for rigid urethane foams, is obtained by reacting propylene oxide with a polyol initiator such as a glycol glycoside in the presence of potassium hydroxide as a catalyst.

Polyether Sulfone. A high-temperature engineering thermoplastic consisting of repeating phenyl groups linked by thermally stable ether and sulfone groups. The material has good transparency and flame resistance, and is one of the lowest smoke-emitting materials available. Both polymer and reinforced grades are available in granule form for extrusion and injection molding. Unreinforced grades are used in high temperature electrical applications, bakery oven windows and medical components. Reinforced grades are used for radomes, structural aircraft and aerospace components, and auto parts.

Polyethylene-Chlorotrifluoroethylene. (E-CTFE Copolymer) A high molecular weight, 1:1 alternating copolymer of ethylene and chlorotrifluoroethylene. Available in pellet and powder form, E-CTFE can be extruded, injection, transfer and compression molded, rotocast and powder coated. It is a strong, highly impact-resistant material that retains useful properties over a broad temperature range. Good electrical properties and chemical resistance make it useful in electrical, chemical plant and corrosion-resistant packaging applications.

Polyethylene Foams. Low density PE foams, weighing around 2 lbs. per cubic foot, are made by mixing a blowing agent with hot, molten polymer under pressure, then releasing the pressure and cooling. High density foams 20 to 30 lbs. per cubic foot and above, are natural polyethylenes containing numerous small, isolated cells filled with inert gas. Pellets impregnated with a foaming agent can be extruded or molded. Cross-linked PE foam is made by blending a peroxide cross-linking agent with the molten compound, then subsequently vulcanizing the molded shapes in a press.

Polyethylene Glycols. Polymers of ethylene glycol with molecular weights ranging from 200 to 6000, ranging from water-clear liquids to hard waxy solids. They are used as plasticizers for cellulose nitrate, as intermediates, and as mold release agents.

Polyethylene Glycol (200) Dibenzoate. $C_6H_5CO(OCH_2CH_2)_xOCOC_6H_5$ A plasticizer compatible with cellulose acetate butyrate, ethyl cellulose, PMMA, polystyrene and vinyls. Its major application is with phenol-formaldehyde resins in laminating applications, to improve flexure and heat distortion characteristics without sacrifice of electrical properties.

Polyethylene Glycol (600) Dibenzoate. A plasticizer with the same structural formula as polyethylene glycol (200) dibenzoate, but with only partial compatibility for the resins given.

Polyethylene Glycol Di-(2-Ethylhexoate). $C_7H_{15}OCOCH_2(CH_2OCH_2)_xCH_2COOC_7H_{15}$ A plasticizer for most cellulosic plastics, PMMA, polystyrene, polyvinyl chloride and vinyl chloride-vinyl acetate copolymers.

Polyethylene Glycol 400 Dilaurate. $C_{11}H_{23}CO(OC_2H_4)_xCOOC_{11}H_{23}$ A plasticizer for cellulose nitrate, PVC and vinyl copolymers.

Polyethylene Glycol Terephthalate. See POLYETHYLENE TEREPHTHALATE.

Polyethylene Naphthalate. A material developed by Teijin Ltd., described as having a naphthalene core chemical structure which includes two additional benzene nuclei. Films of the material resist temperatures to 311 °F and reportedly will receive UL classification as F-grade insulation material.

Polyethylene Oxide. (PEO) Low molecular weight polymers of ethylene oxide are viscous liquids or waxes. Those of high molecular weight are tough, crystalline, ductile thermoplastics which can be processed by molding, extrusion and other conventional techniques. All PEO resins are soluble in water, and thus are used in the form of packaging film for powdered detergents, dyes, insecticides and other household, industrial and agricultural products that are dissolved in water prior to use. The film is heat sealable and permeable to gases.

Polyethylene Propylene Adipate Glycol. (PEPAG) A polymeric diol used in the production of urethane elastomers, marketed by Witco Chemical Co. under the trademark Formrez F 10-91.

Polyethylenes. A family of resins obtained by polymerizing the gas ethylene, C_2H_4. By varying the catalysts and methods of polymerization, properties such as density, melt index, crystallinity, degree of branching and crosslinking, molecular weight and molecular weight distribution can be regulated over wide ranges. Further modifications are obtained by copolymerization, chlorination, and compounding addititives. Low molecular weight polymers of ethylene are fluids used as lubricants; medium weight polymers are waxes miscible with paraffin; and the high molecular weight polymers (over 6000) are the familiar, tough, leathery resins used in the plastics industry in the highest volume of all plastics. Polymers with densities ranging from about .910 to .925 are called *low density* polyethylene; those of densities from .926 to .940 are called *medium density;* and those of densities from .941 to .965 and over are called *high density* polyethylene. The low density types are polymerized at very high pressures and temperatures, and the high density types at relatively low temperatures and pressures. Two other recently-developed types are *extra high molecular weight* (EHMW) materials having molecular weights in the range 150,000 to 1,500,000, and *ultra high molecular weight* (UHMW) materials in the 1,500,000 to 3,000,000 range. The UHMW materials are very difficult to process, but under carefully controlled conditions the EHMW resins can be extruded, blow molded and thermoformed on standard equipment. When fully cross-linked by irradiation or by the use of chemical additives, polyethylene is no longer a thermoplastic. When cured during or after molding, it becomes a true thermoset with good tensile strength, electrical properties and impact strength over a wide range of temperatures.

Polyethylene Terephthalate. (PET, Polyethylene glycol terephthalate) A saturated, thermoplastic polyester resin made by condensing ethylene glycol and terephthalic acid, used for fibers and films (for example, "Mylar" film), and, more recently, for injection molded parts. It is extremely hard, wear-resistant, dimensionally stable, resistant to chemicals, and has good dielectric properties. See also POLYESTERS, SATURATED.

Polyformaldehyde. See PARAFORMALDEHYDE.

Polyformaldehyde Resins. See ACETAL RESINS.

Polyglycidyl Polyepichlorohydrin Resins. A group of epoxy resins derived from epichlorohydrin and hydroxyl compound, possessing flexibility and flame retarding characteristics. They may be cured by themselves, or mixed with conventional epoxy resins to impart their characteristics to laminates.

Polyglycol Distearate. (polyethylene glycol distearate) The distearate ester of polyglycol, used as a plasticizer.

Polyhexafluoropropylene. A FLUOROCARBON RESIN, which see, based on the gas C_3F_6. Hexafluoropropylene can be copolymerized with tetrafluoroethylene to form the FEP family of fluorocarbons.

Polyhexamethylene-Adipamide. See NYLON 6/6.

Polyhexamethylene Sebacamide. See NYLON 6/10.

Polyhydric Alcohols. See POLYOL.

Polyhydroxyether Resins. See PHENOXY RESINS.

Polyimidazopyrrolones. (Pyrrones) Aromatic, heterocyclic polymers which result from the reaction of various tetraacids and tetraamines. Due to their double chain or ladder-like structure, the polymers have outstanding resistance to heat, radiation and chemicals. However, this structure also makes them difficult to process. To overcome this difficulty, pyrrone prepolymers in the form of solutions and salt-like powders have been made available. The powders can be molded under conditions that complete the cyclization or conversion to the ladder-like molecular structure during the molding cycle. The cyclization reaction generates water, which must be removed from the part.

Polyimide Fibers. Polyimides prepared by reaction of diisocyanates with dianhydrides, developed by The Upjohn Company, have been used for fiber spinning by the wet or dry process. They have good thermal-oxidative stability, flame resistance and U.V. stability.

Polyimide Foams. A family of polyimide precursor powders made available in 1969 enables the production of flexible and rigid polyimide foam structures. The powders are poured into molds and heated until sufficient integrity for removal is attained, then subsequently cured at 300 °C.

Polyimides. The first Polyimides were condensation polymers derived from pyromellitic dianhydride and aromatic diamines first developed in 1963 by Du Pont under the name "Polymer SP." Due to rings of four carbon atoms bound tightly together, the material is said to possess a greater resistance to heat than any other unfilled organic material (500 °F continuously, and up to 750 °F intermittently). This first polyimide is a dark brown resin technically classed as a thermoplastic, although it does not melt and cannot be processed by conventional extrusion and injection molding techniques. Parts are made by machining from billets, punching, or by powdered metal techniques. In recent years, addition-type polyimides based on reacting maleic anhydride and 4,4′-methylenedianiline have been developed. They are processible by conventional thermoset transfer and compression molding, film casting and solution fiber techniques. Molding compounds filled with lubricating fillers or fibers produce parts with self-lubrication wear surfaces. Thermoplastic polyimide filled with glass, boron or graphite fibers can be molded into high-strength structural components. Solutions of thermoplastic polyimide are used as laminating varnish to produce radomes, printed circuit boards and other components requiring fire resistance, good electrical properties and strength at high temperatures.

Polyisobutenes. See POLYBUTENES.

Polyisobutylenes. Polymers of the C_4H_8 hydrocarbon isobutylene, for which the IUPAC name is *methylpropene*. Depending on molecular weight, they range from oily liquids to elastomeric solids. The higher molecular weight polymers are used as impact resistance additives for polyethylene and other plastics. The liquid polymers are used as tackifying agents in adhesives.

Polyisocyanates. See ISOCYANATES.

Polyisoprene. Polymers of ISOPRENE, which see. The cis-1,4- type of polyisoprene occurs naturally as the major constituent of natural rubber, and is also produced synthetically. The trans-1,4- type resembles gutta percha, and is used in golf ball covers and shoe soles.

Polyisoprene, Deutero. A polyisoprene in which heavy hydrogen atoms have replaced the ordinary hydrogen atoms. The cis-1,4-deuteropolyisoprene is more elastic than natural rubber.

Polyliner. A perforated longitudinally ribbed sleeve that fits the cylinder of an injection molding machine; used as a replacement for conventional injection cylinder torpedos.

Polymer, Natural. A high molecular weight substance of natural origin, consisting of molecules which are, at least approximately, multiples of low-molecular weight units. Natural polymers are often regarded as organic, but many inorganic materials such as silica, feldspar, glass and porcelain are considered to be entirely or substantially polymeric. Examples of natural organic polymers are natural rubber, gutta percha and many natural textile fibers.

Polymer. (Synthetic) The product of a polymerization reaction. See POLYMERIZATION. The product of polymerization of one monomer is called a *homopolymer, monopolymer* or simply a polymer. When two monomers are polymerized simultaneously the product is called a *copolymer.* The term *terpolymer* is sometimes used for polymerization products of three monomers. However, the term *heteropolymer* is also used for terpolymers as well as for products of more than three monomers. The terms *polymer, resin, high-polymer, macromolecular substance* and *plastic* are often used synonymously, although the latter also refers to compounds containing additives. Note: The definition approved by IUPAC and ISO for polymer is a substance composed of molecules characterized by the multiple repetition of one or more species of atoms or groups of atoms (constitutional units) linked to each other in amounts sufficient to provide a set of properties that do not vary markedly with the addition or removal of one or a few constitutional units.

Polymeric Modifier. A term applied to any polymer which is blended with another polymer to obtain altered characteristics. See also ELASTICIZER, IMPACT MODIFIER.

Polymeric Plasticizers. This term refers to plasticizers with molecular chains much longer than those of monomeric plasticizers which comprise virtually all other classifications of plasticizers. The two main types of polymeric plasticizers are the epoxidized oils of high molecular weight and POLYESTER PLASTICIZERS, which see. Polymeric plasticizers are noted for their permanence, which is due to reduced tendency of the larger molecules to migrate. However, the viscosity, efficiency and low temperature properties of polymeric plasticizers decrease as the molecular weight increases. In cold weather, the higher molecular weight polymerics may be difficult to handle and pump.

Polymeric Polyisocyanate. A generic term for a family of isocyanates derived from aniline-formaldehyde condensation products, used as reactants in the production of urethane foams.

Polymeric Sulfur Nitrides. See SULFUR-NITROGEN POLYMERS.

Polymerizable Plasticizers. See PLASTICIZERS, POLYMERIZABLE.

Polymerization. A chemical reaction in which the molecules of a simple substance (monomer) are linked together to form large molecules whose molecular weight is a multiple of that of the monomer. There are two general types of polymerization reactions, both with many variations: *addition polymerization,* which occurs when reactive monomers unite without forming any other products, and *condensation polymerization,* which occurs by condensation of reactive monomers with the elimination of a simple molecule such as water. Examples of condensation polymers are nylon and phenol-formaldehyde resins. The majority of thermoplastics and some thermosets are made by addition polymerization, types of which processes and related terms are described under the following headings:

ADDITION POLYMERIZATION	GAS-PHASE POLYMERIZATION
ALTERNATING COPOLYMER	GRAFT POLYMER
AUTOACCELERATION	INTERFACIAL POLYMERIZATION
BEAD POLYMERIZATION	IONIC POLYMERIZATION
BRANCHING	ISOTACTIC
BULK POLYMERIZATION	NETWORK POLYMERS
CHAIN TRANSFER AGENT	OXIDATIVE COUPLING
COPOLYMERIZATION	PHOTOPOLYMERIZATION
CONDENSATION POLYMERIZATION	PRECIPITATION POLYMERIZATION
CROSS-LINKING	RADIATION POLYMERIZATION
EMULSION POLYMERIZATION	RANDOM COPOLYMER
FREE RADICAL POLYMERIZATION	REDOX

SOLID STATE POLYMERIZATION   STEREOSPECIFIC POLYMERS
SOLUTION POLYMERIZATION   SUSPENSION POLYMERIZATION
STEREOBLOCK POLYMER   SYNDIOTACTIC
STEREOGRAFT POLYMER   THERMAL POLYMERIZATION
STEREOREGULAR POLYMERS

Polymer Structure. A general term referring to the relative positions, arrangement in space, and freedom of motion of atoms in a polymer molecule. Such structural details have important effects on polymer properties such as the second-order transition temperature, flexibility and tensile strength.

Polymethacrylonitrile. (PMAN) A thermoplastic obtained by the polymerization of methacrylonitrile, a recently developed vinyl monomer containing the nitrile group. The homopolymers have good mechanical strength and high resistance to solvents, acids and alkalis.

Polymethoxy Acetal. (PMAC, 3,5,7,x-polymethoxy dimethyl acetal) High boiling yellowish liquids used as modifiers for phenolic resins, solvents and plasticizers.

Polymethylenic. (adj.) Pertaining to a type of molecular structure consisting of a series of methylene groups (ASTM D 883-65T).

Polymethyl Methacrylate. (PMMA) A prominent member of the family of acrylic resins, made by polymerizing the monomer methyl methacrylate. Two outstanding characteristics of PMMA are its optical clarity (92% light transmission) and unsurpassed resistance to weathering. It also has excellent impact strength, good electrical properties, is tasteless, odorless and non-toxic. PMMA molding powders can be injection molded, extruded and compression molded. The liquid monomer can be cast into rods, sheets, optical lenses, etc. PMMA sheets ae fabricated and thermoformed into a wide variety of products such as airplane canopies, lighting fixtures and outdoor signs. See also ACRYLIC RESINS.

Poly(4-Methylpentene-1). A polyolefin first reported by G. Natta in 1955, but not introduced commercially until 1966 when I.C.I. Ltd. marketed the resin under the tradename "TPX." The monomer, 4-methylpentene-1-, is produced by the dimerization of propylene. Polymerization is conducted with a Ziegler-type catalyst. The polymers are supplied as free-flowing powders or as compounded granules, processable by injection molding, extrusion, and blow molding. Properties of the resins are very low density (0.83), high light transmission (93%, better than many glasses), melting point over 460 °F, rigidity and tensile properties similar to polypropylene, good electrical properties, and high chemical resistance. These properties account for one of the first potential large-scale applications of the material — chemical laboratory apparatus. Later applications included hospital and laboratory ware, extrusion coated paper, high intensity internal lighting fixtures, car body and radiator plugs, textile bobbins, sight glasses, and high frequency electrical components. In 1974 the Japanese firm Mitsui Petrochemical Industries Ltd. completed a plant for producing TPX, claiming it to be the sole manufacturer of TPX in the world. Applications then had spread to include TPX-coated paper for baking cartons, coffee funnels, hair curlers, lampshades and wire coatings. An application reported in 1976 is micro-vials holding from 0.1 to 10 ml. of biochemical and pharmaceutical products, which meet all of the specifications demanded for glass vials.

Poly(Monochloro-p-Xylylene). (parylene C) See PARYLENES.

Polymorphism. The ability of a crystalline substance to exist in two or more forms of crystalline structure.

Polyol. (polyhydric alcohol, polyalcohol) An alcohol having many hydroxyl radicals. In the cellular plastics industry, the term includes compounds containing alcoholic hydroxyl groups such as polyethers, glycols, polyesters and castor oil used as reactants in urethane foams.

Polyolefins. Polymers of relatively simple olefins such as ethylene, propylene, butenes, isoprenes and pentenes; and copolymers and modifications thereof. Polyolefins are most usually processed into end products by extrusion, injection molding, blow molding and rotational molding. Thermoforming, calendering and compression molding processes are used to a lesser degree. An inherent characteristic common to all polyolefins is a non-polar, non-porous inert surface which is non-receptive to adhesives, inks, lacquers etc. without special pretreatment. See ETHYLENE-PROPYLENE RUBBERS, ETHYLENE-VINYL ACETATE COPOLYMERS, IONOMER RESINS, POLYALLOMERS, POLYBUTENES, POLYETHYLENE, POLY(4-METHYLPENTENE-1), POLYPROPYLENE.

Polyorganophosphazenes. Experimental polymers under development in 1975. Phosphorus pentachloride and ammonium chloride are reacted to form cyclic trimer, $(NPCl_2)_3$, or tetramer, $(NPCl_2)_4$, which can be converted to polyorganophosphazines, $(-N=PR_{2-})_n$ where R represents organic side groups. The polymers are being evaluated for hose, gaskets, seals, O-rings, etc. in aviation fuel handling equipment. They appear to have better solvent resistance and low temperature elasticity than siloxane-carborane elastomers, and are lower in cost.

Polyoxamides. Generic name for nylon-type materials made from oxalic acid and diamines.

Polyoxetanes. See CHLORINATED POLYETHER.

Polyoxymethylenes. Linear polymers of the polyoxymethylene glycol type (or of formaldehyde) with the formula $HO(CH_2)_nH$, in which n is above 100. The polymers with n values in the range 100 to 300 are brittle solids used as intermediates. Those with n values above 500 and up to 5000 are ACETAL RESINS, which see.

Polyoxypropylene Glycols. $HO(C_3H_6P)_xH$ Polyethers derived from propylene oxide, used in the production of urethane foams.

Poly-Para-Xylylene. (parylene N) See PARYLENES.

trans-Polypentamer. An elastomer obtained by the polymerization of cyclopentane, using complex catalysts. Its structure is highly linear and of a wide range of molecular weights. Its properties are similar to those of natural rubber and cis-polybutadiene.

Polyphenone. A phenolic-like material under development by Union Carbide, for applications where good color and durability are needed. The natural resin is a light straw color and reportedly, when blended with colorants, a wide range of light colors can be molded with a smooth, deep gloss and high degree of stability after heat or sunlight exposure. The material is said to be resistant to water, chemicals, and most solvents. Electricals at both low and high frequencies are said to be good; physical strength properties are about those of a mineral-filled phenolic; and molding qualities are claimed to be less critical than those of comparable alkyds, melamines, or ureas.

Poly-1,3-Phenylenediamine Isophthalate. A high-temperature fiber, said to be the first developed, marketed by Du Pont in the early 1960's under the designation HT-1, later trademarked Nomex. This fiber resists common flame temperatures around 500 °C for a short time and thus is suitable for fire protective clothing and insulation for motors and transformers.

Polyphenylquinoxalines. (PPQ) A family of high performance thermoplastics (convertible, however to thermosets by rigidizing the linear polymer backbone through reactive latent groups and by crosslinking) which exhibit potential for use and functional and structural resins in applications demanding high chemical and thermal stability. They may be synthesized by many routes. The most attractive method is a copolycondensation reaction between an aromatic bis(o-diamine) powder and a stirred solution or slurry of bis(1,2-dicarbonyl) monomer in an appropriate solvent such as a mixture of m-cresol and xylene. In solution form, the polymers can be used directly for prepreg and adhesive tape formation, film casting, etc. If desired, the polymer can be isolated from solution and used in compression molding.

Polyphenylene Oxide. (PPO) A thermoplastic, linear, non-crystalline polyether obtained by the oxidative polycondensation of 2,6-dimethylphenol in the presence of a copper-amine complex catalyst. These resins have a useful temperature range from less than —275 °F to 375 °F, with intermittent use up to 400 °F possible; excellent electrical properties, unusual resistance to acids and bases, and processbility on conventional extrusion and injection molding equipment. A modified form of polyphenylene oxide resin is marketed by General Electric under the tradename "Noryl." It is based on polyphenylene oxide technology, but is intended for applications not requiring the performance of PPO.

Polyphenylene Sulfide. (PPS) A crystalline polymer having a symmetrical, rigid backbone chain consisting of recurring para-substituted benzene rings and sulfur atoms. A variety of different grades suitable for slurry coating, fluidized bed coating, electrostatic spraying as well as injection and compression molding are offered by Phillips Petroleum Co. under the trade mark Ryton. The polymers exhibit outstanding chemical resistance, thermal stability and fire resistance. The extreme inertness to organic solvents, inorganic salts and bases make for outstanding performance as a corrosion resistant coating suitable for food contact use.

Polyphenylsulphone. A new engineering thermoplastic developed by Union Carbide, named Radel, introduced early in 1977. It is chemically similar to polysulfones, but has higher impact resistance along with good heat resistance, low creep and good electrical properties. It also has good chemical resistance. Its high initial price, $15.00 per pound, makes its potential appliations in areas where other engineering plastics fail to meet specifications.

Polyphosphazene Fluoroelastomers. A family of elastomers developed by the Department of Army primarily for fuel tanks under Arctic conditions, having the typical configuration $(CF_3CH_2O)_2PN(CHF_2C_3F_6CH_2O)_2PN$. The elastomers remain flexible down to —70 °F, lower than any other elastomer previously used for the application.

Polyphosphazenes. A new family of inorganic-base polymers developed by a group at Pennsylvania State University, having phosphorous-nitrogen backbones joined with fluorine or chlorine. Depending on which organic side groups are linked to the backbones, a wide variety of polymers can be made with properties ranging from rigid and flexible thermoplastics to the elastomeric and glass-like thermosets. Early applications as the polymers began to move out of the laboratory stage in 1976 were biomedical devices which outperform the silicones.

Polypropylene. A thermoplastic resin made by polymerizing propylene with suitable catalysts, generally aluminum alkyl and titanium tetrachloride mixed with solvents. Its density (approximately 0.90) is among the lowest of all plastics. Like polyethylene, properties of the polymers vary widely according to molecular weight and method of preparation. The grades used for molding plastics have molecular weights of 40,000 or more, are usually highly crystalline, have good resistance to heat and chemicals, and good electrical properties. Polypropylenes can be modified to gain improved properties by compounding with fillers, e.g. asbestos or glass fibers; by blending with synthetic elastomers, e.g. polyisobutylene; and by copolymerizing with small amounts of other monomers.

Polypropylene Adipate. A plasticizer for vinyl chloride polymers and copolymers.

Polypropylene Glycols. Non-volatile liquids with the general formula $CH_3CHOH(CH_2OCHCH_3)_n\text{-}HC_2OH$. They are similar to the polyethylene glycols, but are more oil-soluble and less water-soluble. They are polyols used in the production of urethane foams, adhesives, coatings and elastomers.

Polypyromellitimides. A family of polymers with enhanced heat resistance developed primarily by Du Pont. They are formed from polyamide carboxylic acids derived by reacting pyromellitic dianhydride with 4,4′-diaminodiphenyl ether. Grades of the polymer are used for forming films, paint components, and a thermoplastic processable under special conditions.

Polysiloxanes. See SILICONES.

Polystyrene. Polymers of styrene (vinyl benzene). The homopolymer is water-white in color, has outstanding electrical properties, good thermal and dimensional stability, and is resistant to staining. However, it is somewhat brittle, and is often copolymerized or blended with other materials to obtain desired properties. High-impact grades are produced by adding rubber or butadiene copolymers. Heat resistance is improved by including some alpha- or methyl styrene copolymers. Copolymerization with methyl methacrylate improves light stability, and copolymerization with acrylonitrile increases resistance to chemicals. Styrene polymers and copolymers possess good flow properties at temperatures safely below degradation ranges, and can be easily extruded, injection molded or compression molded.

Polystyrene Foams. These light-weight foams widely used in packaging and insulation are formed by one of the two basic methods. Extruded foam is made by injecting a volatile liquid such as methyl chloride into molten polystyrene in an extruder. As it emerges from the extruder die, the mass expands to form a low density foam "log" which may be sliced or machined into a large variety of products. In the other basic method, a blowing agent is incorporated in polystyrene beads as they are polymerized. These beads may be molded directly in a closed mold, in which they will expand up to 45 times their original volume; they may be pre-expanded by heating and then molded; or, when combined with nucleating agents, extruded into thick sheets.

Polystyrol. A rarely-used term for POLYSTYRENE, which see.

Polysulfides. A family of sulfur-containing polymers prepared by condensing organic polyhalides with inorganic polysulfides in aqueous suspension. They range from liquids to solid elastomers. The first commercially produced polysulfide was Thiokol A (trademark of Thiokol Chemical Corp.), ethylene tetrasulfide, made from sodium tetrasulfide and ethylene dichloride. This elastomer had outstanding solvent resistance, but its poor physical properties and unpleasant odor limited its use to plasticizing acid-resistant cements. The next development was Thiokol FA, a copolymer of ethylene dichloride and bis-2-chloroethyl formal plasticized with benzothiazyl disulfide. This elastomer is used in hose, printing rolls and gaskets. The newest polysulfide is Thiokol ST, primarily a bis-2-chloroethyl formal polymer. Its principal use is in gas meter diaphragms. Polysulfide products have excellent resistance to oil, solvents, oxygen, ozone, light and weathering. They are also impermeable to gases and vapors.

Polysulfonate Copolyers. (sulfonate-carboxylate copolymers) Transparent, thermoplastics of the polyester family, moldable at 250°—300°C, first announced in 1966. They are formed by reaction of a diphenol, generally bisphenol A, with an aromatic disulfonyl chloride and one of the other disulfonyl chlorides or carboxylic acid chlorides. The copolymers have good electrical and mechanical properties, and excellent stability to hydrolysis and aminolysis.

Polysulfones. A family of sulfur-containing thermoplastics introduced in 1965 by Union Carbide, made by reacting bisphenol A and 4.4′-dichlorodiphenyl sulfone with potassium hydroxide in dimethyl sulfoxide at 130 to 140°C. The structure of the polymer is benzene rings or phenylene units linked by three different chemical groups — a sulfone group, an ether linkage, and an isopropylidene group. Each of these three linking components acts as an internal stabilizer. Polysulfones are characterized by high strength, the highest service temperature of all melt-processable thermoplastics, low creep, good electrical characteristics, transparency, self-extinguishing properties, and resistance to greases, many solvents and chemicals. They may be processed by extrusion, injection molding and blow molding.

Polyterephthalate. See POLYESTERS, SATURATED, and POLYETHYLENE TEREPHTHALATE.

Polyterpene Resins. Thermoplastic resins obtained by polymerizing terpentine or its derivatives such as alpha- or beta-pinene, in the presence of catalysts such as aluminum chloride or mineral acids. The amber colored resins, ranging from solids to viscous liquids, are used as tackifiers, wetting agents and modifiers in the manufacture of adhesives, paints and varnishes, caulking and sealing compounds. They are compatible with natural and synthetic rubbers, polyolefins, alkyd resins, other hydrocarbon resins, and waxes. See also HYDROCARBON PLASTICS.

Polytetrafluoroethylene. (PTFE) The oldest of the fluoroplastic family, discovered in 1938 by Dr. R. J. Plunkett and first marketed under the Du Pont trade name "Teflon." It is made by polymerizing tetrafluoroethylene, C_2F_4, and is available as a powder or an aqueous dispersion. PTFE is characterized by its extreme inertness to chemicals, very high thermal stability, low coefficient of friction, and ability to resist adhesion to almost any material. This last property is evidenced by the popularity of PTFE as a coating for cooking utensils. The high melt viscosity characteristics of PTFE make it difficult to process by conventional extrusion and molding techniques, and therefore the SINTER MOLDING (which see) process is used for most molded articles.

Polytetramethylene Ether Glycol. (PTMEG) A polymeric diol used in the production of urethane elastomers, marketed by Quaker Oats Co. under the trademark QO Polymeg 1000.

Poly(Tetramethylene Terephthalate). (PTMT, Polyterephthalate) A polyester of terephthalic acid and 1,4-butanediol, introduced by Eastman Chemical Products, Inc., in 1971. The polymer is thermoplastic, capable of being injection molded and extruded. The outstanding characteristic is high impact strength. PTMT is expected to compete with acetal and nylon resins in engineering applications. It has been listed by the FDA as an approved non-alcoholic food packaging material with certain extraction limits.

Polythene. The British term for polyethylene.

Polythiazyl. See SULFUR-NITROGEN POLYMERS.

Polytrifluorostyrene. A new, clear thermoplastic announced in December, 1965, by Molecular Research Corp., Cambridge, Mass. It is said to combine the oxidation resistance of PTFE with the mechanical and electrical properties and ease of fabrication of polystyrene.

Polyurethane Elastomers. Condensation polymers made by reacting aromatic diisocyanates with polyols which have an average molecular weight greater than about 750. The aromatic diisocyanates usually employed are toluene diisocyanate (TDI) and diphenylmethane diisocyanate (MDI). The polyol component is either a polyester or a polyether. In the "One Shot" system, the elastomer is prepared directly in the mold which forms the final product. The diisocyanate, polyol and catalyst are mixed and immediately cast into the mold. Prepolymers, in the form of liquids or low melting solids, are used when a longer working life is desired. The prepolymers are mixed with catalysts, heated when necessary, degassed and poured into molds. Also available are millable gums and pellets containing all elements, which have been reacted to a degree which permits further processing by methods used for rubber, including injection molding, compression molding and transfer molding. The polyurethane elastomers most widely used are harder than natural rubber, and possess excellent resistance to abrasion, impact, oils and grease, oxygen, ozone and radiation.

Polyurethane Foams. Although the terms *polyurethane foams* and *urethane foams* are used indiscriminately in the trade magazines and patents, authors of papers in the more academic technical journals usually employ the term URETHANE FOAMS, which see.

Polyurethane/Imide Modified Foam. See URETHANE/IMIDE MODIFIED FOAM.

Polyurethane Resins. (isocyanate resins) A family of resins produced by reacting diisocyanate with organic compounds containing two or more active hydrogen atoms to form polymers having free isocyanate groups. These groups, under the influence of heat or certain catalysts, will react with each other, or with water, glycols, etc., to form a thermosetting material.

Polyurethanes. A large family of polymers with widely varying properties and uses, all based on the reaction product of an organic isocyanate with compounds containing a hydroxyl group. The reaction product of an isocyanate with an alcohol is called a *urethan* according to rules of chemical nomenclature, but the terms *urethane* and *polyurethane* are more widely used in the plastics industry. The types and properties of polyurethanes are so varied that they have been dubbed the "erector set" of the plastics industry. They may be thermosetting or thermoplastic, rigid or soft and flexible, cellular or solid, and properties of any of these types may be varied within wide limits to suit the desired application. Members of this family are described under POLYURETHANE ELASTOMERS, POLYURETHANE RESINS, SPANDEX, URETHANE COATINGS, URETHANE FOAMS, IONIC URETHANES, and INSTANT SET POLYMER.

Polyvinyl Acetal. The general name for resins formed by partially or completely replacing the hydroxyl groups of polyvinyl alcohol with aldehydes, by means of a condensation reaction. Other members of the family are POLYVINYL BUTYRAL and POLYVINYL FORMAL, which see. Sometimes called *polyvinyl acetaldehyde*. Polyvinyl acetal resins are thermoplastics which can be processed by casting, extrusion molding and coating, but their main uses are in adhesives, lacquers, coatings and films.

Polyvinyl Acetate. (PVAc) A colorless, transparent thermoplastic, prepared by the polymerization of vinyl acetate. The homopolymers are available as beads, pearls, powder, solutions, emulsions and latex. The major use is in latex water paints, adhesives, fabric finishes and lacquers. In the plastics industry, the copolymers of vinyl acetate, particularly with vinyl chloride, are of most interest.

Polyvinyl Alcohol. (PVA, PVOH) A water-soluble thermoplastic prepared by partial or complete hydrolysis of polyvinyl acetate. Although it can be extruded and molded, its principal uses are in packaging films, fabric sizes, adhesives, emulsifying agents, etc. The packaging films are impervious to oils, fats and waxes, and have a zero transmission rate for oxygen, nitrogen and helium. Thus, they are often used as barrier coatings for these properties on other thermoplastic films. The water solubility of polyvinyl alcohol films can be regulated to some degree. The "standard" type, made from higher molecular weight polymers and plasticized with glycerine, is only weakly soluble in cold water. The other type, known as CWS or cold water soluble, is made from internally plasticized or lower molecular weight resins. A new application for PVA in the plastics industry, publicized in 1968, is as reinforcing fibers for compression and injection molding compounds. Although lower in tensile strength than glass, PVA fiber reinforcements offer high impact resistance, adhere to plastics so that no coupling agent is needed, do not break under molding stresses, yield a smooth finish on molded parts, and will not scratch polished mold surfaces.

Polyvinyl Butyral. (PVB, polyvinyl butyral acetal) A member of the POLYVINYL ACETAL family, made by reacting polyvinyl alcohol with butyraldehyde, with some unreacted PVA groups retained in the polymer. It is a tough, colorless flexible solid, used primarily in interlayers of laminated safety glass. Other applications include adhesive formulations; base resins for coatings, toners and inks; solutions for rendering fabrics resistant to water, staining and abrasion; and cross-linking with resins such as ureas, phenolics, epoxies, isocyanates, and melamines to improve coating uniformity and adhesion, increase toughness and minimize cratering.

Polyvinyl Carbazole. A thermoplastic resin, brown in color, obtained by reacting acetylene with carbazole. It has excellent electrical properties and good heat and chemical resistance, and is used as an impregnant for paper capacitors.

Polyvinyl Chloride. (PVC) The most important member of the vinyl plastics family, PVC is made by polymerization of vinyl chloride with peroxide catalysts. The pure polymer is hard, brittle and difficult to process, but it becomes flexible when plasticizers are added. It is copolymerized with many monomers and blended with other polymers to obtain a wide variety of properties. PVC resins, especially those copolymerized with minor amounts of vinyl acetate, may be dissolved in volatile solvents to make lacquers, coatings and cast film. A special class of PVC resin of fine particle size, often called *dispersion grade resin,* can be dispersed in liquid plasticizers to form PLASTISOLS, which see. PVC molding compounds can be extruded, injection molded, compression molded, calendered and blow molded to form a huge variety of products, flexible or rigid according to the amount and type of plasticizers incorporated.

Polyvinyl Chloride-Acetate. A copolymer of vinyl chloride and vinyl acetate, usually containing 85 to 97% of vinyl chloride. These copolymers are more flexible and more soluble in solvents than PVC, and are used in solution coatings as well as in most of the processes and applications employing PVC.

Polyvinyl Dichloride. A modified PVC resin was introduced by Goodrich under this name in the mid-1960's. The abbreviations PVD and PVDC were used, the latter of which conflicted with the sometimes-used abbreviation for polyvinylidene chloride. In 1967 this name and the abbreviations were dropped. See CHLORINATED POLYVINYL CHLORIDE.

Polyvinyl Fluoride. (PVF, PVDF) The polymer of vinyl fluoride (fluoroethylene, CH_2CHF), a monomer structurally similar to ethylene because a single fluorine atom replaces one hydrogen atom of the ethylene molecule. This fluorine atom forms a tight bond within the hydrocarbon chain, accounting for properties such as high melting point, chemical inertness and resistance to ultraviolet light. In the form of film, PVF is used for packaging, glazing, and electrical applications. Laminates of PVF film with wood, metal and polyester panels are being used in building construction. Although it cannot be dissolved in ordinary solvents at room temperature, coating solutions can be made by dissolving PVF in hot "latent solvents" such as dimethyl acetamide and the lower boiling phthalate, glycollate and isobutyrate esters. Such solutions are used as linings for rigid metal containers for chemicals and industrial compounds.

Polyvinyl Formal. A member of the POLYVINYL ACETAL family, which see, made by condensing formaldehyde in the presence of polyvinyl alcohol or by the simultaneous hydrolysis and acetylization of polyvinyl acetate. It is used mainly in combination with cresylic phenolics for wire coatings and impregnating, but can also be molded, extruded or cast. It is resistant to greases and oils.

Polyvinyl Halides. A term sometimes used (almost exclusively in patents) for polymers and copolymers of vinyl chloride. Aside from polyvinyl fluoride, which is more similar structurally to polyethylene, no polymers of the other halides (binary compounds of bromine, iodine and astatine) are in commercial existence.

Polyvinylidene Chloride. A thermoplastic polymer of 1,1-dichloroethylene. The homopolymer is of limited commercial value due to difficulty in processing. Copolymers with vinyl chloride (15% or more) are widely used as packaging films under the trade names "Saran," "Velon" and "Cryovac."

Polyvinylidene Fluoride. (VF_2) A member of the fluorohydrocarbon resin family, made by polymerizing the colorless gas $CH_2:CF_2$. It is thermally stable at high temperatures, is stronger and more abrasion resistant than other fluoroplastics, and is easier to process on conventional thermoplastic equipment. Major applications of PVF_2 are in the fields of electrical insulation, pipe and other chemical process equipment, and coatings for industrial and commercial buildings.

Polyvinyl Isobutyl Ether. (PVI) Polymers of vinyl isobutyl ether, available either as a white, opaque elastomer or a viscous liquid depending on the molecular weight. They are used as adhesives, plasticizers, surface coatings, laminating agents and filling compounds in electrical cables.

Polyvinyl Methyl Ether. (PVM) An amber colored balsam-like viscous liquid, soluble in cold water, insoluble in hot water. It is used as a component in pressure-sensitive and hot-melt adhesives for paper, polyethylene and rubber.

Polyvinyl Pyrrolidone. (PVP) A water-soluble polymer prepared by the addition polymerization of N-vinyl-2-pyrrolidone to molecular weights ranging from 10,000 to 360,000. Solutions of the polymer are used as protective colloids and emulsion stabilizers. Films of PVP are clear and hard, but may be plasticized.

Polyvinyl Resins. See VINYL RESINS.

Polyvinyl Stearate. A wax-like polymer of vinyl stearate, of limited use in the plastics industry. However, the monomer is copolymerized with vinyl chloride, etc., acting as an internal plasticizer.

Polywater. In the early 1960's, it was reported that the Russian physicist Boris Derjaguin had discovered a polymeric form of water, formed by condensing ordinary water on the inside of very fine bore quartz capillary tubing. Properties of the polymer, dubbed polywater, were said to be thermal stability up to 500 °C, density 40% greater than ordinary, and solidification to a glass-like state at −40 °C. There were subsequent reports that U.S. scientists had confirmed the existence of polywater. Later, however, it was revealed that the discoverer of "polywater" admitted that the substance he created was actually impurities dissolved from the quartz tubes used in the experiment.

POM. Abbreviation for polyoxymethylene. See ACETAL RESINS, POLYOXYMETHYLENES.

Pony Mixer. See CHANGE CAN MIXER.

Poromeric. A material which has the ability to transmit moisture vapor to some degree while remaining essentially waterproof. The first plastic material of this type was Du Pont's "Corfam," introduced in 1963. Other examples are "Aztran" made by Goodrich, I.C.I.'s "Ortix," and "Clarino" made by Japan's Kurashiki Rayon Co. The term poromeric was coined from micro*porous* and poly*meric*. The wholly plastic poromerics are based on urethane polymers and polyester fibers. The principal application for poromeric materials to date has been shoe uppers.

Porosity. The ratio of the volume of air or void contained within a material to the total volume (solid material plus air or void), expressed as a percentage.

Porous Molds. Molds which are made up of bonded or fused aggregates (powdered metal, coarse pellets, etc.) in such a manner that the resulting mass contains numerous open interstices of regular or irregular size through which either air or liquids may pass through the mass of the mold.

Positive Mold. A compression mold designed to prevent the escape of molding material during the molding cycle.

Post Curing. The process of forming an uncured thermosetting resin article, then completing the curing after the article has been removed from its forming mold or mandrel.

Postforming. A term used in the reinforced plastics industry to denote the heating and reshaping of a fully or partially cured laminate. On cooling, the formed laminate retains the contours and shape of the mold over which it has been formed.

Pot. (n) A chamber to hold and heat molding material for a transfer mold.

Potassium Titanate Fibers. $K_2O.(TiO_2)_n$ wherein n = 4 to 7. Highly refined, single crystals approximately 6 × 0.1 microns, used as reinforcing fibers in thermoplastic composites. The fibers have a melting point of 2500 °F and specific gravity of 3.2. They also act as a white pigment.

Pot Life. See WORKING LIFE.

Pot Plunger. A plunger used to force softened molding material into the closed cavity of a transfer mold.

Pot-Retainer. A plate channeled for heat and used to hold the pot of a transfer mold.

Potting. The process of encasing an article in a resinous mass performed by placing the article in a mold, pouring a liquid resin into the mold to completely surround the article, and curing the resin. The mold is a container which remains attached to the potted article, assisted, when required, by the use of bonding agents. The main difference between potting and ENCAPSULATION, which see, is that in encapsulation the mold is removed from the encapsulated article. These processes are widely used in the electrical industry.

Potting Syrups. See CASTING SYRUPS.

Poundal. The force required to give a mass of one pound an acceleration of one foot per second per second. The SI replacement for the poundal is the newton, obtained by multiplying the value in poundals by 1.382 550 E-01.

Powder Blend. See DRY-BLEND.

Powder Compacts. Molding materials in the form of dry, friable pellets prepared by compacting dry-blended mixtures of resin (usually PVC) with plasticizers and other compounding ingredients. The powder compacts are about as easy to handle and process by extrusion and blow molding as PELLETS, which see, and offer the advantages of low heat history and somewhat lower cost than equivalent materials in the form of fused pellets. See also DRY BLEND.

Powder Density. See BULK DENSITY.

Powdered Plastics. Resins or plastic compounds which have been finely pulverized for use in fluidized bed coating, rotational molding and various sintering techniques.

Powder Molding. A general term encompassing rotational molding, slush molding and centrifugal molding of dry, sinterable powders such as polyethylene, nylon and PVC. The powders are charged into molds which are heated and manipulated according to the process being used, causing the powders to sinter or fuse into a uniform layer against the internal mold wall. See also FLUIDIZED BED COATING, ROTATIONAL MOLDING, CENTRIFUGAL MOLDING, SLUSH MOLDING, SINTER MOLDING.

Power. The rate at which work is done in a unit period of time. The SI-approved unit of power is the watt, which measurement is obtained by Work expressed in joules divided by time in seconds. Other units and equivalents are: 1 horsepower equals 746 watts or 33,000 foot-pounds per minute. Note: for alternating currents, power in watts is equal to EI cos Φ, in which E is the effective value of the electromotive force in Volts, I is the current in amperes, and Φ is the phase angle between the current and the impressed electromotive force.

Power Factor. The ratio of *actual* power being used in an alternating electrical circuit to the power which is *apparently* being drawn from the line. When the load in an a.c. circuit is purely resistive, such as is the case for ovens and incandescent lights, the actual and the apparent power are the same and the power factor is 100%. When the load includes elements such as a.c. motors, generators, transformers and other machinery with coils which introduce inductance, the sine wave of the current lags behind that of the voltage by an angle known as the *phase angle,* which is represented by the Greek symbol θ (theta). This angle ranges from zero (for a purely resistive load) to a theoretical maximum of 90° for a purely inductive load. The cosine of this angle times 100 is the power factor, according to the formula:

$$\% \text{ Power Factor} = \frac{\text{Actual Power}}{\text{Apparent Power}} \times 100$$

$$= \frac{\text{Volts} \times \text{Amperes} \times \text{cosine } \theta}{\text{Volts} \times \text{Amperes}} \times 100$$

$$= \text{cosine } \theta \times 100$$

The difference between actual power and apparent power is known as *reactive power,* which increases as the power factor decreases. Reactive power does no useful work but costs the same as actual power used. Low power factors increase power costs, cause overloading of generators and transformers, and reduce the load handling capability of the plant electrical system. Low power factors due to inductive elements can be improved by introducing capacitors into the circuit, because in a circuit containing only capacitance the sine wave of the current leads that of the voltage — the opposite of an inductive circuit. In a perfect condenser the phase angle is 90°.

PP. Abbreviation for POLYPROPYLENE, which see.

PPI. Abbreviation for POLYMERIC POLYISOCYANATE, which see.

PPO. Registered trademark of General Electric Co. for POLYPHENYLENE OXIDE, which see.

PPOX. Abbreviation for polypropylene oxide.

PPS. Abbreviation for POLYPHENYLENE SULFIDE, which see.

PPSU. Abbreviation for poly(phenylene sulfone).

Precipitation Polymerization. A polymerization reaction in which the polymer being formed is insoluble in its own monomer or in a particular monomer-solvent combination and thus precipitates out as it is formed.

Precure. A partial or full state of cure existing in an elastomer or thermosetting resin prior to its use as an adhesive or in a forming operation.

Precursor. One who or that which precedes and suggests the course of future events. The term is sometimes used by authors of technical articles for a compound or agent that is transformed into another material when added to a substance or mixture.

Predrying. The drying of a resin or molding compound prior to its introduction into a mold. Some plastic compounds are hygroscopic and require this treatment, particularly after storage in a humid atmosphere.

Preform. (n) A compressed shaped mass of plastic material or fibrous reinforcing material or mixtures of both, prepared in advance of a molding operation for convenience in handling or for accuracy in weighing. In the reinforced plastics industry, a preform is a mat of chopped strands bonded together by a resin in approximately the shape of the end product, for use in processes such as matched-die molding. The term also applies to tablets, biscuits and pellets of thermosetting and thermoplastic compounds.

Preform. (v) To make plastic molding powders into pellets, tablets or shapes designed to facilitate filling of cavities in molding processes.

Preform Binder. The resin applied to a preform for a reinforced plastic structure.

Preform Molding. See MATCHED METAL DIE MOLDING.

Pregel. An unintentional extra layer of cured resin on part of the surface of a reinforced plastic. Should not be confused with gel coat.

Preheating Hopper. An extruder feed hopper provided with hot air circulating means to preheat the molding material before it reaches the extruder screw.

Preheat Roll. In extrusion coating, a heated roll installed between the pressure roll and unwind roll, the purpose of which is to heat the substrate before it is coated.

Preimpregnation. The practice of mixing resin and reinforcement before shipping it to the molder. The product of this practice is called a PREPREG, which see.

Premix. (n) A term originally applied to mixtures of polyester resin with sisal or glass fiber reinforcement and fillers, usually prepared by molders just prior to use. The ASTM definition specified that the premix should not be in web or filamentous form. The term premix is now often used for molding compounds of any thermosetting resin mixed with fillers, reinforcements and catalysts. The terms *sheet molding compound* and *bulk molding compound* have been proposed for premixes containing thickening agents such as Group II oxides (e.g. magnesium hydroxide and calcium hydroxide) or thermoplastic polymers, used respectively for sheet molding and injection molding.

Premix Molding. A variation of matched-die molding in which the ingredients, usually chopped roving, resin, pigment, filler and catalyst, are premixed as a "gunk" which can be placed in the mold as accurately weighed charges.

Preplasticization. In injection molding, the technique of premelting molding powders in a separate chamber, then transferring the melt to the injection cylinder. This technique serves to shorten molding cycles.

Pre-Polymer. A polymer of relatively low molecular weight, usually intermediate between that of the monomer or monomers and the final polymer or resin, which may be mixed with compounding additives, and which is capable of being hardened by further polymerization during or after a forming process.

Prepolymer Molding. In the urethane foam industry, this term indicates a system whereby a portion of the polyol is pre-reacted with the isocyanate to form a liquid prepolymer with a viscosity suitable for pumping or metering. This component is supplied to end-users with a second premixed blend of additional polyol, catalyst, blowing agent, etc. When the two components are mixed together, foaming occurs. See also ONE-SHOT MOLDING.

Prepreg. (n) In the reinforced plastics industry, a mat or shaped mass of reinforcing fibers, e.g. glass strands, impregnated with a thermosetting resin advanced in cure only through the B-stage. Such prepregs may be stored until needed for a molding or laminating operation. A prepreg containing a chemical thickening agent is called a MOLD-MAT, which see. The term "prepreg" also covers fabrics such as jute, coconut fiber or rayon yarn impregnated with a thermoplastic resin, e.g. vinyl, ABS or acrylic. The SPI has recently proposed that prepregs in sheet form be called *sheet molding compounds.*

Prepreg Molding. A process similar to MATCHED METAL DIE MOLDING, which see, except that the fibrous mat is pre-impregnated with a thermosetting resin.

Preprinting. In sheet thermoforming, the distorted printing of sheets before they are formed. During forming, the print assumes its proper proportions.

Press Polishing. (planishing) A finishing process used to impart high gloss, improved clarity and mechanical properties to sheets of vinyl, cellulosic and other thermoplastics. The sheets are hot-pressed against thin highly polished metal plates.

Pressure Bag Molding. See BAG MOLDING.

Pressure-Break. As applied to a defect in a laminated plastic, a break apparent in one or more outer sheets of the paper, fabric, or other base visible through the surface layer of resin which covers it. (ASTM D 883-75a).

Pressure Flow. In the metering section of an extruder screw, pressure flow is the relatively backward flow of material down the screw channel caused by pressure in the head. See also DRAG FLOW, LEAKAGE FLOW.

Pressure Forming. A thermoforming process wherein pressure is used to push the sheet to be formed against the mold surface, as opposed to using a vacuum to suck the sheet flat against the mold.

Pressure Pads. Reinforcements of hardened steel distributed around the dead areas in the faces of a mold to help the land absorb the final pressure of closing without collapsing.

Pressure Roll. In extrusion coating, a roll used to apply pressure to consolidate the substrate and the plastic film with which it has been coated.

Pressure-Sensitive Adhesive. An adhesive which develops maximum bonding power by applying only a light pressure.

Primary Plasticizer. A plasticizer that, within reasonable compatibility limits, may be used as the sole plasticizer, is completely compatible with the resin, and is sufficiently permanent to produce a composition that will retain its desired properties under normal service conditions throughout its expected life period. See also PLASTICIZER, SECONDARY PLASTICIZER.

Primer. A coating applied to a substrate to improve the adhesion, gloss or durability of a subsequently applied coating. For example, vinyl copolymer, acrylic, phenolic, epoxy and polyester resins are used as primers for adhering vinyl coatings to metals.

Primrose Chrome. See CHROME YELLOW PIGMENTS.

Primrose Yellows. See LEMON YELLOWS.

Printing on Plastics. Many methods commonly used on paper and other materials are used for printing on plastics, with slight modifications such as the use of special inks. Such processes are letterpress, offset, silk screen, electrostatic and photographic methods. See also ELECTROSTATIC PRINTING, FLEXOGRAPHIC PRINTING, GRAVURE PRINTING, HOT STAMPING, VALLEY PRINTING and SPANISHING. Polyolefins are normally treated before printing in order to make them receptive to inks. See CASING, ELECTRONIC TREATING, FLAME TREATING, ULTRAVIOLET PRINTING.

Progressive Gluing. A method of curing a thermosetting resin adhesive in successive stages by application of heat and pressure between press platens, used for plywood or other laminates which are larger in area than the press platens.

Promoter. A chemical substance which greatly increases the activity of a CATALYST, which see. The promoter may be a feeble catalyst itself.

Proof Resilience. The tensile energy capacity of work required to stretch an elastomer from zero elongation to the breaking point, expressed in foot-pounds per cubic inch of original cross section area.

2-Propanone. See ACETONE.

Propeller Mixers. Devices comprising a rotating shaft with propeller blades attached, used for mixing relatively low viscosity dispersions and holding contents of tanks in suspension.

Propenal. See ACROLEIN.

Propenenitrile. See ACRYLONITRILE.

Propenoic Acid. See ACRYLIC ACID.

Proportional Limit. The greatest stress which a material is capable of sustaining without deviation from proportionality of stress and strain (Hooke's Law). It is expressed in force per unit area, usually pounds per square inch.

n-Propyl Acetate. $C_3H_7OOCCH_3$ A clear, colorless liquid with a pleasant odor, used as a solvent for cellulosics, vinyls, acrylics, polystyrene, phenolics, alkyds and coumarone-indene resins.

Propylene. (propene) $CH_3CH:CH_2$ A colorless gas produced mainly by cracking propane, butane or other refinery off-gases, or by cracking hydrocarbons during the production of ethylene. It is the monomer from which polypropylene is made, and also has many uses as an intermediate.

1,2-Propylene Glycol Monolaurate. $C_{11}H_{23}COOCH_2CHOHCH_3$ A plasticizer for cellulose nitrate, ethyl cellulose, polystyrene, PVC and other vinyls.

1,2-Propylene Glycol Mono-Oleate. $C_{17}H_{33}COOCH_2CHOHCH_3$ A plasticizer for cellulose nitrate and ethyl cellulose.

Propylene Oxide. (1,2-propylene oxide, 1,2-epoxypropane) A low-boiling flammable liquid derived from the intermediate chlorohydrin, produced by reacting chlorine, propylene and water. Propylene oxide is an important intermediate for the manufacture of polyglycols used for urethane foams and resins and polyester resins.

Propylene Plastics. See POLYPROPYLENE.

Propylene-Vinyl Chloride Copolymers. These copolymers, ranging from 2% to 10% by weight in propylene content, provide the end-use property advantage of PVC homopolymers plus the processing advantages attributable to the introduction of stable hydrocarbon structures as end groups. Characteristics of the copolymers are ease of molding or extruding, high thermal stability and low melt viscosity.

n-Propyl Oleate. $C_{17}H_{33}COOC_3H_7$ A plasticizer for cellulose nitrate, ethyl cellulose, polystyrene, and, with limited compatibility, some vinyl and acrylic resins.

Protein Resins. A generic term for resins derived from proteins, constituting CASEIN PLASTICS and ZEIN, which see.

Proton. An elementary particle having a positive charge equivalent to the negative charge of the electron but possessing a mass approximately 1837 times as great as that of the electron. The proton is in effect the positive nucleous of the hydrogen atom.

Prototype Mold. A temporary or experimental mold used to test designs or obtain market reactions. Such a mold is often made from a low-temperature metal casting alloy or from an epoxy resin.

PS. Abbreviation for POLYSTYRENE, which see.

Pseudoplastic Fluid. A pseudoplastic fluid is one whose apparent viscosity or consistency decreases instantaneously with increase in rate of shear; i.e., an intial relatively high resistance to stirring decreases abruptly as the rate of stirring is increased.

PTDQ. Abbreviation for polymerized trimethyl dihydroxyquinoline, an antioxidant used in chemically cross-linked polyethylenes.

PTFE. Abbreviation for POLYTETRAFLUOROETHYLENE, which see.

PTMT. See POLY(TETRAMETHYLENE TEREPHTHALATE).

PU. Abbreviation sometimes used for polyurethane. The British Standards Institution limits the use of this abbreviation to rigid polyurethanes, either of the polyether or of the polyester type.

Pulled Surface. Imperfections in the surface of a laminated plastic ranging from a slight breaking or lifting of its surface in spots to pronounced separation of its surface from its body. (ASTM D 883-75a).

Pull Strength. The bond strength of an adhesive joint, obtained by pulling in a direction perpendicular to the surface of the layer.

Pulp Molding. A process by which a resin-impregnated pulp material is preformed by application of a vacuum and subsequently oven cured or molded. The pulp is first mixed with water and pumped into a tank wherein a mold, usually of wire mesh shaped like the finished article, is positioned. Air is evacuated from the mold to attact the pulp fibers, forming a preformed layer in contact with the mold. The mold is removed from the tank, the deposit stripped off and dried, then the preform is molded to final form by fluid pressure or conventional compression methods.

Pulsed Positive Negative Ion Chemical Mass Spectrometry. See MASS SPECTROMETRY.

Pultrusion. A reinforced plastics technique for continuously producing constant cross-section profiles, both solid and tubular. Strands of reinforcing material are conveyed through a tank of resin, usually polyester but silicone, or epoxy for fewer applications, from which they are pulled through an elongated, heated steel die shaped to impart the desired profile. Heating to both gel and cure the resin is sometimes accomplished entirely within the die length, which can be on the order of 30 inches long. In other variations of the process, preheating of the resin-wet reinforcement is accomplished by dielectric energy prior to entry in the die, or heating may be continued in an oven after emergence from the die. The pultrusion process yields continuous lengths of material with high unidirectional strengths, used for building siding, fishing rods, pipe, golf club shafts, etc.

Pumice. A highly porous igneous rock, used in pulverized form as an abrasive and a filler for plastics.

Pumicing. A finishing method for molded plastics parts, consisting of the rubbing off of traces of tool marks and surface irregularities by means of wet pumice stones.

Punching. Method of producing components, particularly electrical parts, from flat sheets of rigid or laminated plastics by punching out shapes by means of a die and punch.

PUR. Abbreviation for polyurethane. In European literature, the abbreviation PU is more often used.

Purging. In extrusion or injection molding, the cleaning of one color or type of material from the machine by forcing it out with the new color or material to be used in subsequent production, or with another compatible purging material. The operation is usually accomplished more rapidly when the purging material is more viscous than the material being displaced.

Pushback Pins. See RETURN PINS.

Push Up. In the packaging industry, a container bottom with sufficient concavity to prevent rocking of the container when it is filled and placed on a flat surface.

PVA. Abbreviation sometimes used for POLYVINYL ALCOHOL, which see. See also PVAL.

PVAC. Abbreviation for POLYVINYL ACETATE, which see.

PVAL. Abbreviation for polyvinyl alcohol.

PVB. Abbreviation for POLYVINYL BUTYRAL, which see.

PVC. (1) Abbreviation for POLYVINYL CHLORIDE, which see. (2) In the paint industry, abbreviation for *pigment volume concentration.*

PVCA. An abbreviation for copolymers of vinyl chloride and vinyl acetate.

PVD. A rarely-used abbreviation for polyvinyl dichloride. See CHLORINATED POLYVINYL CHLORIDE.

PVDC. (1) Abbreviation for POLYVINYLIDENE CHLORIDE, which see. (2) Abbreviation used briefly in the 1960's for polyvinyl dichloride. See CHLORINATED POLYVINYL CHLORIDE.

PVDF. Abbreviation for POLYVINYLIDENE FLUORIDE, which see.

PVF. Abbreviation for POLYVINYL FLUORIDE, which see. The possibility of confusion exists because this abbreviation has been used in some literature for polyvinyl formal.

PVFM. Abbreviation approved by ASTM for POLYVINYL FORMAL, which see.

PVI. Abbreviation for POLYVINYL ISOBUTYL ETHER, which see.

PVK. Abbreviation for POLYVINYL CARBAZOLE, which see.

PVM. Abbreviation for POLYVINYL METHYL ETHER, which see.

PVOH. Abbreviation for POLYVINYL ALCOHOL, which see. The abbreviation PVA is more commonly used.

PVP. Abbreviation for POLYVINYL PYRROLIDONE, which see.

Pyncometer. A flask for measuring the density of a liquid. It is usually of glass and provided with one or two capillary arms so that it can be filled to a known volume with good precision. A *dilatometer* is a pyncometer equipped with instruments to study density as a function of temperature or time.

Pyranyl Foam. A type of rigid, pour-in-place thermosetting foam similar to a polyurethane foam, but with superior high temperature resistance. It is formed in the same manner as urethane foams, using as the monomer a pyranyl derived from polypropylene by heating and oxidation to form an acrolein dimer, which ultimately forms the pyranyl.

Pyrazolone Red. A metal-free disazo pigment based on a pyrazolone.

Pyrogenic Silica. See FUMED SILICA.

Pyrogram. A chromatogram (see CHROMATOGRAPHY) obtained from the pyrolysis products of a material.

Pyrolysis. The decomposition of a complex organic substance to one of simpler structure by means of heat. Some polymers will depolymerize in the presence of excessive temperatures, either to polymers of lower molecular weight, or, in some cases, back to the monomers from which they were derived.

Pyromellitic Dianhydride. $C_6H_2(C_2O_3)_2$ A curing agent for epoxy resins, and a cross-linking agent for epoxy plasticizers for vinyl resins.

Pyrometer. An instrument for measuring heat. The type most widely used in plastics processing equipment consists of a thermocouple (a pair of wires of dissimilar metals which when heated at their junction produce a feeble electrical e.m.f.) and a milli-voltmeter for measuring the voltage which is proportional to the temperature of the junction.

Pyroxylin. (soluble guncotton) A type of nitrocellulose consisting mainly of cellulose tetranitrate. It is used for making collodion, and has no applications in the plastics industry.

Pyrrones. See POLYIMIDAZOPYRROLONES.

Quadripolymer. A rarely used term for the product of the simultaneous polymerization of four monomers. See POLYMER.

Quantitative Differential Thermal Analysis. Differential thermal analysis (which see) in which the equipment used is designed to produce quantitative results in terms of energy and/or other physical parameters. (ISO)

Quartz. See SILICA.

Quaterpolymer. The IUPAC term for a polymer (which see) derived from four species of monomer.

Quench. (v) A process of shock cooling thermoplastic materials from the molten state.

Quench Bath. The cooling medium used to quench molten thermoplastic materials to the solid state.

Quench-Tank Extrusion. An extrusion process wherein the extrudate is conducted through a water bath for rapid cooling.

Quicklime. See CALCIUM OXIDE.

Quinacridone Pigments. A family of organic pigments based on substituted and unsubstituted forms of linear trans-quinacridones. Colors available include several shades of red, violet, gold, orange, magenta and maroon. These pigments are characterized by good lightfastness, intensity of hue, resistance to bleeding and chemical attack, good transparency and heat resistance.

Rad. The unit of energy absorbed by a material from ionizing radiation. One Rad is equal to 100 ergs per gram. See also ROENTGEN.

Radiation Polymerization. A polymerization reaction initiated by exposure to radiation such as gamma rays rather than by means of chemical catalysts.

Radiation Processing. See IRRADIATION.

Radical. A group of atoms, existing in a molecule, which remains unchanged through many chemical reactions. A typical organic radical is C_2H_5, the ethyl radical.

Radio Frequency. A frequency within the range of radio transmission, e.g. from about 15,000 to 10^{11} cycles per second.

Radio Frequency Heating. See DIELECTRIC HEATING.

Radio Frequency (R.F.) Preheating. A method of preheating used for molding materials to facilitate the molding operation or reduce the molding cycle. The frequencies most commonly used are between 10 and 100 Mc/sec.

Radio Frequency Welding. A method of welding thermoplastics using a radio frequency field to apply the necessary heat. Also known as *high frequency welding.*

Ram. The press member that enters the cavity block and exerts pressure on the molding compound, designated by its position in the assembly as the top force or bottom force.

Ram Extruder. See EXTRUDER, RAM.

Ramie. A natural vegetable fiber obtained from stems of the hemp *Boehmeria nivea,* used as a reinforcement.

Ram Injection Molding. See INJECTION MOLDING.

Ram Travel. The distance the injection ram moves in filling the mold in injection or transfer molding.

Random Copolymer. A copolymer consisting of alternating segments of two monomeric units of random lengths, including single molecules. A random copolymer usually results from the copolymerization of two monomers in the presence of a free-radical initiator, for example the rubbery copolymer of ethylene and propylene.

Rankine Temperature. The Rankine temperature scale is the absolute Fahrenheit scale, that is the sum of absolute zero on the Fahreinheit scale (—459.69) and the Fahrenheit temperature.

Rayon. The definition established for rayon by the Federal Trade Commission on December 11, 1951 is "Generic name for a manufactured fiber composed of regenerated cellulose, as well as manufactured fibers composed of regenerated cellulose in which substituents have replaced not more than 15% of the hydrogens of the hydroxyl groups." Prior to this date, back to 1924 when the name rayon was first used (inspired by its sheen having the brilliance of a ray of sunlight), the term was used for all man-made fibers derived from cellulose, including cellulose acetate and cellulose triacetate. Rayon is the oldest of the synthetic fibers, having been produced commercially in 1855. All methods of producing rayon are based on treating fibrous forms of cellulose to make them soluble, extruding the material through small holes of a spinneret, then converting the filaments into solid cellulose. Most rayon fibers are produced from the intermediate VISCOSE, which see. See also CUPRAMMONIUM RAYON.

Reaction Injection Molding. (RIM) This term is usually applied to the process of injection molding of urethane reactants in which the two primary constituents, isocyanate and polyol, are pumped by a metering device into a mixing head from which the mixed reactants are rapidly injected into a closed mold. The injection pressure is much lower than in conventional injection molding of solid thermoplastics, enabling the use of inexpensive, light-weight molds. However, the mixing head is a high-pressure impingement mixer in which pressures may reach 2000 to 3000 psi. One mixing head may be used to feed up to 10 separate molding presses. One of the largest applications of RIM is in the production of exterior automotive parts such as body panels and bumpers. The term LIQUID INJECTION MOLDING (LIM) is usually applied to the similar process of molding other thermosetting resins such as polyesters, epoxies, silicones, alkyds and DAP. The terms *Liquid Reaction Molding (LRM)* and *High Pressure Injection Molding (HPIM)* are also sometimes used for one or both processes.

Reactive Plasticizers. See PLASTICIZERS, POLYMERIZABLE.

Reagent Resistance. (chemical resistance) The ability of a plastic to withstand exposure to acids, alkalis, solvents and other chemicals.

Ream. (n) (1) Layers of unhomogeneous material parallel to the surface in a transparent or translucent plastic. (ASTM D 883). (2) 500 Sheets.

Reciprocating Screw Injection Molding. In this process an extruder is modified to permit lengthwise movement of the screw by a hydraulic cylinder located at the rear of the screw. The plastic material is advanced and plasticated by the screw as in conventional extrusion while the cylinder is retracted, until an injection shot is to be made. Screw rotation is then stopped, and the hydraulic cylinder pushes a charge into the injection mold. The advantage of the process is that extremely high clamping pressures do not have to be maintained throughout the whole cycle as in conventional injection molding.

Recycle. (n) See REGRIND.

Red 2B Pigments. (permanent red 2B, pigment red 48). The calcium, strontium or barium precipitations of the coupling from diazotized 2-chloro-4-aminotoluene-5-sulfonic acid with 2-hydroxy-2-naphthoic acid. These pigments have high tinctorial strength and good resistance to bleeding and high temperatures, but are poor in chemical resistance and fair in light resistance. Colors range from orange red to rubine.

Redox. A contraction of the term "oxidation-reduction." A *redox catalyst* is one entering into an oxidation-reduction reaction. A polymerization initiator comprising a mixture of a peroxide and a reducing agent is called a *redox initiator.* Polymers formed by such reactions are sometimes called *redox polymers.*

Reduced Viscosity. Ratio of the specific viscosity to the concentration. Reduced viscosity is a measure of the specific capacity of the polymer to increase the relative viscosity. The IUPAC term for reduced viscosity is viscosity number. (ASTM D 1243-60). See also DILUTE SOLUTION VISCOSITY.

Reduction. (1) Any process which increases the proportion of hydrogen or base-forming elements or radicals in a compound. (2) The gaining of electrons by an atom, ion, or element, thereby reducing the positive valence of that which gained the electron.

Re-Entrant Mold. A mold containing an undercut which tends to resist withdrawal of the molded object.

Reflectivity. (Coefficient of reflection) The ratio of light reflected from a surface to the total incident light. The coefficient may refer to diffuse or to specular reflection. In general it varies with the angle of incidence and with the wavelength of the light.

Refractive Index. See INDEX OF REFRACTION.

Refractivity. The index of refraction minus 1. Specific refractivity is given by $\dfrac{n-1}{d}$ where n is the index of refraction and d is the density.

Regenerated Cellulose. A transparent cellulosic plastic made by mixing cellulose xanthate with a dilute sodium hydroxide solution to form a viscose, extruding the viscose into film, sheeting or fiber form, then treating the extrudate with acid to effect regeneration. In fiber form, the material is called RAYON (which see). The term CELLOPHANE (which see) is used for films and sheets.

Regrind. (n) Waste material such as sprues, runners, excess parison material and reject parts from injection molding, blow molding and extrusion operations, which has been reclaimed by shredding or granulating. Regrind is usually mixed with virgin compound at a predetermined percentage for remolding.

Regular Block. A block that can be described by only one species of constitutional repeating unit in a single sequential arrangement (IUPAC).

Regular Polymer. A polymer whose molecules can be described by only one species of constitutional repeating unit in a single sequential arrangement. (IUPAC).

Regulator. With respect to a polymerization reaction, a substance used in small quantities to control the molecular weight during polymerization.

Reinforced Molding Compound. Compound supplied by raw material producer in the form of ready-to-use materials, as distinguished from PREMIX, which see.

Reinforced Plastic. (RP) A plastic composition in which fibrous reinforcements are imbedded, with strength properties greatly superior to those of the base resin. The reinforcements are usually fibers, rovings, fabrics or mats of glass, asbestos, metals, paper, sisal, cotton or nylon. Resins most commonly used are polyesters, phenolics, aminos, silicones and epoxies. The term *reinforced plastic* includes some forms of LAMINATE (which see) and molded parts in which the reinforcements are not in laminated form. When the resin is a thermoplastic, the term "reinforced thermoplastic" is often used. Methods of forming reinforced plastics articles are defined under AXIAL WINDING, BAG MOLDING, CONTACT PRESSURE MOLDING, CENTRIFUGAL CASTING, CERAPLASTICS, FIBERFIL MOLDING, FILAMENT WINDING, FURAN PREPREGS, IMPREGNATION, LAMINATE, LAY-UP MOLDING, LAP WINDING, LOW PRESSURE LAMINATES, MATCHED METAL DIE MOLDING, PREFORM, PREMIX, PREPREG, PREPREG MOLDING, PULTRUSION, REVERSE HELICAL WINDING, SHEET MOLDING COMPOUND, SLURRY PREFORMING, and SPRAY-UP.

Reinforced Thermoplastics. (RTP) Reinforced structures in which the bonding resin is a thermoplastic rather than a thermoset. In recent years, applications for RTP have grown rapidly, mainly with nylon, polycarbonates, acetal resins and polystyrene. The tensile strength of a thermoplastic can be at least doubled by the addition of glass reinforcement, and deformation under load is greatly decreased. Unlike the thermosetting reinforced plastics, RTP compounds can be pelleted and used in conventional injection molding equipment.

Reinforcement. A strong, inert fibrous material incorporated in a plastic mass to improve its physical properties. Typical reinforcements are ASBESTOS, BORON FIBERS, CARBON FIBERS, CERAMIC FIBERS, FLOCK, GLASS FIBER REINFORCEMENTS, GRAPHITE, JUTE, SISAL and WHISKERS, all of which see. Others sometimes used are chopped paper, macerated fabrics, and synthetic fibers. To be effective, a reinforcement must form a strong adhesive bond with the resin used, for which purpose adhesion promoting substances known as coupling agents are often applied to the fibers. The function of a reinforcement is similar in some respects to that of a FILLER, which see, the primary difference being that a reinforcement markedly improves tensile and flexural strength, whereas a filler does not.

Reinforcing Pigments. Pigments which also serve to improve the properties of the finished product. An example is carbon black.

Relative Density. The ratio of the absolute density of a substance at a stipulated temperature to the absolute density of water at 3.98 °C (its maximum value). See also DENSITY, APPARENT DENSITY and SPECIFIC GRAVITY.

Relative Humidity. The ratio of the quantity of water vapor present in the atmosphere to the quantity which would saturate it at the existing temperature. It is also the ratio of the pressure of water vapor present to the pressure of saturated water vapor at the same temperature. An excess of relative humidity can cause "blushing" in the painting of plastics as well as metals.

Relative Viscosity. Ratio of the kinematic viscosity of a specified solution of the polymer to the kinematic viscosity of the pure solvent. The IUPAC term for relative viscosity is *viscosity ratio*. (ASTM D 1243-60). See also DILUTE SOLUTION VISCOSITY.

Relaxation. A decrease in stress under sustained constant strain, or creep and rupture under constant load.

Relaxation Time. The time required for a stress under a sustained constant strain to diminish a stated fraction of its initial value.

Release Agent. See PARTING AGENT.

Release Paper. A layer of paper which can be separated from the surface of a plastic article to which it has been applied or against which the plastic article has been formed. The term applies to papers used as interleaves between plastic sheets, temporary backings for pressure-sensitive adhesives, and papers used as temporary carriers in film and foam casting processes. The release papers may be smooth or embossed, to impart any desired texture to the film or foam cast against them. They may also be preprinted with an ink that is transferred to the cast film.

Relief Angle. In a mold, e.g. for blow molding or injection molding, the relief angle is the angle between the narrow pinch-off land and the cutaway portion adjacent to the pinch-off land.

Reluctance. The property of a magnetic circuit which determines the total magnetic flux in the circuit when a given magnetomotive force is applied; the resistance of a magnetized body to a magnetic flux.

Rennet Casein. A type of casein precipitated from milk by means of rennet, the dried extract of stomach secretions containing the enzyme rennin. See also CASEIN and CASEIN PLASTICS.

Residual Monomer. The unpolymerized monomer that remains incorporated in a polymer after the polymerization reaction is completed. Polystyrene and vinyls are examples of resins in which residual monomer can be found, although manufacturers of vinyls have taken steps to nearly eliminate residual monomer from their resins during recent years.

Resilience. The degree to which a body can quickly resume its original shape after removal of a deforming stress. It can be expressed as the ratio of energy returned, upon recovery from deformation, to the work input required to produce deformation.

Resin. See RESINS, NATURAL; RESIN, SYNTHETIC.

Resin Applicator. In filament winding, a device which deposits the liquid resin onto the reinforcement band.

Resin-Bonded Laminate. See LAMINATE, REINFORCED PLASTIC.

Resinoid. A term sometimes used for a thermosetting resin, either in its initial temporarily fusible state or in the final cured state.

Resin-Pocket. An apparent accumulation of excess resin in a small localized area between laminations in laminated plastics, visible on cut edges or molded surfaces. (ASTM D 883-65T).

Resin-Rich Area. A region in a reinforced plastic article in which there is an objectionable excess of resin.

Resins, Natural. When certain trees and plants are wounded, either by natural accident or by tapping, they exude liquids that harden or partially harden upon exposure to air to form soft resinous or balsamlike products. Deposits of such secretions undergo chemical changes, oxidation and polymerization when buried for long periods, forming solid or semi-solid viscous products that are soluble in oils and organic liquids but insoluble in water. Such water-insoluble resins are generally known as the natural resins, sometimes called varnish resins. The water-soluble secretions are known as gums or essential oils. Examples of natural resins are accroides, amber, Canada balsam, Congo copal, dammar, elemi, Kauri copal and sandarac. They are used in varnishes and lacquers, and also as modifiers for plastics. See also RESIN, SYNTHETIC.

Resin-Starved Area. An area of a reinforced plastic article which has an insufficient amount of resin to wet out the reinforcement completely, evidenced by low gloss, dry spots or fiber show. The condition may be caused by improper wetting or impregnation, or by excessive molding pressure.

Resin Streak. A long, narrow surface imperfection on the surface of a laminated plastic caused by a local excess of resin.

Resin, Synthetic. The term *resin* is defined by ASTM (D 883-75a) as a solid or pseudosolid material often of high molecular weight, which exhibits a tendency by flow when subjected to stress, usually has a softening or melting range, and usually fractures conchoidally. A note added to this ASTM definition explains that in a broad sense, the term is used to designate any polymer that is a basic material for plastics. However, common uses of the term in the plastics industry do not always conform to this definition. The term *resin* is also used for uncured fluid thermosetting materials, some chemically modified natural resins, and is often used synonymously with the terms *plastic* and *polymer*.

Resistance. (Electrical) The ohm remains as the SI unit of resistance, the electrical resistance between two points of a conductor when a constant difference of potential of one volt, applied between these two points, produces in this conductor a current of one ampere, this conductor not being the source of any electromotive force.

Resistance Welding. See THERMOBANDE WELDING.

Resistivity. (specific resistance) A proportionality factor characteristic of different substances, equal to the resistance of a specified volume (usually a cubic centimeter) or area (usually a square centimeter) of the substance. See also SURFACE RESISTIVITY, VOLUME RESISTIVITY.

Resite. A term sometimes used for a thermosetting resin in the fully cured or C-stage.

Resitol. See RESOLITE.

Resole. (resol) A thermosetting resin composition in its unformed and uncured state, but containing all of the necessary materials for hardening upon heating. Also called *A-stage resin.*

Resolite. (resitol, B-stage) A thermosetting resin in an intermediate, partially cured form.

Resonance. In chemistry, the moving of electrons from one atom of a molecule or ion to another atom of the same molecular or ion. Thus, given atoms may remain in a fixed spatial arrangement with their electrons arranged so as to simultaneously satisfy two or more classical structural formulas.

Resorcinol. $C_6H_4(OH)_2$ (resorcin, meta-dihydroxybenzene, 3-hydroxyphenol) A highly reactive phenol, which when reacted with formaldehyde produces resins suitable for cold-setting adhesives.

Resorcinol Monobenzoate. A white, crystalline solid used as an ultraviolet screening agent in plastics. It is particularly useful in applications requiring a high degree of transparency, and can be used with cellulosics, vinyls and certain polyesters.

Restricted Gate. A very small orifice between runner and cavity in an injection or transfer mold. When the piece is ejected this gate breaks cleanly, simplifying separation of runner from piece.

Restrictor Ring. A ring-shaped part protruding from the torpedo surface which provides increase of pressure in the mold to improve, e.g., welding of two streams.

Retainer Plate. In injection molding, a plate which reinforces the cavity block against the injection pressure, and also serves as an anchor for the cavities, ejector pins, guide pins and bushings. The retainer plate is usually cored for circulating water or steam for cooling and heating.

Retaining Pin. A pin on which an insert is placed and located prior to molding. Also sometimes used for *guide pin* or *dowel pin.*

Retarder. See INHIBITOR.

Reticulated Urethane Foams. Very low density urethane foams characterized by a three-dimensional skeletal structure of strands with few or no membranes between the strands, containing up to 97% or more of void space. They are made by treating an open-cell foam structure with a dilute aqueous sodium hydroxide solution under controlled conditions so that the thin membranes are dissolved, leaving the strands substantially unaffected. Ultrasonic vibrating is sometimes used to assist the solution process. These foams are used in filters for air-conditioners, automobile carburetors, air cleaning systems; and in acoustical panels, humidifiers and various household products.

Return Pins. Pins which return the ejector mechanism to molding position.

Reverse-Flighted Screw. A type of extruder screw with left hand flights on one end and right hand flights on the other end, so that material can be fed at both ends of the barrel and extruded from the center.

Reverse Helical Winding. In filament winding, a pattern in which a continuous helix is laid down, reversing direction at the polar ends. It is contrasted to biaxial, compact or sequential winding in that the fibers cross each other at definite equators, the number depending on the helix. The minimum number of crossover regions would be three.

Reverse Impact Test. A test for sheet material in which one side of the specimen is struck by a pendulum or falling object and the reverse side is inspected for damage.

Reverse-Roll Coating. A method of coating wherein the coating material is premetered between a pair of rolls, one of which deposits the coating on a substrate. The thickness of the coating is controlled by the gap between the rollers and also by the speed of rotation of the coating roll.

Reworked Material. Scrap plastic parts such as runners, flash and reject parts that have been reclaimed for reprocessing.

Reynold's Number. A ratio used to determine whether the flow of a viscous fluid through a pipe is streamlined or turbulent. The formula is DUP/u, wherein D is the inside pipe diameter, U is the average velocity of flow, P is density, and u is the viscosity. Values below about 2100 represent streamlined flow, and values above 3000 denote turbulent flow.

Reworked Material. (Thermoplastic) A plastic material that has been reprocessed, after having been previously processed by molding, extrusion, etc., in a fabricator's plant. Note: In many specifications the use of reworked material is limited to clean plastic that meets the specifications for the virgin material, and yields a produce essentially equal in quality to one made from only virgin material. (ASTM D 883-75a). Another use for the term is for excess material such as sprues, runners and blow molding parisons which have been re-ground for blending with virgin material for re-use in the fabricator's plant.

RF Heating. See DIELECTRIC HEATING.

RHE. In the old cgs system, rhe is the reciprocal of the poise. It is the unit of fluidity.

Rheology. The study of flow. The term rheology, derived from the Greek *rheos* meaning "something flowing," was proposed by Bingham in 1929 for the rapidly growing science of studying flow and deformation properties of materials, including gases, liquids and solids, in terms of stress, strain and time. See also CONSISTENCY, DILATANCY, NEWTONIAN FLOW, THIXOTROPY, VISCOELASTICITY and YIELD VALUE.

Rheometer. (plastometer) An instrument for determining the flow properties of a thermoplastic material, usually of high viscosity or in the molten condition. The most commonly used type is the EXTRUSION RHEOMETER, which see. Instruments for measuring flow properties of fluids are more often called VISCOMETERS, which see.

Rheopecticity. (rheopexy) The opposite of thixatropy. The viscosity of rheopectic materials increases with time under a constantly applied stress, and decreases upon removal of the stress.

Rhodamines. A class of organic dyes which exhibit bright orange to red fluorescent colors when viewed under ultraviolet light. They can be incorporated in vinyls as indicators of the completeness of fusion, since their shades of color vary with the degree of fusion of the vinyl compound when exposed to UV light.

Ribbon Blenders. Mixing devices comprising helical ribbon-shaped blades rotating close to the edge of a u-shaped vessel. They are used for relatively high viscosity fluids and dry-blends such as PVC calendering and extrusion compounds.

Ricinus Oil. See CASTOR OIL.

Ridge Forming. See THERMOFORMING.

Rigidity, Modulus Of. The slope of the linear portion of the initial stress-strain curve.

Rigid Plastic. For purposes of general classification, a plastic that has a modulus of elasticity either in flexure or in tension greater than 700 MPa (7000 kgf/cm^2 or 100,000 psi) at 23 °C and 50% relative humidity when tested in accordance with ASTM methods D 747, D 790, D 638 or D 882. (ASTM D 883-75a). This ASTM criterion is not always observed in the literature, especially with respect to vinyls in which impact strengths and other properties can vary widely while elastic modulus remains fairly constant. A proposed system for vinyls places the elastic modulus of rigids at 200,000 psi and above, semi-rigids at from 60,000 to 200,000, and flexibles at below 60,000 psi.

Rigidsol. A coined term for a plastisol which forms an article of very high durometer hardness. Such hardness is obtained by compounding techniques which permit the use of relatively small amounts of plasticizer, and/or by the incorporation of monomers which serve as diluents at room temperatures but which cross-link or polymerize upon heating.

RIM. Abbreviation for REACTION INJECTION MOLDING, which see.

Ring Gate. An annular opening for entrance of material into the cavity of an injection or transfer mold.

Rings. A polymeric structure resulting from the reaction of one end of a molecule with the other end, forming a ring structure that may be compared to a snake biting its own tail. The stability of ring molecules formed from a carbon-carbon chain is generally greatest in 5 to 6-membered rings, and least in 9 to 11-membered rings. The probability of ring formation decreases rapidly as the length of the molecule increases, so that the presence of a few small rings in a polymer chain is usually insignificant.

Rocker. A colloquial term for a blown container which is defective by reason of a bulged or deformed bottom causing the container to rock when placed upright on a flat surface.

Rockwell Hardness. The hardness of a material expressed as a number derived from the net increase in depth of impression as the load on an indentor is increased from a fixed minor amount to a major load and then returned to the minor load. As specified in ASTM D 785, the minor load is fixed at 10 kg. Various scales, depending on the diameter of the ball indentor and the major load, are used. In order of increasing hardness these are the R, L, M, E and K scales. See also SCRATCH HARDNESS, INDENTATION HARDNESS.

Rodent Resistance. The ability of a plastic to withstand or repel attacks by rodents. Some plastics require additives to prevent rodents from chewing objects such as cable insulation. One example of such an additive is TETRAMETHYLTHIURAM DISULFIDE, which see.

Roentgen. (r) The international unit of quantity or dose for X-rays or gamma rays, equal to the quantity of X- or gamma rays which will produce as a result of ionization one electrostatic unit of electricity of either sign in 1 cc of dry air at 0 °C and standard atmospheric pressure. However, the use of the roentgen unit has been extended to include particle radiation such as alpha and beta particles and neutrons. In such cases the term *roentgen equivalent physical* abbreviation *rep,* is used and defined as the quantity of particle radiation which upon absorption in 1 gram of body tissue is accompanied by the gain of 93 ergs of energy. See also RAD. In the new SI, the roentgen is expressed as 2.58 E-04 coulombs per kilogram.

Roller Coating. (roll coating) The process of coating substrates with fluid resins, solutions or dispersions by contacting the substrate with a roller on which the fluid material is spread. The process is often used to apply a contrasting color on raised lettering or markings. See also KISS ROLL COATING, REVERSE ROLL COATING and GRAVURE COATING.

Roll Leaf Stamping. See HOT STAMPING.

Roll Mill. An apparatus for admixing a plastic material with compounding ingredients, comprising two rolls placed in close relationship to one another. The rolls turn at different speeds to produce a shearing action to the materials being compounded. Mixing plows and slitting knives are sometimes provided to work the stock across the rolls, aiding the uniform dispersion of additives.

Rossi-Peakes Tester. An instrument for measuring the temperature at which a given amount of a molding powder will flow through a standard orifice in a prescribed period of time under a prescribed pressure.

Rotary Molding. A term sometimes used to denote a type of injection, transfer, compression or blow molding utilizing a plurality of mold cavities mounted on a rotating table or dial. Not to be confused with *rotational molding.*

Rotary Vane Feeders. Devices for conveying and metering dry materials, comprising a cylindrical housing containing a shaft with blades or flutes attached, rotating at a rate selected to feed at a desired rate.

Rotating Spreader. A type of injection torpedo which consists of a finned torpedo rotated by a shaft extending through a tubular cross-section injection ram behind it.

Rotational Casting. The process of forming hollow articles from fluid materials by rotating a mold containing a charge of the material about one or more axes at relatively low speeds until the charge is distributed on the inner mold walls by gravitational forces and hardened by heating, cooling or curing. Rotation about one axis is suitable for cylindrical objects. Rotation about two axes and/or rocking motions are employed for completely closed articles. The process dates back to 1855, when a British patent was granted for the rotational casting of hollow articles from molten metals. Vinyl plastisols were first used in the process in 1947 by Claude Delacoste, a French toy manufacturer. The process of rotational casting of plastisols comprises placing a measured charge of plastisol in the bottom half of an opened mold, closing the mold, rotating the mold about one or more axes in the presence of heat until the charge has been distributed and fused against the mold walls, cooling the mold until the deposit is of sufficient strength, opening the mold and removing the article. See also CENTRIFUGAL CASTING, ROTATIONAL MOLDING.

Rotational Injection Molding. A modified injection molding process applicable to hollow, symmetrical articles such as cups and beakers, in which the male half of the mold is rotated during the molding cycle until the material has hardened to a predetermined degree. The rotation produces multiaxial molecular orientation and increased crystallinity of polymers such as polystyrene, resulting in improved toughness and stress-craze resistance.

Rotational Molding. The preferred term for a variation of the rotational casting process utilizing dry, finely divided sinterable powders, such as polyethylene rather than fluid materials. The powders are first sintered, then fused against the mold walls. This variation of the process dates back to 1947, when a British patent issued to I.C.I. Ltd. for the rotational molding of powders such as polyethylene and PVC.

Rotomolding. A contraction of the term *rotational molding,* sometimes used indiscriminately for both the processes of ROTATIONAL CASTING and ROTATIONAL MOLDING, which see.

Roving. A form of fibrous glass used in reinforced plastics comprising from 8 to 120 (usually 60) single filaments or strands gathered together in a bundle, and usually treated with a coupling agent to promote adhesion of the glass to the plastic. When the strands are twisted together, the term *spun roving* is used. Roving is used in continuous lengths for filament winding, chopped into short lengths for use in reinforced plastic molding compounds, and woven into skeins or mats for use in laminates. See also REINFORCEMENT.

RP. Abbreviation for REINFORCED PLASTIC, which see.

RTP. Abbreviation for REINFORCED THERMOPLASTICS, which see.

RTV. Abbreviation for *room temperature vulcanizing,* a characteristic of some elastomers which do not require heating to cure.

Rubber Hydrochloride. A non-flammable thermoplastic material obtained by treating a solution of rubber with anhydrous hydrogen chloride under pressure at low temperatures. The packaging film "Pliofilm" is an example.

Rubber, Natural. (India rubber, caoutchouc) A polymer consisting essentially of cis-1,4-polyisoprene, obtained from the sap (latex) of certain trees and plants, usually the *Hevea Brasiliensis* tree. The material is shipped from plantations in one of two primary forms: Latex, usually preserved with ammonia and centrifuged to remove part of the water; or sheets made by milling the coagulum from the latex.

Rubber Plate Printing. A marking method sometimes employed for intricate parts such as molded terminal blocks. Numbers, instructions or part names are stamped with conventional rubber stamps or printing plates.

Rubber Plunger Molding. A variation of matched-die molding, employing a deformable rubber plunger and a heated metal female mold. The process enables the use of high fiber loadings.

Rubbers, Synthetic. See ELASTOMER.

Rubber, Synthetic Natural. An awkward term sometimes used for the elastomers which most nearly resemble natural rubber, such as cis-1,4-polyisoprene.

Rubber Transition. (rubbery transition, gamma transition) See GLASS TRANSITION.

Runner. In an injection or transfer mold, the feed channel, usually of circular cross section, that connects the sprue with the cavity gate. The term is also used for the plastic piece formed in this channel.

Runnerless Injection Molding. A molding process in which the runners are insulated from the cavities and kept hot, so that the molded parts are ejected with only small gates attached. See also HOT RUNNER MOLD.

Runnerless Injection Molding. (Thermosets) See COLD RUNNER INJECTION MOLDING.

Runner System. This term is sometimes used for the entire mold feeding system, including sprues, runners and gates, in injection or transfer molding.

Rutile. One of the crystalline forms of TITANIUM DIOXIDE, which see.

SABRA. Abbreviation for Surface Activation Beneath Reaction Adhesives, a method of bonding plastics such as polyolefins and Teflon which are not normally receptive to adhesives without pretreatment. The method consists of mechanically abrasion of the surfaces to be joined to roughen their outer layers, scission of bonds with creation of free radicals, and further reaction with primers in the liquid, vapor or gaseous phase. An adhesive such as an epoxy is then applied.

Sag. In blow molding, the local reduction in diameter of the parison, or "necking down," caused by gravity. It is usually greatest on the portion nearest to the die, and increases as the parison becomes longer. In a thermoforming operation, sag is the downward bulge in the molten sheet.

SAIB. Abbreviation for SUCROSE ACETATE ISO-BUTYRATE, which see.

Salt. Any substance which yields ions, other than hydrogen or hydroxyl ions. A salt is obtained by displacing the hydrogen of an acid by a metal.

SAN. Abbreviation for STYRENE-ACRYLONITRILE COPOLYMERS, which see.

Sanding. A finishing process employing abrasive belts or discs, sometimes used on thermosetting resins parts to remove heavy flash or projections, or to produce radii or bevels which cannot be formed during molding.

Sand Mill. An apparatus used for preparing pigment dispersions, consisting of a vertical cylinder with a centrally mounted agitator shaft on which are mounted several flat, annular disc impellers. The mill is charged with natural sands or high-silica ceramic beads as the grinding media. The pigment slurry is pumped into the bottom of the mill and becomes mixed with the grinding media. As the mixture is forced upward through the mill it passes through several zones of agitation and finally flows through a screen at the top which retains the grinding media.

Sandwich. A term sometimes employed for a laminate comprising at least three layers; for example, a cellular plastic core sandwiched between two layers of other material. See also LAMINATE.

Sandwich Heating. A method of heating a thermoplastic sheet prior to forming which consists of heating both sides of the sheet simultaneously.

Saran. Generic name for thermoplastics consisting of polymers of vinylidene chloride or copolymers of same with lesser amounts of other unsaturated compounds.

Saran Fiber. Generic name for a manufactured fiber in which the fiber-forming substance is any long chain synthetic polymer composed of at least 80% by weight of vinylidene chloride units ($-CH_2CCl_2-$). (Federal Trade Commission)

Saturated Compounds. Compounds whose available atomic valence bonds are attached to other atoms of the same compound, and thus cannot add on elements or other compounds.

Saturated Polyester. See POLYESTERS, SATURATED.

Saturators. Machines designed to impregnate paper, fabrics and the like with resins. The web to be saturated is conveyed by rollers through a pan containing a solution of the resin, then through metering devices such as squeeze rolls, scraper blades or suction elements which control the amount of resin retained.

Saybolt Viscosity. The time in seconds required to fill a 60 cc flask with a liquid specimen preheated to a standard temperature through an orifice of specified diameter.

SB. Abbreviation for copolymers of styrene and butadiene.

SBR. Abbreviation for STYRENE-BUTADIENE RUBBERS, which see.

Scarf Joint. A joint made by cutting away similar angular segments on two pieces to be joined, with the cut areas fitted together.

Scintillometer. An instrument which detects radiation by emitting flashes of light.

Scleroscope. An instrument for measuring impact resilience by dropping a ram with a flattened cone tip from a specified height onto the specimen, then noting the height of rebound.

"Scotch Tape" Test. A method for evaluating the adhesion of a lacquer or paint to a plastic substrate. Pressure-sensitive adhesive tape is applied to an area of the painted plastic article, which is sometimes cross-hatched with scratched lines. Adhesion is considered to be adequate if no paint is pulled off by the tape when it is removed.

Scrap. All products of a processing operation which are not present in the primary finished articles. This includes flash, runners, sprues, excess parison, and reject articles. Scrap from thermosetting molding operations is generally not reusable. That from most thermoplastic operations can usually be reclaimed for reuse in the molders own plant or for sale to a commercial reclaimer.

Scrap Grinders. See GRANULATORS.

Scrapless Thermoforming. See THERMOFORMING.

Scratch Hardness. The resistance of a material to scratching by another material. The test most often employed for plastics is the Bierbaum test, in which the specimen is moved laterally on the stage of a microscope under a loaded diamond point. The standard load is 3 grams. The width of the scratch is measured by a screw micrometer eyepiece, and the hardness value is expressed as the load in kilograms divided by the square of the width of the scratch in millimeters. Other tests employ pencils of different hardnesses (Kohinoor Value); scratching with various minerals (see MOHS VALUE); and abrasion by falling carborundum particles.

Screen Pack. See EXTRUDER SCREEN PACK.

Screen Process Printing. (silk screen printing) A printing process widely used on plastic bottles and other articles, employing as a stencil a taut woven fabric secured in a frame, the fabric coated in selected areas with a masking material which is not affected by the ink being used. The stencil fabric is commonly called a "silk screen" even though silk is rarely used today. Nylon is most often used, and screens of copper, stainless steel and many other materials are suitable. The screen is placed above the part to be decorated, and a flexible squeegee forces ink through the openings in the screen onto the surface of the plastic article.

Screw. In extrusion, the shaft provided with helical grooves which conveys the material from the hopper outlet through the barrel and forces it out through the die. See also EXTRUDER SCREW.

Screw Extruder. See EXTRUDER.

Screw Injection Molding. See INJECTION MOLDING.

Screw-Piston Injection Molding. See INJECTION MOLDING.

Screw Plasticating Injection Molding. See INJECTION MOLDING.

Scrim. A low cost, nonwoven, open mesh reinforcing fabric made from continous filament yarn.

Sealing. See HEAT SEALING.

Sebacic Acid. $COOH(CH_2)_8COOH$ (sebacylic acid, octanedicarboxylic acid) White leaflets derived from butadiene or castor oil, used as an intermediate in the production of plasticizers, alkyd resin and certain types of nylon.

Secant Modulus. The ratio of total stress to corresponding strain at any specific point on the stress-strain curve. It is expressed in force per unit area, usually pounds per square inch, and reported together with the specified stress or strain.

Secondary Plasticizer. (extender plasticizer) A plasticizer which is less compatible with a given resin than is a PRIMARY PLASTICIZER, which see, and thus would exude or cause surface tackiness if used in excess of a certain concentration. Secondary plasticizers are used in conjunction with primary plasticizers to reduce cost or to obtain improvement in electrical or low temperature properties.

Second Order Transition. See GLASS TRANSITION.

Second-Surface Decorating. A decorating process in which the decoration is applied to the back of a transparent plastic part so that it is visible through the part, but is not exposed.

Seebeck Effect. The electrical phenomenon responsible for the action of a THERMOCOU-PLE, which see. If a circuit consists of two metals, one junction being hotter than the other, a current flows in the circuit. The intensity and direction of the flow depend on the metals used and on the temperature of the junctions.

Segregation. A close succession of parallel, rather narrow and sharply defined, wavy lines of color on the surface of a plastic differing in shade from surrounding areas, and creating the impression that components of the plastic have separated. (ASTM D 883-65T).

Self-Extinguishing. A somewhat loosely-used term describing the ability of a material to cease burning once the source of flame has been removed. PVC, vinyl chloride-acetate copolymers, polyvinylidene chloride, nylon and casein plastics are examples of self-extinguishing materials. See also FLAME RETARDANTS.

Semi-Automatic Molding Machine. A machine in which only part of the operation is controlled by the direct action of a human. The automatic part of the operation is controlled by the machine according to a predetermined program.

Semipositive Mold. A mold with a plunger which is loosely fitting within the cavity as the mold begins to close, allowing excess material to escape as flash, but which becomes tightly fitting as the mold closes completely, thus exerting full clamping pressure on the material. The semipositive mold combines the advantage of the free flow of material inherent in a flash mold with that of high material density obtained with a positive mold.

Semirigid Plastic. For purposes of general classification, a plastic that has a modulus of elasticity either in flexure or in tension of between 700 and 7000 Kg per sq cm (10,000 and 100,000 psi) at 23 °C and 50 percent relative humidity when tested in accordance with . . . ASTM D 747, . . . D 790, . . . D 638 . . . or D 882. (ASTM D 883-75a).

Sequential Arrangement. The arrangement of head and tail linkages of constitutional units in a polymer chain. (ISO)

Sequential Winding. See BIAXIAL WINDING.

Sequestering Agents. Agents which prevent metallic ions from precipitating from solutions by means of reactions which normally would cause precipitation in the absence of a sequestering agent. See also CHELATE.

Serpentine. A type of asbestos containing CHRYSOTILE, which see.

Set. (n) Strain remaining after complete release of the force producing deformation. (ASTM D 883-75a).

Set. (v) To convert into a fixed or hardened state by chemical or physical action, such as condensation, polymerization, oxidation, vulcanization, gelation, hydration, or evaporation of volatile constituents. See also CURE, PERMANENT SET.

S-Glass. A magnesia-alumina-silicate glass, especially designed to provide very high tensile strength. See also GLASS FIBER REINFORCEMENTS.

Shark Skin. A surface of irregularity of a container in the form of finely spaced sharp ridges caused by a relaxation effect of the melt at the die exit.

Shaw Pot. A name used in the early years of the industry for the original thermosetting transfer molding machine. It consisted of a conventional hydraulic press with a pot suspended above the mold. Material was charged into the pot and forced into the mold by closing of the press.

Shear. An action or stress resulting from applied forces which causes or tends to cause two contiguous parts of a body to slide relative to each other in a direction parallel to their plane of contact.

Shear Modulus. The ratio of shearing stress to shearing strain within the proportional limit of a material.

Shear Rate. The overall velocity over the cross section of a channel with which molten or fluid layers are gliding along each other or along the wall in laminar flow. Shear rate is expressed in reciprocal seconds, derived from the relationship

$$\frac{\text{velocity}}{\text{clearance}} = \frac{\text{cm/sec.}}{\text{cm.}} = \text{sec.}-1.$$

Shear Strength. The maximum load required to shear the specimen in such a manner that the moving portion has completely cleared the stationary portion.

Shear Stress. The stress developing in a polymer melt when the layers in a cross section are gliding along each other or along the wall of the channel (in laminar flow).

Sheet. (sheeting) Sheets are distinguished from films in the plastics and packaging industry only according to their thicknesses. A web under 10 mils (.010″) thick is usually called a film, whereas one 10 mils and over in thickness is usually called a sheet. Sheeting is most commonly made by extrusion, casting and calendering.

Sheeter Lines. Parallel scratches or projecting ridges distributed over considerable area of a plastic sheet such as might be produced during a slicing operation. (ASTM D 883-65T).

Sheet Forming. See THERMOFORMING.

Sheet Molding Compound. (SMC) A fiber glass reinforced thermosetting compound in sheet form, usually rolled into coils interleaved with plastic film to prevent autoadhesion. This term has been designated by the SPI to replace the term *prepreg,* which was deemed to be confusing and insufficiently descriptive. SMC can be molded into complex shapes with little scrap, and is low in cost.

Sheet Train. The entire assembly necessary to produce sheet which includes extruder, die, polish rolls, conveyor, draw rolls, cutter and stacker.

Shelf Life. (storage life) The length of time over which a product will remain suitable for its intended use during storage under specific conditions. The term is applied to finished products as well as to raw materials. See also WORKING LIFE.

Shell Cup Viscometer. A stainless steel cup of 23 milliliter capacity which drains through a one-inch long capillary at the bottom. There are four sizes of Shell Cups, differing only in diameter of the capillary tubes. The entire cup is submerged in the sample, then raised and held above the surface. The time in seconds from the moment the top of the cup emerges from the sample until the first break in the stream from the capillary orificce is the measure of kinematic viscosity.

Shell Flour. A filler obtained by grinding shells of walnuts, coconuts, pecans or peanuts, used primarily in thermosetting molding compounds and as extenders for adhesives.

Shell Molding. In the foundry industry, a process of casting metal objects in thin molds made from sand or a ceramic powder mixed with a thermosetting resin. Some authors have misused the term by applying it to plastics processes such as dipping and slush casting.

Shish Kebab. A term derived from the Turkish *shish* (skewer) and *kebab* (roast beef). However, in the plastics industry the term has been applied to polymeric structure in which the random coil chain of an amorphous polymer (shish) has been interlaced with crystalline segments (kabobs) produced by straining the polymer in the molten condition or in solution. The resulting lumpy chain structure so resembles the edible delicacy that its name has been applied to the polymer structure resulting from stress-induced crystallization.

Shoe. See CHASE.

Shore Hardness. See INDENTATION HARDNESS.

Short. In reinforced plastics, an imperfection caused by an absence of surface film in some areas, or by lighter unfused particles of material showing through a covering surface film, accompanied possibly by thin-skinned blisters.

Short Beam Shear Strength. The inter-laminar shear strength of a parallel fiber reinforced plastic material as determined by three-point flexural loading of a short segment cut from a ring-type specimen.

Short Shot. In injection molding, failure to fill the mold completely. It results in voids in the article, unfused particles showing through a surface covering, or possibly thin-skinned blisters.

Shortstopper. A term used for an agent added to a polymerization reaction mixture to inhibit or terminate polymerization.

Shot. (1) One complete cycle of a molding machine. (2) The yield from one complete molding cycle, including the molded part, cull, runner system and flash. See also SPRAY.

Shot Capacity. The maximum weight of material that can be delivered to an injection mold by one stroke of the ram. In the case of screw injection molding machines, slippage of material may occur in the screw flights and thus affect calculations of shot capacity based on swept volume or cubic inch displacement.

Shrinkage Allowance. The dimensional allowance which must be made in molds to compensate for shrinkage of the plastic compound on cooling. The ASTM method for determining shrinkage from mold dimensions is D 955-73. This method does not provide for the measurement of shrinkages that may occur as molded materials age, e.g. after the first 48 hours after removal from the mold.

Shrinkage Pool. An irregular, slightly depressed area on the surface of a molding caused by uneven shrinkage before complete hardening is attained.

Shrink Film. A term sometimes used for pre-stretched or oriented film used in SHRINK PACKAGING, which see.

Shrink Fixture. See COOLING FIXTURE.

Shrink Mark. A shallow depression or dimple on the surface of an injection molded article caused by local internal shrinkage after the gate seals, or by a short shot.

Shrink Packaging. A method of wrapping articles utilizing pre-stretched (oriented) films which are heated to cause them to shrink tightly around the articles. First, the article is placed in a loose envelope of two layers of film, usually in the form of a V-folded strip. This envelope is heat sealed around the edges and detached form the strip, both of which operations can be performed by an L-shaped thermal impulse sealer and cutter. The package is then conveyed through a hot air oven or other heating device to shrink the film.

Shrink Tunnel. An oven in the form of a tunnel mounted over or containing a continuous conveyor belt, used to shrink oriented films in the shrink packaging process.

SI. (1) Abbreviation for silicone plastics. See SILICONES. (2) Abbreviation for "International System of Units," derived from the French name, Le Systeme International d'Unites. The system is a modern version of the MKSA (meter, kilogram, second, ampere) system published by an international treaty organization which is attempting to have the system adopted throughout the world. Although not officially enforced in the U.S., on December 23, 1975 President Ford signed the "Metric Conversion Act" which established a board to coordinate the "voluntary" conversion to the SI metric system in the U.S. Subsequently, the National Bureau of Standards published "Guidelines for the Metric System of Weights and Measures" for use in the U.S. It recognizes that certain units which are not part of the SI are used so widely that it is impractical to abandon them. In this Dictionary, certain SI units likely to be encountered in the plastics industry are given, along with the old units and conversion factors. It should be noted that exponential factors for multiples and fractions are given in the SI as a number greater than one and less than ten with six or less decimal places. This number is followed by an asterisk (*) after the sixth decimal place if all subsequent digits are zero — otherwise the asterisk is omitted. Following the asterisk, if any, is the letter E indicating "exponent," a plus or minus symbol, and two digits which indicate the power of 10 by which the number must be multiplied to obtain the correct value. For example, 3.234 000*E + 03 is 3.234 $\times$ 10³. Although not required by law, several technical organizations such as ASTM and the American Chemical Society have stipulated that papers and definitions submitted to them for publication must give SI units. The ASTM Standard for Metric Practice is given in publication No. E 380-76ᵉ. It also recognizes that some units outside the SI should remain in use in the U.S.

Siamese Blow. A colloquial term applied to the process of blow molding two or more objects or parts of objects in a single blowing mold, then cutting them apart.

Side Bars. Loose pieces used to carry one or more molding pins, and operated from outside the mold.

Side Draw Pins. Projections used to core a hole in a direction other than the line of closing of a mold, and which must be withdrawn before the part is ejected from the mold.

Siemens. (S) The new SI-approved term for electrical conductance, expressed by dividing amperes by volts. It is equivalent to the old term *mho,* the reciprocal of resistance.

Silane Coupling Agents. Silanes (compounds of silicon and hydrogen of the formula Si_nH_{2n+2}) and other monomeric silicon compounds which have the ability to bond inorganic materials such as glass, mineral fillers, metals and metallic oxides to organic resins. The adhesion mechanism is due to two groups in the silane structure. The $Si(OR_3)$ portion reacts with the inorganic reinforcement, while the organofunctional (vinyl-, amino-, epoxy- etc.) group reacts with the resin. The coupling agent may be applied to the inorganic materials (e.g. glass fibers) as a pre-treatment and/or added to the resin. Examples of silane coupling agents are:

N-beta-(AMINOETHYL)-gamma-AMINOPROPYLTRIMETHOXY SILANE,
gamma-AMINOPROPYLTRIETHOXY SILANE,
BIS(beta-HYDROXYETHYL)-gamma-AMINOPROPYLTRIETHOXY SILANE,
beta-(3,4-EPOXYCYCLOHEXYL)ETHYLTRIMETHOXY SILANE,
gamma-GLYCIDOXYPROPYLTRIMETHOXY SILANE,
gamma-METHACRYLOXYPROPYLTRIMETHOXY SILANE,
SULFONYLAZIDOSILANES,
VINYL TRICHLOROSILANE,
VINYL TRIETHOXYSILANE, and
VINYL-TRIS(beta-METHOXYETHYL) SILANE.

A new class of silane coupling agents is known as *silyl peroxides,* represented by the general formula $R'_m R''_{(4-n-m)} Si(OOR)_n$. A typical member of this family is Vinyltris-(t-butylperoxy)silane. The coupling mechanism of the silyl peroxides, effected by heat only, is free radical in nature. The conventional silanes require an external free radical source and couple via an ionic mechanism initiated by hydrolysis.

Silgan. Tradename of SWS, a subsidiary of Stauffer Chemical Co., for a line of room-temperature vulcanizing silicone elastomers reinforced with polymeric organic fibers which are chemically grafted to the silicone network during the manufacturing process. SILGAN is used for potting electrical components, gaskets formed from the liquid compound and molded parts.

Silica. SiO_2 (silicon dioxide) A substance occurring naturally as quartz, sand, flint, chalcedony, opal, agate, etc. In powdered form it is used as a filler, especially in phenolic compounds for ablative nose cones of rockets. Synthetic silicas, made from sodium silicate or heating silicon compounds are also available. See also FUMED SILICA.

Silicone Foams. Foams based on fluid silicone resins are made by mixing the resins with a catalyst and blowing agent, pouring the mixture into molds, and curing at room temperature for about 10 hours on at elevated temperatures for shorter periods. Silicone foam sponge is made by mixing unvulcanized silicone rubber with a blowing agent and heating at vulcanizing temperature.

Silicone-Polycarbonate Copolymers. Introduced in 1969 by the General Electric Company, these thermoplastic copolymers vary from strong elastomers to rigid "engineering plastics" depending on composition. They can be extruded at temperatures from 550 to 625 °F, or cast or molded into optically clear films. An initial application has been for permselective membranes which are heat-sealable and ten times more permeable to oxygen than non-silicone polymers.

Silicone Rubber. A synthetic rubber made by vulcanizing a silicone elastomer gum such as dimethyl silicone. A free-radical generating catalyst such as penzoyl peroxide is usually used as the vulcanizing agent. The tensile strength of unreinforced silicone rubber is only about 50 psi. Higher tensile strengths are attained by adding reinforcing fillers such as finely-divided or fumed silica, or by putting crystallizing segments such as silphenylene into the polymer. See also SILICONES.

Silicones. A family of semi-organic polymers comprising chains of alternating silicon and oxygen atoms, modified with various organic groups attached to the silicon atoms. Depending on the nature of the attached organic groups and the extent of crosslinking between the molecules the polymers may be fluids ranging in viscosity from under 1 to over 1 million centistokes, elastomers, or solid resins. The earliest silicones were dimethyl polysiloxanes, made by treating silicon derived from sand with methyl chloride in the presence of a catalyst to form a chlorosilane, hydrolyzing this chlorosilane to form a cyclic trimer of siloxane, then polymerizing the siloxane to form a dimethyl polysiloxane. Although this type of silicone is still in widespread use, many modifications have been made such as by the incorporation of phenyl groups, halogen atoms, alkyds, epoxides, polyesters and other organic compounds containing OH groups. The silicone fluids are used as lubricants, mold release agents, heat transfer fluids and water-repellant coatings. The elastomers, often called silicone rubbers and reinforced with inorganic fillers or fibers, are vulcanizable and offer superior resistance to high temperatures and weathering. The silicone resins, possessing good electrical propertiess and strength at high temperature, are widely used for encapsulating and potting electrical components and in reinforced laminates. Solvent solutions are also available for use in coating and varnishes.

Silk Screen Printing. See SCREEN PROCESS PRINTING.

Siloxanes. See SILICONES.

Silver Spray Process. (chemical spray process) A metallizing process based on the glass mirror art. The plastic article is prepared by cleaning and lacquering as in vacuum metallizing, then this lacquer coat is sensitized in an acidic salt solution such as a mixture of sulphuric acid, potassium dichromate and water. A silver-forming solution, e.g. silver nitrate and an aldehyde, is sprayed on the article, usually with a two-nozzle spray gun so that the components are separated until they reach the surface. A final topcoat of protective lacquer is applied over the silver.

Silver Streaking. See SPLAY MARKS.

Silyl Peroxides. See SILANE COUPLING AGENTS.

Single Circuit Winding. A winding in which the filament path makes a complete traverse of the chamber, after which the following traverse lies immediately adjacent to the previous one.

Single-Stage Resin. (single-step resin) See RESOL.

Sinking. (mold making process) See HOBBING.

Sink Mark. See SHRINK MARK.

Sinter Coating. A coating process in which the article to be coated is preheated to sintering temperature and immersed in a plastic powder, then is withdrawn and heated to a higher temperature to fuse the sintered coating adhering to the article. See also FLUIDIZED BED COATING.

Sintering. The welding together of powdered plastic particles at temperatures just below the melting or fusion point. The particles are fused (sintered) together to form a relatively strong mass, but the mass as a whole does not melt.

Sinter Molding. The process of compacting thermoplastic particles under pressure at temperatures below their melting point until the particles become sintered together, often followed by further heating and/or post forming. Porous nylon bearings capable of absorbing lubricants are made by this method, and some fluorocarbon resin parts are most economically made by sinter molding.

Sisal. The fiber obtained from the leaves of agave plants, most commonly the *agave sisalana* which is native to America. Sisal is sometimes used in short, chopped lengths as a reinforcement in thermosetting molding compounds, imparting moderate impact resistance.

Sizing. (v) The process of applying a material to a surface to fill pores and thus reduce the absorption of the subsequently applied adhesive or coating or to otherwise modify the surface. Also, the surface treatment applied to glass fibers used in reinforced plastics. The material used is sometimes called *size*.

Skin. (n) A relatively dense layer at the surface of a cellular polymeric material. (ASTM D 883-75a).

Skin Packaging. A variation of the thermoforming process in which the article to be packaged serves as the mold. The article is usually placed on a printed card prepared with an adhesive coating or mechanical surface treatment to seal the plastic film to the card. See also BLISTER PACKAGING.

Slate. A fine grained metamorphic rock of varied composition, used in powdered form as a filler, especially in flooring compounds.

Sleeve Ejector. A bushing-type knockout.

Slip. With reference to adhesives, slip is the ability to move or position the adherends after an adhesive has been applied to the surfaces.

Slip Agent. A modifier that acts as an internal lubricant which exudes to the surface of a plastic during and immediately after processing. In other words, a non-visible coating blooms to the surface to provide the necessary lubricity to reduce the coefficient of friction and thereby improve slip characteristics. See also ANTIBLOCKING AGENTS.

Slip Forming. (slip ring forming) A variation of the process of THERMOFORMING, which see, employing a sheet clamping frame provided with tensioned pressure pads which permit the plastic sheet to slip inwards as the part is being formed. This controlled slippage contributes to more uniform wall thickness of the formed article.

Slip-Plane. Plane within transparent plastic visible in reflected light because of poor welding and shrinkage on cooling. (ASTM D 883-65T).

Slitting. The conversion of a given width of plastic film or sheeting to several smaller widths by means of knives. The operation can be performed as material emerges from a production unit such as a calender, film casting unit or an extruder ("In-line slitting"); by unwinding, slitting, then rewinding of rolls; or by slitting of rolls without rewinding ("roll slicing"). Slitting knives may be actual flat-blade knives or razor blades, or circular knives.

Slot Extrusion. A method of extruding film sheet in which the molten thermoplastic compound is forced through a straight slot.

Slug Molding. A process for making thin walled containers of 200- to 400-cc capacity. Slugs from an extruder are fed into a metering head, which delivers slugs of uniform weight into cylindrical bushings. From the bushings the slugs are propelled into a single cavity mold by a high speed ram that passes through the bushing. Finished parts are removed by a mechanical arm and transferred to an air conveyor that takes them to an automatic stacking unit. No secondary trimming operations are required.

Slurry Preforming. Method of preparing reinforced plastics preforms by wet processing techniques similar to those used in the pulp molding industry. For example, glass fibers suspended in water are passed through a screen which passes the water but retains the fibers in the form of a mat.

Slush Casting. A method of forming hollow objects, widely used for doll parts and squeeze toys, in which a fluid plastic mixture, usually vinyl plastisol, is poured into a hollow mold provided with an opening until the mold is full. Heat, applied to the mold before and/or after filling, causes a layer of material to gel against the inner mold wall. After the layer has reached the desired thickness, the excess fluid material is poured out, and additional heat is applied to fuse the layer. After cooling, the article is stripped from the mold. Molds for slush casting are thin-walled for rapid heat transfer. Electroformed copper molds or aluminum castings are most often used.

Slush Molding. The preferred term for the process similar to slush casting but employing dry, sinterable powders.

SMA. Abbreviation for copolymers of styrene and maleic anhydride.

SMS. Abbreviation for copolymers of styrene and alpha-methylstyrene.

(SN)$_x$. Abbreviation for SULFUR-NITROGEN POLYMERS, which see.

Snap-Back Forming. See VACUUM SNAP-BACK FORMING.

Soaps, Metallic. Products derived by reacting fatty acids with metals, widely used as stabilizers for plastics. The fatty acids commonly used are the lauric, stearic, ricinoleic naphthenic octoic or 2-ethylhexoic, rosin and tall oil. Typical metals are aluminum, barium, calcium, cadmium, copper, iron, lead, magnesium, tin and zinc.

Sodium Aluminum Hydroxycarbonate. $NaAl(OH)_2CO_3$ Tradenamed Dawsonite by Alcoa, this material is produced in the form of microfiber crystals useful for upgrading physical properties of thermoplastics. In PVC compounds it also acts as a smoke suppressant and HCl scavenger. During 1976, the material was made only in pilot plant quantities.

Sodium Borohydride. $NaBH_4$ A white crystalline powder, used as a blowing agent for foamed plastics such as rigid PVC and polystyrene, and for elastomers. The material decomposes at room temperature in the presence of water and an acidic medium, releasing hydrogen.

Sodium Stearate. $NaOOCC_{17}H_{35}$ A white powder used as a non-toxic stabilizer.

Softening Range. The range of temperature in which a plastic changes from a rigid to a soft state. Note: Actual values will depend on the method of test. Sometimes referred to as softening point. (ASTM D 883-65T). The foregoing definition would apply to thermoplastics which are normally rigid at room temperature, and possibly to thermosetting materials which depolymerize or disintegrate at the test temperatures. For normally flexible thermoplastics, a more suitable definition of softening range is the range of temperature in which a plastic exhibits a rather sudden and substantial decrease in hardness. See also VICAT SOFTENING POINT, BALL AND RING TEST.

Solid Casting. The process of forming solid articles by pouring a fluid resin or dispersion into an open mold, causing the material to solidify by curing or heating and cooling, then removing the formed article.

Solid-Phase Forming. This term includes the shaping of plastic sheets or billets into three-dimensional articles either at room temperature (See COLD FORMING) or at higher temperatures up to the softening or melting range (see WARM FORGING) by processes such as those used in the metal working industry. These processes include forging with closed metal dies, rubber pad forming (using one metal die and a rubber pad as the matching die), diaphragm forming, stamping, drawing, cold heading, threading and coining. An advantageous characteristic of the solid-phase forming processes is that working of the material below its melt temperature orients it so that thin sections are stronger than the heavier sections. In thermoforming processes employing temperatures above the melting range of the material, thin sections tend to be weaker than heavier sections. Among the materials suitable for at least some of the solid-phase forming processes are ABS, acetals, cellulosics, polyolefins, polycarbonates, polyphenylene oxides and polysulfones. Brittle materials such as acrylics and polystyrene cannot be formed by solid-phase processes.

Solid-State Polymerization. A chain-growth polymerization reaction initiated by exposing to ionizing radiation a crystalline solid monomeric substance. A large number of olefin and cyclic solid monomers have been so polymerized, the crystalline monomer converting directly to the polymer with no obvious change in appearance of the solid.

Solprene. Trademark of Phillips Chemical Co. for a line of solution-polymerized copolymers of butadiene and styrene that differ from conventional copolymers that have the two monomers in a linear block configuration. The Solprenes have a radial-teleblock structure in which several polybutadiene chain segments extend from a central hub, with a polystyrene block attached to the outward end of each segment. This structure forms an elastomeric network without chemical crosslinks having high strength, resilience and coefficient of friction. The polymers, dubbed "Plastomers," process much like thermoplastics but perform like elastomers.

Solute. That constituent of a solution which is considered to be dissolved in the solvent. The solvent is usually present in larger amount than the solute.

Solution. A homogeneous mixture of two or more components, such as a resin completely dissolved in a liquid, which forms more or less spontaneously, will not settle, and has no fixed proportions of the components. Solutions are used in the plastics industry as coatings, for film casting and for spinning fibers. The term also covers a gas dissolved in another gas, and a liquid in a liquid.

Solution Casting. See FILM CASTING.

Solution Coating. Any coating process employing a solvent solution of a resin, as opposed to a dispersion, hot-melt or uncured thermosetting system. See also SPREAD COATING.

Solution Polymerization. A polymerization process in which the monomer or mixture of monomers and the polymerization initiators are dissolved in a non-monomeric liquid solvent or diluent at the beginning of the polymerization reaction. The liquid is usually also a solvent for the resulting polymer or copolymer. The solution process is most advantageous when the polymers are to be used for coatings, lacquers or adhesives. Vinyl acetate, olefins, styrene and methyl methacrylate are the monomers most often employed.

Solvation. The process of swelling, gelling, or solution of a resin by a solvent or plasticizer as a result of mutual attraction.

Solvency. Solvent action, or strength of solvent action.

Solvent. That constituent of a solution which is present in larger amount; or, the constituent which is liquid in the pure state, in the case of solutions of solids or gases in liquids.

Solvent Bonding. See SOLVENT WELDING.

Solvent Casting. A process for forming thermoplastic articles by dipping a male mold in a solution or dispersion of the resin and drawing off the solvent to leave a layer of plastic film adhering to the mold.

Solvent Cement. See ADHESIVES.

Solvent Polishing. A method for improving the gloss of thermoplastic articles by immersion in or spraying with a solvent which will dissolve surface irregularities, followed by evaporation of the solvent. The method is used primarily for cellulosics, for which dipping is suitable. Plastics which are subject to crazing such as polystyrene are usually sprayed with the solvent.

Solvent Resistance. The ability of a plastic material to withstand exposure to a solvent. Plastics vary widely in their resistance to specific solvents.

Solvents. Broadly defined as substances with the ability to dissolve other substances, solvents are used by the plastics industry in three ways. As intermediates, solvents are used in the production of many monomers and resins. In plastics processing, solvents are used in etching, welding, polishing and making laminates. Finally, solvents are widely used in adhesives, printing inks and surface coatings for plastics as well as those based on plastics and used on other materials. The major types of solvents used in all of these applications are alcohols, esters, glycol ethers, ketones, aliphatic hydrocarbons, chlorinated hydrocarbons, and nitroparaffins.

Solvent Welding. (solvent bonding, solvent cementing) The process of joining articles made of thermoplastic resins by applying a solvent capable of softening the surfaces to be joined, and pressing the softened surfaces together. Adhesion is attained by means of evaporation of the solvent, absorption of the solvent into adjacent material and/or polymerization of the solvent cement. Plastics which may be joined by this method include ABS, acrylics, cellulosics, polycarbonates, polystyrenes and vinyls.

Solvesso. Trademark for a group of aromatic petroleum derivatives, used as diluents in PVC organosols.

Somel. Du Pont's trademark for a series of polyolefin-type thermoplastic elastomers with properties midway between those of flexible plastics and conventional rubbers. They can be processed on short-cycle plastics equipment, and require no plasticizing or curing.

Soot-Chamber Test. A test for evaluating the relative effectiveness of anti-static agents in finished products such as blown polyethylene bottles. The specimens are placed in a closed cabinet and conditioned under temperature and relative humidity approximating retail shelf conditions in Winter in the Midwestern U.S. states. Then, soot is generated in an adjacent chamber by burning filter paper soaked in toluene and circulated throughout the test chamber by a fan. The specimens are then inspected visually and assigned ratings from 1 to 10 to indicate their relative cleanliness.

Sorption. The binding of one substance to another by any mechanism, such as adsorption, absorption or persorption.

Soybean Meal. The product of grinding soybean residue after extraction of its oil, sometimes treated with formaldehyde to reduce moisture absorption. It is used as a filler, often in conjunction with wood flour, in thermosetting resins.

Soybean Oil. A pale yellow oil extracted from soybeans, used in epoxidized form as plasticizer-stabilizers for vinyl resins.

Spandex. Generic name for a manufactured fiber in which the fiber-forming substance is a long chain synthetic polymer comprised of at least 85% of a segmented polyurethane. (Federal Trade Commission). These fibers are used in garments to enhance elasticity.

Spangles. See GLITTER.

Spanishing. A printing process similar to valley printing. Ink is deposited on the bottoms and sides of depressed areas of an embossed plastic film.

SPE. Abbreviation for Society of Plastics Engineers.

Specific Adhesion. Adhesion between two surfaces which are held together by valence forces of the same type as those which give rise to cohesion, as opposed to mechanical adhesion in which the adhesive holds the parts together by interlocking action.

Specific Gravity. The ratio of the weight of a given volume of a substance to that of an equal volume of water at the same temperature. The temperature selected varies among industries, 20° and 23 °C being the usual standards. In analytical work when corrections are made for the effects of air buoyancy, the term *absolute specific gravity* is used. The term *apparent specific gravity* is used to denote the specific gravity of a porous solid when the volume used in the calculations is considered to exclude the permeable voids. The term *bulk specific gravity* denotes specific gravity measurements in which volume of a solid includes both the permeable and impermeable voids. The ASTM test for specific gravity and density of plastics by displacement is D 792-66, found in Part 35 of the 1976 Book of ASTM Standards. See also DENSITY, BULK FACTOR.

Specific Heat. The ratio of the thermal capacity of a substance to that of water at 15 °C; the amount of heat required to raise a specified mass by one unit of a specified mass by one unit of a specified temperature, usually expressed as Btu/lb/ °F or cal/g/ °C.

Specific Inductive Capacity. See PERMITTIVITY.

Specific Insulation Resistance. See VOLUME RESISTIVITY.

Specific Viscosity. The specific viscosity of a polymer solution of known concentration is equal to the relative viscosity of the same solution minus one. it represents the increase in viscosity that may be attributed by the polymeric solute. See also DILUTE SOLUTION VISCOSITY.

Specific Volume. The volume of a unit of weight of a material; the reciprocal of density. Specific volume is expressed in cubic feet per pound, gallons per pound, or milliliters per gram.

Spectrograph. An instrument for observing and recording a spectrum photographically.

Spectrophotometer. An instrument for measuring the brightness of the various portions of spectra. One useful application of this instrument is in the formulation of colorants to match a given sample under all types of illumination. The instrument produces a curve representing the amounts of light energy the specimen will absorb over a wide range of wave lengths. Matching of this curve assures that the developed compound will look like the specimen under any lighting condition. Another important application is in the field of quantitative chemical analysis. See also COLORIMETER.

Spectroscope. An instrument for producing and observing a spectrum visually.

Spectroscopy. (1) The study of spectra by means of instruments such as spectroscopes, spectrographs, and spectrophotometers. (2) The study of electromagnetic waves which are absorbed or emitted by substances when excited by an arc, a spark, X-rays or magnetic fields. Each element emits light of characteristic wave lengths, by which minute quantities can be detected. See also NUCLEAR MAGNETIC RESONANCE SPECTROSCOPY, ELECTRON SPIN RESONANCE SPECTROSCOPY.

Specular Gloss. The luminous fractional reflectance of a specimen at the specular direction. The ASTM method for measuring specular gloss is D 523-67.

Specular Transmittance. The transmittance value obtained when the measured transmitted flux includes only that transmitted in essentially the same direction as the incident flux. (ASTM D 883-75a).

Spew Groove. See FLASH GROOVE.

Spew Line. See PARTING LINE.

Spherulite. A rounded aggregate of radiating crystals with a fibrous appearance. Spherulites are present in most crystalline plastics. They originate from a nucleus such as a particle of contaminant, catalyst residue, or a chance fluctuation in density. They may grow through stages: first needles, then bundles and sheaf-like aggregates, and finally the spherulites. Spherulites may range in diameter from a few tenths of a micron to several millimeters.

SPI. Abbreviation for Society of the Plastics Industry.

Spider. (1) In a molding press, that part of an ejector mechanism which operates the ejector pins. (2) In extrusion, a term used to denote the membranes supporting a mandrel within the head and die assembly. (3) In rotational casting, the gridwork of metallic members supporting cavities in a multi-cavity mold.

Spider Gating. An injection mold gating system in which the cavities are fed by runners radiating from a central sprue.

Spider Lines. In blow molding, vertical marks on the parison or molded part caused by improper welding of several melt flow fronts formed by the legs with which the torpedo is fixed in the extruder head.

Spin Dyeing. (mass dyeing, dope dyeing) The process of coloring fibers or yarns by incorporating pigments or dyes in the material prior to spinning, either during or after polymerization of the material.

Spinneret. A type of extrusion die, e.g. a metal plate with many tiny holes, through which a plastic melt or solution is forced to make fine fibers and filaments. Conventionally, the spinneret holes are round and thus produce fibers of circular cross section. A recent trend has been toward the use of orifices of varied shapes designed to decrease the fiber-bundle density, giving warmth, moisture permeability and enhanced dye receptivity to the textile fabric. Filaments emerging from the spinneret may be hardened by cooling in air, water, etc., or by chemical action.

Spinning. The process of forming synthetic fibers by extruding polymers. There are three main variations of the process: *melt spinning, dry spinning* and *wet spinning*. All employ extrusion nozzles with from one to many thousands of tiny orifices, called jets or spinnerets. In melt spinning, the polymer compound is heated to melt temperature. In both wet and dry spinning the polymer is dissolved in a solvent prior to extrusion. In dry spinning the extrudate is subjected to a hot atmosphere which removes the solvent by evaporation. In wet spinning the jet or spinneret is immersed in a liquid, which either diffuses the solvent or reacts with the fiber composition. The spinning operation is often followed by stretching to orient the polymer molecules.

Spin Welding. (friction welding) A process for joining thermoplastic articles of circular cross section by rotating one part in contact with the other until sufficient heat is generated by friction to cause a melt at the interface, which solidifies under pressure when rotation is stopped to weld the articles together. The process can be performed manually in a drill press with suitable chucks to hold the parts, or can be automated by adding devices for feeding, timing, controlling stroke and pressure of the press, and ejection. See also FRICTION WELDING.

Spiral Flow Test. A method for determining the flow properties of a thermoplastic or thermosetting resin based on the distance it will flow under controlled conditions of pressure and temperature along a spiral runner of constant cross section. The test is usually performed with a transfer molding press and a test mold into which the material is fed at the center of the spiral cavity.

Spiral Mold Cooling. A method of cooling injection molds or similar molds wherein the cooling medium flows through a spiral cavity in the body of a mold. In injection molds, the cooling medium is introduced at the center of the spiral, near the sprue section, as more heat is localized in this section.

Splash. A term used for small pit-like surface defects caused by excessive water in injection molding resins.

Splay Marks. Scars or surface defects on injection moldings caused by the high velocity injection of a stream of molten material into the mold ahead of the normally advancing material front. The prematurely-injected material, especially in the case of crystalline polymers with a sharp freezing point, cool and solidify before the mold cavity is completely filled. These defects occur most frequently in the gate area, but may be washed into other areas of the cavity. Remedies are increasing the mold temperature, local heating of the gate area, reduction of ram speed, and increasing the gate area. Sometimes called *silver streaking*. Other types of defects are sometimes called splay marks, e.g. those resulting from gases or voids in the polymer melt, short-shots, or residual monomer in the resin.

Split Ring Mold. A mold in which a split cavity block is assembled in a chase to permit the forming of undercuts in a molded piece. These parts are ejected from the mold and then separated from the piece.

Spray. (n) A complete impression of an injection mold, including the molded parts with their gates and runners attached.

Spray-And-Wipe Painting. (fill-in marking) A decorating process for articles with depressed letters, figures or designs. A lacquer or an enamel is applied by spraying either the entire surface or a restricted area around the depression, then the excess wet paint is removed by buffing or wiping the raised areas.

Spray Coating. The application of a plastic coating to a substrate by means of a spray gun. The process is used for coating any material with a plastic, and for coating plastics for decorative purposes. In the latter application, masks are usually employed to apply the coating only to selected areas. See also ELECTROSTATIC SPRAY COATING, FLAME SPRAY COATING, PLASMA SPRAY COATING.

Spray Drier. A heated cylinder or conical bottomed container onto which a resin emulsion is sprayed and from which the dried resin is removed by scraping or an air blast. Spray drying is employed for emulsion-polymerized vinyl resins, and for amino and phenolic resins.

Sprayed Metal Molds. Molds made by spraying molten metal onto a master form to obtain a shell of desired thickness, which may be subsequently backed up with plaster, cement, casting, resin, etc. Such molds are used most frequently in sheet forming processes.

Spray Molding. See SPRAY-UP.

Spray-Up. A general term covering several processes using a spray gun. In reinforced plastics, the term applies to the simultaneous spraying of resin and chopped reinforcing fibers onto the mold or mandrel. In the foamed plastics field, the term refers to the spraying of fast-reacting polyurethane or epoxy resin systems onto a surface where they react to foam and cure. In both processes, resins and catalysts are usually sprayed through separate nozzles so that they become mixed externally, thus avoiding pot life problems in the spray equipment and tanks.

Spread. (n) In the adhesives industry, the quantity of adhesive applied to a unit area of a material to be adhered to another. It is usually expressed in pounds per thousand square feet of joint area.

Spread Coating. A process for coating fabrics, sheet metals and the like with fluid dispersions such as vinyl plastisol. The substrate is supported on a carrier, and the fluid material is applied to it just ahead of a blade or "doctor knife" which regulates the thickness of the coating. The deposit is then heated to fuse the coating to the substrate, often followed by embossing to impart the desired texture.

Spreader. A streamlined metal block placed in the path of flow of the plastic material in the heating cylinder of extruders and injection molding machines to spread it into thin layers, thus forcing it into intimate contact with the heating areas.

Spring Box Mold. A type of compression mold equipped with a spacing fork which prevents the loss of bottom-loaded inserts or fine details, and which is removed after partial compression.

Sprue. In an injection or transfer mold, the main feed channel that connects the mold filling orifice with the runners leading to each cavity gate. The term is also used for the piece of plastic material formed in this channel.

Sprue Bushing. A hardened steel insert in an injection mold which contains the tapered sprue hole and has a suitable seat for the nozzle of the injection cylinder. Sometimes called an *adapter*.

Sprue-Ejector Pin. See SPRUE-PULLER.

Sprue Gate. A passageway through which molten resin flows from the nozzle to the mold cavity.

Sprue Lock. In injection molding, a portion of the plastic composition which is held in the cold slug well by an undercut, used to pull the sprue out of the bushing as the mold is opened. The sprue lock itself is pushed out of the mold by an ejector pin. When the undercut occurs on the cavity block retainer plate, this pin is called the sprue ejector pin.

Sprue-Puller. A pin having a Z-shaped slot undercut in its end, by means of which it serves to pull the sprue out of the sprue bushing.

Spunbonded Sheet. A sheet structure resembling paper or fabric, made by heat-sealing webs of randomly arranged, continuous thermoplastic fibers. The materials were introduced by Du - Pont in 1968 under the tradenames "Tyvek" (for polyethylene), "Typar" (polypropylene) and "Reemay" (polyester). Favorable characteristics are non-directional tensile strength, high tear resistance, good flex life and puncture resistance. Suggested applications are for wall coverings, book covers, tags and labels, packages, industrial clothing, electrical insulation, shoe components and filter media.

Spun Roving. A heavy, low-cost glass fiber strand consisting of filaments that are continuous but doubled back on each other.

Spur. A term sometimes used for the piece of plastic formed in the sprue of an injection or transfer mold.

Squeegee. A soft, flexible roll or blade used in wiping operations, especially in SCREEN PROCESS PRINTING, which see.

Squeeze Molding. A process for making prototypes of sheet molding compounds with inexpensive tooling and very low molding pressure, to develop designs for parts that will be injection molded or die-cast from metal for production runs. An epoxy two-piece mold is prepared, details such as ribs, gussets and bosses are positioned, then the mold is filled with reinforced SMC under 20 to 30 psi until cured.

SRP. Abbreviation approved by ASTM for styrene-rubber plastics.

SS. Abbreviation for single stage. See RESOL.

Stabilizer. An agent used in compounding some plastics to assist in maintaining the physical and chemical properties of the compounded materials at suitable values throughout the processing and service life of the material and/or the parts made therefrom. An *emulsion stabilizer* serves to keep emulsions, suspensions and the like from separating. A *viscosity stabilizer* is used in vinyl dispersions to retard viscosity increase on aging. An agent used primarily to protect plastics and rubbers from deterioration by oxidation is usually called an ANTIOXIDANT, which see. The remaining, and most important, types of stabilizers are those which protect plastics from the effects of heat and light. Such effects are evidenced by (1) a change of color, progressing from a slight yellowing to blackening; (2) a progressive decrease in mechanical properties; (3) a decrease in electrical properties; and (4) undesirable surface conditions such as blisters and spew or exudation of ingredients rendered incompatible by heat or light. Stabilizers that function primarily by absorbing U.V. light are described under ULTRAVIOLET STABILIZERS. Thousands of compounds have been proposed as heat stabilizers and as combination heat and light stabilizers for various plastics. The principle classes of such compounds are: (1) Group II metal salts of organic acids (primarily the barium, cadmium and zinc salts of fatty acids and phenols, the most important group); (2) ORGANOTIN STABILIZERS, which see; (3) EPOXY STABILIZERS, which see; (4) salts of mineral acids, e.g. carbonates, sulphates, silicates, phosphites and phosphates; (5) other organic compounds of metals and metalloids, e.g. alcoholates and mercaptides. Such heat stabilizers are used nearly exclusively with vinyl resins.

Stain Resistance. The ability of a plastic material to resist staining caused by traffic, spilled foods, waxing compounds, grease deposits, and exposure to any other staining agents. In the case of plasticized PVC, the most severe staining is caused by shoe polish, ink, tobacco smoke, lipstick, nail polish, ketchup and mustard. The degree of staining can be reduced to some extent by use of certain plasticizers, e.g. butyl benzyl phthalate, diethylene glycol dibenzoate, dipropylene glycol dibenzoate and 2,2,4-trimenthyl-1,3-pentanediol monoisobutyrate benzoate.

Staking. A term sometimes used for the process of forming a head on a protruding portion of a plastic article for the purpose of holding a surrounding part in place. Ultrasonic heating has recently been found to be advantageous for heating the protrusion in this process.

Stalk. (n) A European term for SPRUE, which see.

Stamping. See DIE CUTTING.

Stannous 2-Ethylhexoate. (stannous octoate) A polymerization catalyst for urethane foams.

Staple Fibers. Short, spinnable fibers of lengths between ½ and 5 inches.

Starch, Permanent. An aqueous emulsion of a synthetic resin for application to fabrics, which upon ironing become stiff as if starched.

Starved Area. See RESIN-STARVED AREA.

Statcoulomb. The unit of electric charge in the old metric system. In the new SI, the stat-coulomb is expressed as 3.335 640 E-10 coulombs.

Static Eliminators. Mechanical devices for removing electrical static charges from plastics articles by creating an ionized atmosphere in close proximity to the surface which neutralizes the static charges. Types of static eliminators include static bars, ionizing blowers and air guns, and radioactive elements. All except the latter operate on the principle that a high-voltage discharge from the applicator to ground creates an ionized atmosphere. A recently developed device employs ceramic microspheres containing nuclear matter that emits alpha particles. A layer of the microspheres is bonded to a substrate with a resinous binder, and the laminate is installed in an aluminum housing. The device ionizes air to neutralize charges on film.

Stationary Platen. In an injection molding machine, the large front plate to which the front plate of the mold is secured. This platen does not normally move.

Steam Molding. A process for molding parts of pre-expanded beads of polystyrene, using steam as a source of heat to further expand the blowing agent in the material. The steam in most cases is contacted intimately with the beads directly or may be used indirectly to heat mold surfaces which are in contact with the beads.

Steam Plate. See FORCE PLATE.

Stearyl Methacrylate. A group name for CH_2:$C(CH_3)COO(CH_2)_nCH_3$, in which n is from 13 to 17. It is a polymerizable monomer for acrylic plastics.

Steel Rule Die. A sharp-edged shaped knife used for DIE CUTTING, which see.

Stereoblock. A regular block that can be described by one species of stereorepeating unit in a single sequential arrangement. (IUPAC)

Stereoblock Polymer. A polymer whose molecules consist of stereoblocks connected linearly. (IUPAC) See also STEREOSPECIFIC POLYMERS.

Stereograft Polymer. A polymer consisting of chains at an atactic polymer grafted to chains of an isotactic polymer. For example, atactic polystyrene can be grafted to isotactic polystyrene under suitable conditions.

Stereoregular Polymer. A regular polymer whose molecules can be described by only one species of stereorepeating unit in a single sequential arrangement. A stereoregular polymer is always a tactic polymer, but a tactic polymer need not have every site of stereoisomerism defined. (ISO) See also ISOTACTIC, SYNDIOTACTIC.

Stereorepeating Unit. A configurational unit repeating unit having defined configuration at all sites of stereoisomerism in the main chain of a polymer molecule. (IUPAC)

Stereoselective Polymerization. Polymerization in which a polymer molecule is formed from a mixture of stereoisomeric monomer molecules by incorporation of only one stereospecific species. (ISO)

Stereospecific. (adj.) Implies a specific or definite order of arrangement of molecules in space. This ordered regularity of the molecules in contrast to the branched or random arrangement found in other plastics permits close packing of the molecules and lead to high crystallinity (i.e., as in polypropylene).

Stereospecific Polymers. Polymers whose molecular chains are arranged in STEREOSPECIFIC (which see) form. There are five types of stereospecificity, namely the cis, trans-, isotactic-, syndiotactic and tritactic structures. These structures are most commonly produced in polyolefins, by means of Ziegler and Natta catalysts derived from a transition metal halide and a metal alkyl, etc.

Stiffness. The capacity of a material to resist elastic displacement under stress.

Stir-In Resin. (dispersion resin, paste resin) A vinyl resin which does not require grinding or extremely high shear mixing to effect dispersion in a plasticizer to form a plastisol or organosol.

Stitching. (stitch welding) The progressive welding of thermoplastic materials by successive applications of two small mechanically operated electrodes, connected to the output terminals of a radio frequency generator, using a mechanism similar to that of a normal sewing machine.

Stoddard Solvent. A petroleum distillate comprising 44% naphthenes, 39.8% paraffins and 16.2% aromatics, used as a diluent in PVC organosols.

Stoichiometric. Pertaining to a mixture of chemical reactants, each element of which is present in the exact amount necessary to complete a reaction without an excess of any reactant. For example, each ingredient of a urethane foam formula should be present in a stoichiometric quantity in order to assure a consistently uniform product.

Stoke. The c.g.s. unit of kinematic viscosity. It is obtained by dividing the absolute viscosity of a fluid by the density of the fluid. A centistoke is one one hundredth of a stoke. In the new SI, the stoke is to be expressed as meters ^{2} per second, the conversion factor being one stoke times $1.000\ 000{*}E{-}04$ equals one m^2/s.

Storage Life. See SHELF LIFE, WORKING LIFE.

Stormer Viscometer. An instrument consisting of a small paddle operated through a rotor which is actuated by the force of falling weights of various sizes. The paddle is sized to fit in a one pint, friction top paint can. The weight in grams which produces 100 revolutions in 30 seconds is the viscosity in Krebs units. Use of the Stormer Viscometer is described in ASTM D 562.

Strain. In tensile testing, the ratio of the elongation to the gage length of the test specimen, that is, the change in length per unit of original length. The term is also used in a broader sense to denote a dimensionless number that characterizes the change in dimensions of an object during a deformation or flow process.

Strain Gages. Small metallic grid elements which can be adhered to the surface of a plastic article to measure the deformation occurring in the plastic immediately below each gage. The deformation causes a change in electrical resistance of the metallic grid proportional to the amount of deformation, which difference is measured with a sensitive galvanometer.

Strain Relaxation. See CREEP.

Strands. Primary bundles of continuous filaments (or slivers) combined in a single compact unit without twist. The number of filaments in a strand is usually 52, 102 or 204.

Stress. The force producing or tending to produce deformation in a body measured by the force applied per unit area. In the old Cgs system the unit of stress is one dyne per square centimeter. In the new SI, the unit of stress is the pascal, equal to 10 dynes per square centimeter.

Stress Concentration. The magnification of the level of an applied stress in the region of a notch, void or inclusion.

Stress Corrosion. The preferential attack of areas under stress in a corrosive environment, when the environment alone would not have caused corrosion.

Stress-Crack. (n) External or internal cracks in a plastic caused by tensile stresses less than that of its short-time mechanical strength. Note: The development of such cracks is frequently accelerated by the environment to which the plastic is exposed. The stresses which cause cracking may be present internally or externally or may be combinations of these stresses. (ASTM D 883-75a).

Stress Relaxation. The decay of stress at a constant strain. If a strained plastic is held under stress and the recovery of the strain is prevented, the chain segments of the molecules will make an effort to realign themselves in order to lower the free energy of the system, often by means of breakage of covalent bonds. The elastic and retarded strains which generally recover upon removal of stress are converted into non-recoverable strains when stress is maintained and chain segments rearrange themselves. The strains induced in processing many thermoplastics at elevated temperatures do not completely recover before cooling and become frozen into the material. At ambient temperatures these strains recover slowly, retarded by internal stresses. The decrease in stress is often plotted against time. The greater the relaxation achieved the better will be the dimensional stability of the finished product.

Stress Rupture. The sudden, complete failure of a plastic specimen held under a definite constant load for a given period of time at a specific temperature. Loads may be applied by tensile, bending, flexural, biaxial or hydrostatic methods.

Stress-Strain Diagram. The curve plotting the applied stress on a test specimen in tension versus the corresponding strain. The test is usually carried out at a constant rate of elongation or strain.

Stress Wrinkles. Distortions in the face of a laminate caused by uneven web tensions, slowness of adhesive setting, selective absorption of the adherends, or by reaction of the adherends with materials in the adhesive.

Stretch Blow Molding. See BLOW MOLDING.

Stretch-Film Wrapping. A packaging process used mainly to wrap pelletized multiple-package loads. Films of PVC or polyethylene are stretched over the load by one of several methods, including winding with film wider than the load, spiral winding with narrower film, or pushing the load against a tensioned web. The film is capable of stretching from 10 to 30%, and maintains tension on the load elements.

Stretch Forming. A plastic sheet forming technique in which the heated thermoplastic sheet is stretched over a mold and subsequently cooled. See also DRAPE FORMING, THERMO-FORMING.

Stretching. See ORIENTATION.

Striae. Surface or internal thread-like inhomogeneities in transparent plastic. (ASTM D 883-65T).

Striation. In blow molding, a rippling of thick parisons caused by a local orientation effect in the melt imparted by the spider legs.

Strippable Coatings. Temporary coatings applied to finished articles to protect them from abrasion or corrosion during shipment and storage, which can be removed when desired without damage to the substrate. Vinyl plastisols, applied by dipping, spraying or roll coating, are often used for this purpose.

Stripper-Plate. A plate that strips a molded piece from core pins or force plugs. The plate is actuated by the opening of the mold.

Stripping Fork. A tool, usually of brass or laminated sheet, used to remove articles from the mold. Also called *comb*.

Structural Foam. A term originally used for cellular thermoplastic articles with integral solid skins having high strength-to-weight ratios, but now sometimes also used for high density cellular plastics which are strong enough for structural applications. In the original Union Carbide process called structural foam molding, pellets of resin containing a blowing agent are fed into an extruder provided with an accumulator, where the melt is maintained above the foaming temperature but at a pressure high enough to prevent foaming. A piston in the accumulator cylinder forces a measured charge of molten resin into the mold, the volume of the charge being only about half of the mold volume, but the expanding gas rapidly swells the charge to completely fill the cavity. As the foam flows through the mold the cells in contact with the mold surfaces collapse to form a solid skin. Parts produced by this process are from three to four times as rigid as injection moldings of the same weight. Another process (the Engelit process, licensed by Phillips Petroleum) employs a heated turntable to melt the resin, which is then scraped into the hopper of an extruder. The blowing agent and any desired pigment are introduced into the extruder barrel just before the material enters the die, and the flow of material into the mold is closed off while the mold is only partially filled as in the Union Carbide process. In recent years so many variations have been developed that the process cannot be defined as a basic method. Most variations employ a form of injection molding but extrusion methods have been developed also. The trend in the art is toward elimination of swirls, by various methods of relieving pressure in the mold after the initial skin is formed. For example, in the "TM Process" developed by the Institute for Metal Science & Technology at the Bulgarian Academy of Sciences, prior to introduction of material into the mold, the empty mold is pressurized with a gas. The initial charge of foamable material fills the mold and forms an un-formed skin on the interior surface of the mold. At this time, when the skin has attained the desired thickness, the excess pressure is relieved to allow the material in the interior of the part to foam. The excess volume of expanded material escapes through the sprue and is directed towards a vertical cylinder with a plunger, from which it is returned under pressure into the mold at the end of the next molding cycle. This TM Process is said to avoid surface swirls and non-uniform cell problems encountered in earlier methods of structural foam molding.

Styrenated Alkyds. See ALKYD MOLDING COMPOUNDS.

Styrene. $C_6H_5CH:CH_2$ (vinyl benzene, phenylethylene, cinnamene, cinnamol) A colorless liquid produced from the catalytic dehydrogenation of ethylbenzene. Styrene monomer is easily polymerized by exposure to light, heat or a peroxide catalyst.

Styrene-Acrylonitrile Copolymers. (SAN) Copolymers of about 70% styrene and 30% acrylonitrile, with higher strength, rigidity and chemical resistance than straight polystyrene. They may be blended with butadiene, either as a terpolymer or by grafting onto the butadiene, to make ABS resins.

Styrene-Butadiene Rubbers. (SBR, Buna-S, GR-S) A group of widely-used synthetic rubbers comprising about 3 parts of butadiene copolymerized with 1 part of styrene with many modifications yielding a large variety of properties. The copolymers are first prepared as latices, in which form they are sometimes used. The latices can be coagulated to produce crumb-like particles resembling natural crepe rubber.

Styrene-Butadiene Thermoplastics. A new family of true elastomers that behave like thermoplastics in processing, introduced in 1965 by Shell Chemical Co. under the trade name "Thermolastic." They are believed to be linear block copolymers of styrene and butadiene, produced by lithium-catalyzed solution polymerization, with a sandwich molecular structure comprising a long polybutadiene center surrounded by shorter polystyrene ends. The materials are marketed as fully compounded pellets ready for processing by conventional extrusion, blow molding, injection molding and thermoforming processes.

Styrene Plastics. See POLYSTYRENE.

Styrene-Rubber Plastics. Polystyrene (which see) compounds containing rubber modifiers to improve impact resistance. Also called *high impact polystyrenes.*

Styrol. The name given to styrene by the chemist who first observed the monomer in 1839. The name was changed to styrene by German researchers in about 1925.

Submarine Gate. (tunnel gate) A type of edge gate where the opening from the runner into the mold is located below the parting line or mold surface as opposed to conventional edge gating where the opening is machined into the surface of the mold. With submarine gates, the item is broken from the runner system on ejection from the mold. See also GATE.

Sucrose Acetate Iso-Butyrate. (SAIB) A modifying extender for lacquers and finishes based on plastics resins such as butyrate, acrylic, alkyd and polyester. Properties such as viscosity, density, solids content and coating performance can be adjusted by partially replacing the resin with SAIB.

Sucrose Benzoate. $C_{12}H_{14}O_3(C_6H_5COO)_7$ A plasticizer for polystyrene, cellulosics and some vinyls.

Sucrose Octaacetate. $C_{12}H_{14}O_3(OCOCH_3)_8$ A plasticizer for cellulosic and polyvinyl acetate resins.

Sulfide Staining. A discoloration of a plastic caused by the reaction of one of its constituents with a sulfide in a liquid, solid or gas to which the plastic article is exposed. Stabilizers based on salts of lead, cadmium, antimony, copper or other metals sometimes react with external sulfides to form a staining metallic sulfide.

Sulfonate-Carboxylate Copolymers. See POLYSULFONATE COPOLYMERS.

Sulfonylazidosilanes. A new family of organofunctional coupling agents that, in contrast to conventional silane coupling agents, enter into direct chemical reaction with organic polymers. They function by insertion into carbon-hydrogen bonds which avoids generation of free radicals and degradation of radical-sensitive polymers such as polypropylene, polyisobutylene and polystyrene.

Sulfonyldianiline. $(C_6H_4NH_2)_2SO_2$ A curing agent for epoxy resins.

Sulfur-Nitrogen Polymers. $(SN)_x$; (Polymeric sulfur nitrides, polythiazyl) First synthesized in 1910 but remaining little more than a laboratory curiosity until recently, this covalent polymer has been re-studied recently and has been found to have the optical, physical and electrical properties of a metal. It is formed by the epitaxial growth between the two solid phases, one crystal forming a preferred position for the deposition of a second crystal. $(SN)_x$ crystals are malleable and can be flattened into thin sheets or fibers under pressure, having an electrical conductivity similar to that of mercury.

Sunlight Resistance. See LIGHT RESISTANCE.

Surface Conductance. The direct current conductance between two electrodes in contact with a specimen of solid insulating material when the current is passing only through a thin film of moisture on the surface of the specimen.

Surface Mat. A thin mat of fine fibers used primarily to produce a smooth surface on a reinforced plastic. (ASTM D 883-75a).

Surface Pins. See EJECTOR RETURN PINS.

Surface Resistance. The surface resistance between two electrodes in contact with a material is the ratio of the voltage applied to the electrodes to that portion of the current between them which flows through the surface layers.

Surface Resistivity. The ratio of the potential gradient parallel to the current along the surface of a material to the current per unit width of the surface. Surface resistivity is numerically equal to the surface resistance between opposite sides of a square of any size when the current flow is uniform.

Surface Tension. Two fluids in contact exhibit phenomena, due to molecular attractions, which appear to arise from a tension in the surface of separation. It may be expressed as dynes per cm or as ergs per sq. cm. One method of measuring surface tension is by means of a capillary tube. If a liquid of density d rises a height h in a tube of internal radius, r, the surface tension is equal to rhdg/2. The result will be in dynes per cm if r and h are in cm, d in grams per cm^3 and g in cm per sec^2. In the new SI, surface tension is to be expressed in Newton meters, the conversion factor being one dyne per centimeter times 1×10^{-7} equals one Newton meter.

Surface Treating. Any method of treating a surface to render it more receptive to adhesives, paints, inks, lacquers or to other surfaces in laminating processes. See CORONA DISCHARGE TREATMENT, FLAME TREATING, CASING, ELECTRONIC TREATING, PLASMA ETCHING, ION PLATING, IRRADIATION.

Surfacing Mat. A very thin mat, usually 7 to 20 mils thick, of highly filamentized glass fiber used primarily to produce a smooth surface on a reinforced plastic laminate.

Surfactant. A widely used contraction of *surface active agent,* a compound that alters surface tension of a liquid in which it is dissolved.

Surging. In extrusion, an unstable pressure buildup leading to variable output and waviness of the surface of the extrudate. In extreme cases, the flow of extrudate may even cease momentarily at intervals. Surging may be caused by pulsations in motor load, instability of the flux point within the length of the screw, improper type or adjustment of barrel or die temperatures, improper location of thermocouples, and poor mixing in the screw.

Surlyn A. See IONOMER.

Suspension. A liquid medium with fine particles of any solid dispersed therin. The particles are called the *disperse phase,* and the suspending medium is called the *continuous phase.* When the particles do not settle out and are small enough to pass through ordinary filters, the suspension is called a *colloid* or *colloidal suspension.* In the plastics field, a suspension is essentially synonymous with dispersion. See also DISPERSION, EMULSION.

Suspension Polymerization. (pearl-, bead-, granular polymerization) A polymerization process in which the monomer or mixture of monomers is dispersed by mechanical agitation in a second liquid phase, usually water, in which both the monomer and polymer are essentially insoluble. The monomer droplets are polymerized while maintained in dispersion by continuous agitation. Polymerization initiators and catalysts used in the process are generally soluble in the monomer. According to the type of monomer, emulsifier, protective colloid and other modifiers used the resulting polymers may be in the form of pearls, beads, soft spheres or irregular granules, which are easily separated from the suspending medium when agitation is ceased. Suspension polymerization is used primarily for PVC, PVAc, PMMA, PTFE and polystyrene.

Sweating. Exudation of small drops of liquid, usually a plasticizer or softener, on the surface of a plastic part.

Swirl. A term applied to visual and tactile surface roughness sometimes obtained in the structural foam molding process. It results from high-speed fill causing surface wrinkling as the polymer melt flows along the wall of the mold. The condition can be eliminated by measures such as increasing the fill time and mold temperature. Viscosity reduction may also help.

Sylvic Acid. See ABIETIC ACID.

Syndiotactic. Derived from the Greek words *syndio,* meaning "every other" and *tatto,* meaning "to put in order," the term syndiotactic is sometimes used to denote a polymer structure in which monomer units attached to the polymer backbone alternate in a-b-a-b- fashion on one side of the backbone and, if present on the other side, are arranged in b-a-b-a fashion. See also ISOTACTIC.

Syndiotactic Polymer. A regular polymer whose molecules can be described by alternation of configurational base units that are enantiomeric. Note: In a syndiotactic polymer the configurational repeating unit consists of two configurational base units that are enantiomeric. (IUPAC, ISO)

Syneresis. The contraction of a gel upon standing, usually accompanied by the separation of a liquid.

Synergism. A phenomenon wherein the effect of a combination of two additives is greater than the effect that could be expected from the known performance of each additive used singly. For example, some stabilizers for plastics have a mutually reinforcing effect when used together, and thus are called synergistic.

Syntactic. (adj.) A term derived from the Greek word *syntaxis,* meaning orderly arrangement.

Syntactic Foams. A term applied to composites of tiny, hollow spheres and a resin or plastic material. The spheres are usually of glass, although phenolic microspheres were used in the early years of the art. The resin most widely used is epoxy, followed by polyesters, phenolics and PVC. Syntactic foams of the most usual type, glass microspheres in a binder of high-strength thermosetting resin, are made by mixing the spheres with the fluid resin, its curing agent and other additives, to form a fluid mass that can be cast into molds, trowelled onto a surface or incorporated into laminates. After forming, the mass is cured by heating. These foams are characterized by low density, ranging from 36 to 42 pounds per cubic foot; and very high compressive strength. Their first applications were for deep-submergence buoys capable of withstanding ocean depths of 20,000 feet. When both gas bubbles and hollow glass spheres are used in the same mixture, the resulting composite has been called a *diafoam.*

Syrups. See CASTING SYRUPS.

Tab Gate. A small removable tab of approximately the same thickness as the molded item, usually located perpendicularly to the item. The tab is used as a site for edge gate location, usually on items with large flat areas. See also GATE.

TAC. Abbreviation for TRIALLYL CYANURATE, which see.

Tack. The stickiness of an adhesive, measureable as the force required to separate an adherend from it by viscous or plastic flow of the adhesive.

Tackifier. A substance such as a rosin ester which is added to synthetic resins or elastomeric adhesives to improve the initial and extended tack range of the deposited adhesive film.

Tack Range. The period of time in which an adhesive will remain in the tacky-dry condition after application to an adherend, under specified conditions of temperature and humidity.

Tactic Block. A regular block that can be described by only one species of configurational repeating unit in a single sequential arrangement. (IUPAC)

Tactic Block Polymer. A polymer whose molecules consist of tactic blocks connected linearly. (IUPAC)

Tacticity. (1) The orderliness of the succession of configurational repeating units in the main chain of a polymer molecule. (IUPAC, ISO). (2) Any type of regular or symmetrical molecular arrangement in a polymer structure, as opposed to random positioning of substituent groups along a polymer backbone. See also STEREOSPECIFIC.

Tactic Polymer. A regular polymer whose molecules can be described by only one species of configurational repeating unit in a single sequential arrangement. (IUPAC, ISO).

Take-Off. The mechanism for drawing extruded or calendered material away from the extruder or calender. The most common form of take-off is a pair of endless caterpillar belts with resilient grip pads conforming to the section being extruded, driven at a speed synchronized with that of the extrudate.

Talc. $Mg_3Si_4O_{10}(OH)_2$ (steatite, talcum, mineral graphite) A natural hydrous magnesium silicate, used infrequently as a filler.

Tandem Extruders. See EXTRUDER, TANDEM.

Tangent Modulus. The slope of the line at any point on a static stress-strain curve expressed in psi per unit strain. This slope is the tangent modulus at that point in shear, extension, or compression as the case may be.

TBT. Abbreviation for TETRABUTYL TITANATE, which see.

TCEF. Abbreviation for trichloroethyl phosphate, a plasticizer.

TCP. Abbreviation for TRICRESYL PHOSPHATE, which see.

TDI. Abbreviation for toluene diisocyanate, an 80-20 mixture of 2,4- and 2,6-toluene diisocyanate isomers. See also DI-ISOCYANATES.

T-Die. A center-fed, slot extrusion die for film which, in combination with the die adapter, resembles an inverted T.

Teflon® . Du Pont's trademark covering all of its fluorocarbon resins, including PTFE, FEP, and various copolymers.

Telomer. An addition polymer, usually of low molecular weight, in which the growth of molecules is terminated by a radical-supplying chain transfer agent. The term is also used synonymously with *oligomer,* meaning simply a polymer with a very few (two to ten) repeating units.

Tenacity. A term used in yarn and textile manufacturing to denote the strength of a yarn or filament of given size. Numerically it is the grams of breaking force per denier unit of yarn or filament size. (gpd) The yarn is usually pulled at the rate of 12 inches per minute.

Tensile Heat Distortion Temperature. See HEAT DISTORTION POINT.

Tensile Impact Test. A test similar to the Izod test, except that the specimen is clamped in a fixture attached to the swinging pendulum and is ruptured by tensile stresses as it strikes an anvil. The test is described in ASTM D 1822-61T.

Tensile Modulus. See MODULUS OF ELASTICITY.

Tensile Product. The product of tensile strength and elongation at break, usually divided by 10,000; or the product of tensile strength and gage length plus deformation at break. The latter definition is an approximation (assuming no volumetric change) to actual rupture stress.

Tensile Strength. The maximum tensile stress sustained by the specimen during a tension test. The result is usually expressed in pounds per square inch, the area being that of the original specimen at the point of rupture rather than the reduced area after break. The ASTM test for tensile properties of plastics is D 638-72, found in Part 35 of the 1976 Annual Book of ASTM Standards. Note: In the new SI, results are to be expressed in kilograms per square centimeters of area.

Tentering. The process of stretching, the name being derived from the tenter frame used by housewives to stretch curtains and the like on frames with pins. In the plastics field, the term is used in connection with film ORIENTATION (which see) when the orientation is performed by tenter-like conveyors moving the film through heating chambers.

TEP. Abbreviation for TRIETHYL PHOSPHATE, which see.

Tera-. (T) The SI-approved prefix for a multiplication factor to replace 10^{12}.

Terephthaldehyde Resins. See POLYESTERS, SATURATED.

Terephthalic Acid. $C_6H_4(COOH)_2$ (TPA, paraphthalic acid, benzene-para-dicarboxylic acid)

White crystals or powder, used in the production of alkyd resins and polyethylene terephthalate.

Terpene Resins. See POLYTERPENE RESINS.

Terpolymer. The product of simultaneous polymerization of three different monomers, or of the grafting of one monomer to the copolymer of two different monomers. An example of a terpolymer is ABS resin, derived from acrylonitrile, butadiene and styrene.

Terra Alba. $CaSO_4.2H_2O$ A finely powdered form of gypsum, used as a filler.

Terra Ponderosa. See BARIUM SULPHATE.

TESLA. The SI unit of magnetic flux density given by a magnetic flux of one weber per square meter.

Tetrabasic Lead Fumarate. $4PbO.PbC_2H_2(COO)_2.2H_2O$ A creamy-white powder used as a heat stabilizer for vinyl phonograph records, electrical grade plastisols and insulation. It is also used as a curing agent for chlorosulfonated polyethylene.

Tetrabromobisphenol A. $C_{15}H_{12}Br_4O_2$ An off-white crystalline solid, used as a flame retardant in epoxy resins, polyesters and polycarbonates.

Tetrabromophthalic Anhydride. A reactive intermediate containing 69% bromine, used as a flame retardant. It is sold by Michigan Chemical as Firemaster PHT4.

2,2′,6,6′-Tetrabromo-3-3′,5,5′-Tetramethyl-4,4′-Dihydroxydiphenyl. (TTB) A recently developed aromatic brominated flame retardant synthesized easily by a two-step process from 2,6-dimethylphenol. The unusual chemical structure of TTB enables its use as both a reactive and additive flame retardant. It has been used in high-impact polystyrenes.

Tetrabutyl Titanate. $Ti(OC_4H_9)_4$ (TBT, butyl titanate, titanium butylate) A catalyst for condensation and cross-linking reactions, also used to improve the adhesion of plastics compounds to metals.

Tetrachlorobisphenol A. $C_{15}H_{12}Cl_4O_2$ A monomer for flame retardant epoxy, polyester and polycarbonate resins.

Tetrachloromethane. See CARBON TETRACHLORIDE.

Tetraethylene Glycol Dicaprylate. $(C_7H_{15}COOCH_2CH_2OCH_2CH_2)_2O$ A plasticizer for vinyl chloride polymers and copolymers.

Tetraethylene Glycol Di(2-Ethylhexoate). A secondary plasticizer for vinyl resins and a primary plasticizer for cellulose plastics and synthetic rubbers. In vinyls, it is used as 15 to 30 percent of the total plasticizer to impart good low temperature flexibility. In nitrocellulose lacquers, it improves the cold-check resistance.

Tetraethylene Glycol Monostearate. $C_{17}H_{35}COO(CH_2CH_2O)_4H$ A plasticizer for ethyl cellulose and cellulose nitrate.

Tetrafluoroethylene. $F_2C:CF_2$ (TFE, perfluoroethylene) A colorless gas derived by passing chlorodifluoromethane through a heated tube, used as the monomer for polytetrafluoroethylene resins.

Tetrahydrofuran. C_4H_8O (THF, tetramethylene oxide) A colorless liquid obtained by the catalytic hydrogenation of furan. In addition to its many uses as an industrial intermediate, THF is a powerful solvent for PVC and polyvinylidene chloride. Its presence in relatively small amounts increases the "bite" of vinyl printing inks, lacquers and adhesives. In recent years, THF has been polymerized to polytetramethylene ether glycol for use in the production of polyurethanes.

Tetrahydrofurfuryl Oleate. $C_{17}H_{35}COOCH_2.OC_4H_7$ A plasticizer for polystyrene, cellulosic plastics, acrylic and vinyl resins. In vinyls it is used as a secondary plasticizer with low temperature resistance, and as a lubricant in stiff or highly filled calendering and extrusion compounds.

Tetramer. A molecule formed by uniting four different simple molecules.

1,2,4,5-Tetramethylbenzene. See DURENE.

1,1,3,3-Tetramethylbutyl Peroxy-2-Ethyl-Hexanoate. An organic peroxide catalyst marketed by Lucidol Div., Pennwalt Corp., under the trade mark Lupersol 259. It is superior to benzoyl peroxide as a catalyst for polyesters and, being a liquid, is easier to handle.

Tetramethylethylenediamine. (TMEDA) A colorless, anhydrous liquid used as a catalyst for urethane foams, coatings and elastomers, and as a curing agent for epoxy resins.

5,5′-Tetramethylene Di(1,3,4-Dioxazol-2-One). (Adipodinitrile carbonate, ADNC) A white crystalline solid, capable of being reacted with diols and polyols to form light-stable urethane coatings, elastomers and foams.

Tetramethylene Glycol. See 1,4-BUTYLENE GLYCOL.

Tetramethylthiuram Disulfide. $[(CH_3)_2NCS]_2S_2$ A white crystalline powder used as a fungicide, bacteriostat and rodent repellent in vinyl compounds.

Tex. A basic unit of linear density of a fiber; the weight in grams of a fiber one kilometer in length. See also CUT, DENIER, GREX NUMBER.

TFE. Abbreviation for TETRAFLUOROETHYLENE, which see.

Tg. A symbol for the temperature of GLASS TRANSITION, which see.

TGA. Abbreviation for THERMOGRAVIMETRIC ANALYSIS, which see.

"Therimage." A trademark for a decorating process for plastics which transfers the image of a label or decoration to the object under influence of heat and light pressure.

Thermal Black. See CARBON BLACK.

Thermal Capacity. The quantity of heat necessary to produce a unit change of temperature in a unit mass of a substance. In the old cgs system, the unit is calories per degree Centigrade. In the new SI, the unit is to be the joule per kilogram/ °K, the conversion factor being one g/ °C times $4.186\ 800*E+03 = J/Kg/ °K$.

Thermal Conductivity. The rate at which heat is transferred by conduction through a unit cross-sectional area of a material when a temperature gradient exists perpendicular to the area. The coefficient of thermal conductivity, sometimes called the k-factor, is expressed as the quantity of heat that passes through a unit cube of the substance in a given unit of time when the difference in temperature of the two faces is 1°. The ASTM test for thermal conductivity of dry specimens of insulating, building and other materials is C 111-71, "Thermal Conductivity of Materials by means of the Guarded Hot Plate."

Thermal Decomposition. Decomposition resulting from action by heat. It occurs at a temperature at which some components of the material are separating or reacting together, with a modification of the macro- or microstructure.

Thermal Diffusivity. A measure of the rate at which a temperature disturbance at one point in a body travels to another point. It is expressed by the relationship K/dC_p, where k is the coefficient of thermal conductivity, d is the density, and C_p is the specific heat at constant pressure.

Thermal Expansion Coefficient. See COEFFICIENT OF THERMAL EXPANSION.

Thermal Fluids. Heat-stable, non-corrosive liquids such as oils and glycols which are used in heat transfer equipment. Examples of applications in the plastics industry are jacketed molds for rotational casting, heating of calenders, and maintenance of temperature in storage tanks.

Thermal Gravimetric Analysis. See THERMOGRAVIMETRIC ANALYSIS.

Thermal Impulse Sealing. See IMPULSE SEALING.

Thermally Foamed Plastic. A cellular plastic produced by applying heat to effect gaseous decomposition or volatilization of a constituent. (ASTM D 883-65T).

Thermal Polymerization. A polymerization process performed solely by heat, in the absence of a catalyst. Monomers such as styrene and methyl and methacrylate are examples of those which can be thermally polymerized.

Thermal Sealing. (thermal heat sealing) The method of bonding two or more layers of plastics by pressing them between heated dies or tools which are maintained at a relatively constant temperature. See also HEAT SEALING.

Thermal Stability. See HEAT STABILITY.

Thermal Stress Cracking. (TSC) Crazing and cracking of some thermoplastic resins which results from overexposure to elevated temperatures.

Thermionic Emission. Electron or ion emission due to the temperature of the emitter. The rate of emission increases rapidly with an increase of temperature of the emitter. It is also very sensitive to the state of the surface.

Thermobande Welding. A variation of the hot plate welding method. A metallic tape acting as a resistance element is adhered to the material to be welded. Low voltage is applied to heat the material to softening temperature.

Thermocompression Bonding. The joining together of two materials without an intermediate material by the application of pressure and heat in the absence of an electrical current.

Thermocouple. A pair of two dissimilar metal wires welded together at one end, which when heated at the welded junction generates a feeble electrical current through a circuit connected to the opposite ends of the wires. The current strength varies according to the temperature, and thus can be measured with a millivoltmeter calibrated in degrees of temperature. The complete system comprising a thermocouple and the detecting instrument is called a *pyrometer*. This type of pyrometer is widely used for indicating and controlling temperatures of extrusion and molding machines in the plastic industry.

Thermodynamic Temperature. See KELVIN.

Thermoelasticity. (n) Rubber-like elasticity (exhibited) by a rigid plastic resulting from an increase in temperature. Note: Retention of the desired shapes may be achieved by cooling in place after forming. In the case of thermosetting materials, prolonged heating may be necessary to effect cure in place. (ASTM D 883-65T).

Thermoform. The product which results from a thermoforming operation.

Thermoforming. The process of forming a thermoplastic sheet into a three-dimensional shape by clamping the sheet in a frame, heating it to render it soft and flowable, then applying differential pressure to make the sheet conform to the shape of a mold or die positioned below the frame. When the pressure is applied entirely by vacuum, the process is called *vacuum forming*. When air pressure is employed to partially preform the sheet prior to application of vacuum the process becomes *air-assist vacuum forming*. In another variation, mechanical pressure is applied by a plug to partially preform the sheet (*plug assist forming*). In the *drape forming* modification, the softened sheet is lowered to drape over the high points of a male mold prior to application of vacuum. Still other modifications are *plug-and-ring-forming* (using a plug as the male mold and a ring matching the outside contour of the finished article); *ridge forming* (the plug is replaced with a skeleton frame); *slip forming* or *air slip forming* (the sheet is held in pressure pads which permit it to slip as forming progresses); and *bubble forming* (the sheet is blown by air into a blister, then pushed into a mold by means of a plug). The term thermoforming also includes methods employing only mechanical pressure, such as *matched mold forming,* in which the hot sheet is formed between registered male and female molds. Corrugated sheets are produced by this method. A recent variation of the process, introduced in 1977, is known as *scrapless thermoforming.* Designed to reduce the amount of scrap inherent in previous thermoforming methods, the process starts with a blank, usually square, of the material to be thermoforming. This blank is forged in a press to a circular preform, which can be thermoformed to a bowl or other shape without scrap. See also BLISTER PACKAGING and SKIN PACKAGING.

Thermogram. A curve plotting weight loss of a specimen against temperature. See THERMOGRAVIMETRIC ANALYSIS.

Thermographic Transfer Process. A modification of the hot stamping process wherein the design to be transferred is first printed on a film, from which it is transferred to the plastic part by means of heat and pressure.

Thermogravimetric Analysis. (TGA) A testing procedure in which changes in weight of a specimen are recorded as the specimen is progressively heated. A typical apparatus consists of an analytical balance supporting a platinum crucible for the specimen, the crucible situated in an electric furnace, and means for plotting the weight change as a function of temperature. Some TGA tests are conducted in air, others in controlled, successive atmospheres such as nitrogen for a first stage, followed by air for a second stage. Thermogravimetric curves thus obtained (thermograms) provide useful information regarding polymerization reactions, the efficiencies of stabilizers and activators, the thermal stability of final materials, and direct analysis. TGA is also used for evaluating the heat performance of plasticizers and stabilizers in PVC.

Thermolastic. A trade name of Shell Chemical Co. See STYRENE-BUTADIENE THERMOPLASTICS.

Thermoplastics. Resins or plastics compounds which in their final state as finished articles are capable of being repeatedly softened by increase of temperature and hardened by decrease of temperature by means of physical changes. Examples of thermoplastics are the following resins: ABS, acetal, acrylic, cellulosic, chlorinated polyether, diallyl phthalate, fluorocarbons, phenoxy, polyamides (nylons), polycarbonate, polyethylene, polypropylene, polystyrene, some types of polyurethanes, PVC and other vinyls.

Thermoplastic Elastomers. The family of polymers that resemble elastomers in that they can be repeatedly stretched without distortion of the unstressed part shape, but are true thermoplastics and thus do not require curing or vulcanization as do rubber-like elastomers. See ELASTOMERS for examples.

Thermoplastic Polyesters. See POLYBUTYLENE TEREPHTHALATE, POLYTETRAMETHYLENE TEREPHTHALATE.

Thermoplastic Rubbers. See ELASTOMER.

Thermosetting Plastics. (thermosets) Resins or plastic compounds which in their final state as finished articles are substantially infusible and insoluble. Thermosetting resins are often liquids at some stage in their manufacture or processing, which are cured by heat, catalysis or other chemical means. After being fully cured, thermosets cannot be resoftened by heat. Some plastics which are normally thermoplastic can be made thermosetting by means of crosslinking with other materials. Examples of thermosetting plastics are alkyd, allyl, amino, epoxy, furane, phenolic, polyacrylic esters, polyesters, protein and silicone resins.

Theta Solvent. A solvent which performs in an ideal manner (activity coefficient $= 1$) in dilute solution measurements of molecular weight.

Theta State. A term introduced by Flory to describe the condition in a polymer solution in which there is little interaction between the molecules of the solvent and those of the polymer, and in which the polymer molecules exist as statistical coils.

Theta Temperature. (Flory temperature) With respect to molecular interactions in dilute polymer solutions, theta temperature is the temperature at which the second virial coefficient disappears. That is, the temperature at which the coiled plymer molecules expand and tend to their full contour lengths and become rod shaped.

THF. Abbreviation for TETRAHYDROFURAN, which see.

Thickening Agents. (anti-sag agents) Substances which increase the viscosity of and/or impart thixotropy to fluid dispersions or solutions. They are used widely in coatings and paints to prevent their flowing or slumping while setting to their final form. Examples of thickening agents are bentonite, calcium carbonates with high oil absorption, clays, chrysotile asbestos, hydrated siliceous minerals, magnesium oxide, soaps, stearates, and special organic waxes.

Thickness Gaging. Many calendered, extruded and cast products must be measured in thickness during their manufacture in order to adjust machines to maintain the thickness within specified tolerances. The simplest methods, called *contact gaging,* use elements such as calipers, micrometers and rolls which physically touch the product being measured. The more advanced methods, classified as *non-contact gaging,* yield continuous data which can be fed back to the machine to make the necessary adjustments automatically. Non-contact gaging devices employ nuclear radiation such as X-rays and Beta rays (see BETA GAGE), infrared radiation, air nozzles with means for measuring back-pressure which varies with product thickness, electrical capacitance sensors, and optical devices employing beams of light.

Thin-Layer Chromatography. See CHROMATOGRAPHY.

Thinner. A fluid substance incapable of dissolving a resin but which can partly substitute for a solvent and, at the same time, reduce the viscosity of a paint, varnish, lacquer or adhesive. See also DILUENT.

2,2′-Thiobis(4-Tert-Octylphenolato)-n-Butyl Amine Nickel. (Cyasorb UV 1084) A UV absorber used in polyolefins for items such as agricultural film wherein good weatherability is required.

Thiokol. Trademark of Thiokol Chemical Company for a line of polysulfides and similar materials. See POLYSULFIDES.

Thiourea-Formaldehyde Resin. A member of the amino plastics family, made by condensation of thiourea ($[NH_2]_2CS$) with formaldehyde.

Thixotropic Agents. See THICKENING AGENTS.

Thixotropic Fluids. See THIXOTROPHY.

Thixotropy. A flow characteristic evidenced by a decrease in viscosity of a fluid when it is stirred at a constant or increasing rate of shear. When the stirring or shearing is discontinued, the apparent viscosity of the fluid gradually increases back to the original value. Changes in both directions are dependent on time as well as shear.

Thread Plug. The part of a mold that shapes an internal thread and must be unscrewed from the finished piece.

Three Plate Mold. An injection mold with an intermediate, movable plate which permits center or offset gating of each cavity.

Threshold Limit Values. Parts of vapor or gas per million parts of air by volume (PPM), or milligrams of particulate material per cubic meter or air (Mg/cu.m.) to which workers may be exposed under certain conditions. Threshold limit values for many materials are published by the American Conference of Governmental Industrial Hygienists (ACGIH). However, it is stipulated that these limits are intended for use by persons trained in the field of industrial hygiene.

Throwing. A textile term referring to the act of imparting twist to a yarn, especially while plying and twisting together a number of yarns; a throwster.

Tie Bars. In plastic molding presses, bars which provide structural rigidity to the clamping mechanism often used to guide platen movement.

Tin Stabilizers. See ORGANOTIN STABILIZERS.

TIOTM. Abbreviation for TRI-ISO-OCTYL TRIMELLITATE, which see.

Titanate Couplers. A family of monoalkoxy titanates developed by Kenrich Petrochemicals, Inc. They are especially useful in conjunction with mineral-type fillers and flame retardants, where they are more effective than silane coupling agents because they have three pendant organic functional groups, compared to one for silanes. The titanate couplers act as plasticizers to enable much higher filler loadings and/or to achieve better flow.

Titanium Dioxide. TiO_2 (titanic anhydride, titanic acid anhydride, titanic oxide, titanium white, titania) A white powder available in two crystalline forms; the *anatase* and *rutile* types. Both are widely used as opacifying pigments in thermosets and thermoplastics, used alone when whites are desired or in conjunction with other pigments when tints are desired. They are essentially chemically inert, light-fast, resistant to migration, and resistant to heat. The rutile form has the higher refractive index (2.75, versus 2.55 for the anatase form), and thus has the greater opacifying power.

Titanium Trichloride. $TiCl_3$ A catalyst for polymerizing olefins.

TMDI. Abbreviation for 2,2,4-TRIMETHYL-1,6-HEXANE DIISOCYANATE, which see.

TOF. Abbreviation for TRIOCTYL PHOSPHATE, which see.

Toggle Action. A mechanism which exerts pressure developed by the application of force on a knee joint. It is used as a method of closing presses and also serves to apply pressure at the same time.

Toluene. $CH_3C_6H_5$ (toluol, methylbenzene, methylbenzol) A colorless, flammable liquid with a benzene-like odor, used as a solvent for the cellulosics, vinyl organosols and other resins. Toluene is also used as an intermediate for polyurethanes and polyesters.

Toluene-2,4-Diamine. (TDA, meta-tolylene-diamine) $CH_3C_6H_3(NH_2)_2$ A colorless, crystalline material used in the production of toluene diisocyanate, a key material in the manufacture of urethanes.

Toluene-2,4-Diisocyanate. (2,4-toluene diisocyanate, meta-toluene diisocyanate, TDI) $CH_3C_6H_3(NCO)_2$ A water-white to pale yellow colored liquid with a sharp, pungent odor, produced by reacting 2,4-diaminotoluene with phosgene. It reacts with water to produce carbon dioxide. It is widely used in the production of urethane foams and elastomers, but due to its toxicity must be handled carefully to keep its concentration in air to which workmen are exposed to safe limits. See also DIISOCYANATES.

para-Toluenesulfonamide. (PTSA) $CH_3C_6H_4SO_2NH_2$ White leaflets, existing also in the ortho form, used as solid plasticizers for ethyl cellulose, polyvinyl acetate and rigid PVC.

p-Toluene Sulfonyl Hydrazide. A blowing agent marketed by Uniroyal under the trademark Celogen TSH. It is similar to benzene sulfonyl hydrazide except that the melting point and decomposition temperatures are higher.

p-Toluene Sulfonyl Semicarbazide. A blowing agent marketed by Uniroyal under the tradename "Celogen RA." Its relatively high decomposition temperature (235 °C) makes it useful for plastics that are processed at high temperatures, such as high density polyethylene, polypropylene, rigid PVC, ABS resins, polycarbonates and nylons.

2,4-Tolylene Diisocyanate. British spelling for TOLUENE-2,4-DIISOCYANATE, which see.

Toner. A coloring agent in which all pigments and vehicles are organic. See also LAKE.

Top Blowing. A blow molding process in which air is injected into the parison at the top of the mold.

Torpedo. (spreader) A streamlined metal block placed in the path of flow of the plastic material in the heating cylinder of extruders and injection molding machines, to spread it into thin layers, thus forcing it into intimate contact with the heating areas.

Torr. A unit of force per unit area used mainly in the field of vacuum technology. In the new SI system the Torr is to be replaced by the Pascal, the conversion factor being one Torr (expressed in mm of Hg at 0 °C) is equal to 1.33322×10^2 Pascals.

Torsion. Stress caused by twisting a material.

Torsional Braid Analysis. A method of performing torsional tests on small amounts of materials in states in which they cannot support their own weight, e.g. liquid thermosetting resins. A glass braid is impregnated with a solution of the material to be tested. After evaporation of the solvent, the impregnated braid is used as a specimen in an apparatus which measures motion of the oscillating braid as it is being heated at a programmed rate in a controlled atmosphere.

Torsional Tests. Tests for determining the stiffness properties of plastics, based on measuring the torque required to twist a specimen to a predetermined degree of arc. One such method is described in ASTM D 1043-51.

Tortuosity Factor. The distance a molecule must travel to pass through a film divided by the thickness of the film.

TOTM. Abbreviation for TRIOCTYL TRIMELLITATE, which see.

Toughness. A term with a wide variety of meanings, no single precise mechanical definition being generally recognized. Toughness generally implies a lack of brittleness; having very great elongation to break accompanied by high tensile strength. One proposed definition for toughness is the energy required to break a material, equal to the area under the stress-strain curve.

Tow. A textile term for a bundle of untwisted fibers. The term has been used in the plastics industry for graphite fibers used in reinforced plastics.

Toxicity. (Poisonousness) Although most pure resins and polymers are relatively nontoxic, compounding additives such as stabilizers, colorants and plasticizers must be carefully selected when products are to be used for food packaging or other applications involving body contact. See NON-TOXIC MATERIALS for a source of information about the toxicity status of resins and additives.

TPA. Abbreviation for TEREPHTHALIC ACID, which see.

TPP. Abbreviation for TRIPHENYL PHOSPHATE, which see.

TPR. Registered trade mark of Uniroyal, Inc. for a family of thermoplastic rubbers based mainly on ethylene and propylene. There are four basic grades of TPR ranging in hardness from 65 to 90 Shore A. The materials are processed by conventional thermoplastic techniques such as injection molding and extrusion. Products have the properties of vulcanized rubber.

TPS. Abbreviation used by British Standards Institution for "toughened polystyrene," equivalent to high impact polystyrene.

TR. Abbreviation used by British Plastics Institution for thio rubber. See POLYSULFIDES.

Tracking. A phenomenon wherein a high voltage source current creates a leakage or fault path across the surface of an insulating material by slowly but steadily forming a carbonized path appearing as a thin, wiry line between the electrodes.

Trade Molder. The British term for CUSTOM MOLDER, which see.

Trans-. A prefix denoting an isomer in which certain atoms or groups are located on opposite sides of a plane.

Transfer Coating. A process for coating fabrics such as knits, which are extremely difficult to coat directly by conventional spread coating methods. In a typical version of the process a layer of plastisol is cast against a silicone-treated release paper. This first layer will become the top coat or wear layer in the final product. After gelling of the first layer a second coating of urethane solution is applied to serve as an adhesive tie layer for the fabric substrate. The composite is finally heated and pressed together. Many variations of this typical version are possible, e.g. using foam vinyl or urethane, embossing, etc.

Transfer Molding. A molding process used mainly for thermosetting resins and vulcanizable elastomers. The molding material, which may be preheated, is placed in an open pot at the top of a closed mold. The area of this pot is about 15% larger than the projected area of all cavities and runners in the mold. A plunger is placed in the pot above the material. Pressure applied by a press platen to the plunger forces the molding material into the gates, runners and cavities of the heated mold. Following a heating cycle during which the material is cured or vulcanized, the press is opened and the parts are ejected. In a variation called *plunger molding,* the plunger is more a part of the press rather than of the mold, and pressure is applied to the plunger by an auxiliary hydraulic ram. Plunger molding develops more frictional heat so that molding cycles are generally shorter than those of true transfer molding.

Transfer Molding Pressure. The pressure applied to the cross-sectional area of the material pot or cylinder, expressed in pounds per square inch. (ASTM D 883-65T).

Transition Section. In an extruder, the section of the screw that contains material in both the solid and molten state. The transition can be regulated from gradual to rapid by screw design. Experiments with polyethylene have indicated that a gradual transition screw provides greater uniformity of output and melt temperature, particularly at higher temperatures, pressures and melt indices.

Transition Temperature. The temperature at which a polymer changes from (or to) a viscous or rubbery condition to (or from) a hard and relatively brittle one. See also GLASS TRANSITION.

Transparent Plastics. Eight basic types of plastics are generally regarded as being permanently transparent, although others possess near-transparency, at least for a limited period. These are: acrylics (the foremost in usage), cellulosics, allyl diglycol carbonates, some nylons, polycarbonates, polysulfones, styrenics, and polyphenyl-sulfones. The ASTM test for Haze and Luminous Transmittance is D 1003-61, found in Part 35 of the 1976 Book of Standards.

Transuranic Elements. All elements of atomic numbers above 92. All of these are radioactive and are products of artificial nuclear changes. All are also called members of the *actinide group.*

Treater. Equipment for preparing resin-impregnated reinforcements including means for delivery of a continuous web or strand to a resin tank, controlling the amount of resin pickup, drying and/or partially curing the resin, and rewinding the impregnated reinforcement.

Tremolite. $Ca_2Mg_5Si_8O_{22}(OH)_2$ A variety of silicate mineral similar to and sometimes sold as fibrous talc. It can be used in many applications in place of asbestos as a filler.

Triacetate. A generic term for fibers of cellulose acetate in which at least 92% of the hydroxyl groups are acetylated.

Triacetin. See GLYCEROL TRIACETATE.

Triallyl Cyanurate. $(CH_2:CHCH_2OC)_3N_3$ (TAC) A colorless liquid or solid, highly reactive, used in copolymerizations with vinyl-type monomers to form resins of the ALLYL RESIN family. It is also used in the cross linking of unsaturated polyesters.

Triallyl Phosphate. (TAP) A monomer that can be copolymerized with methyl methacrylate to produce flame-retardant copolymers.

Triaryl Phosphate. A synthetic ester-type plasticizer derived from isopropylphenol feedstock, developed by FMC Corp. and sold under the trademark Kronitex 100. It is useful as a flame retarding plasticizer in vinyl plastisol.

Triazine Resins. A new class of thermosetting polymers designated as NCNS resins, prepared from primary and secondary bis-cyanamides with pendant aryl sulfonyl groups. The bis-cyanamides are reacted together in solutions to form soluble prepolymers by an addition polymerization reaction. By refluxing these solutions, stable laminating varnishes are obtained. Evaporation of solvents from the solutions yields the prepolymer in powder form for molding. Laminates prepared with NCNS resins have good mechanical strength at high temperatures and are relatively fire-retardant.

Tribasic. Pertaining to acids or salts which have three displaceable hydrogen atoms per molecule. Such substances having one displaceable H atom are called *monobasic,* and those with two are called *dibasic.*

Tribasic Lead Maleate. A yellowish-white crystalline powder, used as an effective heat stabilizer in vinyls and as a curing agent for chlorosulfonated polyethylene.

Tribasic Lead Sulphate. $3PbO.PbSo_4.H_2O$ A heat stabilizer, especially for vinyl electrical insulation compounds. It is very effective, has good electrical properties, and produces no gassing.

Tributoxyethyl Phosphate. $(C_4H_9OC_2H_4O)_3PO$ A primary plasticizer for cellulosics, acrylic and vinyl resins, imparting low temperature flexibility and flame retardance. In vinyl plastisols, small amounts of tributoxyethyl phosphate will markedly reduce viscosity. However, when used alone or in high percentages, this plasticizer causes very rapid gelation.

Tri-n-Butyl Aconitate. $C_3H_3(COOC_4H_9)_3$ A combination plasticizer and stabilizer for polyvinylidene chloride and synthetic rubbers.

Tributyl Borate. $(C_4H_9)_3BO_3$ (butyl borate) A colorless liquid, used as an anti-blocking agent for plastic films and sheets.

Tri-n-Butyl Citrate. $C_3H_5O(COOC_4H_9)_3$ A non-toxic plasticizer for most thermoplastics, including cellulosics, polystyrene and vinyls.

Tributyl Phosphate. $(C_4H_9O)_3PO$ (TBP) A colorless liquid used as a primary plasticizer and solvent for nitrocellulose, cellulose acetate, chlorinated rubber and, in special applications, for vinyl resins. Its relatively low boiling point limits its use as a plasticizer for vinyls.

Tri-n-Butyl Phosphine. $(CH_3CH_2CH_2CH_2)_3P$ A curing agent for epoxy resins, and a catalyst for vinyl and isocyanate polymerization.

1,1,1-Trichloroethane. (methyl chloroform) CH_3CCl_3 A nonflammable chlorinated solvent used in adhesives. A stabilized form, sold under Dow Chemical's trademark Chlorothene NU, is used to form resin-based adhesiess that are nonflammable by milling the resin with the solvent.

Trichlorofluoromethane. CCl_3F (Freon 11) A blowing agent for foam plastics, e.g. urethanes and polystyrene. Also a refrigerant and propellant for aerosols.

N-(Trichloromethylthio) Phthalimide. A bacteriostatic agent used in vinyl fabrics for hospitals and households. It stops the growth of bacteria such as staphylococcus aureus that cause pink staining in white and pastel colored PVC.

Trichloromonofluoromethane. A chemically inert blowing agent (Allied Chemical's trademark Genetron 11) used in conjunction with water in flexible urethane formulations to control foam density and load bearing properties.

Trichlorotrifluoroethane. CCl_2FCClF_2 (Freon 113) A colorless, nearly odorless fluorocarbon solvent with a boiling point around $118\,°F$, used as a blowing agent in integral skin urethane foam compositions and as an intermediate in the production of chlorotrifluoroethylene.

Tricresyl Phosophate. $(CH_3C_6H_4O)PO$ (TCP, tritolyl phosphate) One of the earliest commercially used plasticizers for PVC, also suitable for cellulosics, alkyds and polystyrene. Along with other phosphate plasticizers it imparts flame resistance and fungus resistance, even when used in small amounts such as 5% of the total plasticizer content. TCP and cresyl diphenyl phosphate are the two most widely used phosphate plasticizers for this purpose.

Tricresyl Phosphite. $CH_3C_6H_4O)_3P$ A colorless liquid used as a flame retardant plasticizer and stabilizer for thermoplastics.

Tricyclohexyl Citrate. $(C_6H_{11}OCOCH_2)_2C(OH)COOC_6H_{11}$ A non-toxic plasticizer for cellulose nitrate, polystyrene, acrylic and vinyl resins.

Tridecyl Phosphite. $(C_{10}H_{21}O)_3P$ A colorless liquid used as a stabilizer for polyolefins and PVC.

Tri-Dimethylphenyl Phosphate. $[(CH_3)_2C_6H_3O]_3PO$ (tri-xylenyl phosphate, TPX) A plasticizer for cellulosics and vinyl compounds in which a low-gravity electrical grade flame retardant plasticizer is required.

Triethyl Aluminum. $(C_2H_5)_3Al$ (aluminum triethyl, ATE) A catalyst for the polymerization of olefins.

Triethyl Citrate. $C_3H_5O(COOC_2H_5)_3$ (ethyl citrate) An ester of citric acid, used as a plasticizer for many thermoplastics including vinyls, cellulosics and polystyrene. It has been FDA approved for use in food packaging.

Triethylene Diamine. The most widely-used amine-type catalyst for urethane foams, elastomers and coatings. It is soluble in water and polyols.

Triethylene Glycol Diacetate. $(CH_2OCH_2CH_2OCOCH_3)_2$ A plasticizer for cellulosic plastics and some acrylic resins.

Triethylene Glycol Dibenzoate. $C_6H_5CO(OCH_2CH_2)_xOCOC_6H_5$ A secondary plasticizer, partially compatible with all common thermoplastics. In most resin systems it has a tendency to crystallize and bloom at higher concentrations, which property may be used to advantage for anti-blocking purposes.

Triethylene Glycol Dicaprylate. $(CH_2OCH_2CH_2OCOC_7H_{15})_2$ (triethylene glycol dioctate) A plasticizer for cellulose nitrate, ethyl cellulose and vinyls, with good low temperature properties.

Triethylene Glycol Di-Caprylate-Caprate. A plasticizer for cellulose nitrate, ethyl cellulose, and vinyl resins.

Triethylene Glycol Di-(2-Ethylbutyrate). $C_5H_{11}OCOCH_2(CH_2OCH_2)_2CH_2COOC_5H_{11}$ A plasticizer for cellulosic, acrylic and vinyl resins. It is most widely used as a plasticizer for the polyvinyl butyral interlayer in safety glass.

Triethylene Glycol Di-(2-Ethylhexoate). $C_7H_{15}CO(OC_2H_4)_3OCOC_7H_{15}$ A plasticizer for cellulosic plastics, PMMA, polyvinyl chloride and vinyl chloride-vinyl acetate copolymers. In vinyls it is usually used as a secondary plasticizer at 10 to 25 percent of the total plasticizer to impart low temperature flexibility.

Triethylene Glycol Dipelargonate. $(CH_2OCH_2)_2(CH_2OCOC_8H_{17})_2$ A plasticizer for PVC, other vinyl resins, cellulose nitrate, and ethyl cellulose.

Triethylene Glycol Dipropionate. $(CH_2OCH_2CH_2OCOC_2H_5)_2$ A plasticizer for cellulosic plastics and PMMA.

Triethylenetetramine. (TETA) $NH_2(C_2H_4NH)_2C_2H_4NH_2$ A viscous, yellowish liquid, used as a curing agent for epoxy resins.

Tri(2-Ethylhexyl) Citrate. A plasticizer for polyvinyl chloride, used especially when non-toxicity is desired.

Tri(2-Ethylhexyl) Phosphate. (TOF). See TRIOCTYL PHOSPHATE.

Triethyl Phosphate. $(C_2H_5)_3PO_4$ (TEP) A flame-retardant plasticizer for cellulosics, acrylics, some vinyl polymers, and unsaturated polyesters.

Trihydrazide Triazine. A chemical blowing agent which decomposes at 480 to 500 °F yielding about 175 cc per gram of gas consisting mainly of ammonia and nitrogen. It is used with polypropylene, ABS, nylon and other high temperature plastics.

Tri-Iso-Decyl Trimellitate. $C_6H_3(COOC_{10}H_{21})_3$ A plasticizer for PVC.

Tri-Iso-Octyl Trimellitate. (TIOTM) A plasticizer for cellulosic and vinyl plastics with very low volatility, high resistance to soapy water extraction, and essentially no marring effect on lacquered surfaces.

Trimer. A molecule formed by the union of three molecules of a monomer. See POLYMER.

Trimethyl Borate. $(CH_3O)_3B$ (methyl borate, trimethoxyborine) A colorless liquid used as a flame retardant for plastics.

Trimethylene Glycol Di-p-Aminobenzoate. A diamine compound proposed in 1976 by the Polaroid Corp. as a substitute for MOCA in the curing of polyurethanes. It is said to be very stable to cleavage in the melt and highly soluble in a variety of coating solvents.

2,2,4-Trimethyl-1,6-Hexane Diisocyanate. (TMDI)

$$\begin{array}{cc} CH_3 & CH_3 \\ | & | \\ OCNCH_2C-CH_2-CH-CH_2CH_2NCO \\ | \\ CH_3 \end{array}$$

An aliphatic isocyanate used in the production of polyurethanes.

2,2,4-Trimethyl-1,3-Pentanediol. (trimethylpentanediol) One of the principal glycols used in making polyester resins, alkyd resins and polyester plasticizers.

2,2,4-Trimethyl-1,3-Pentanediol Monoisobutyrate Benzoate. A plasticizer for PVC imparting good stain resistance.

Trimethylolethane Tribenzoate. A solid plasticizer for rigid PVC.

2,2,4-Trimethyl-1,3-Pentanediol. (TMPD®) A moderately high-molecular weight glycol containing one primary and one secondary hydroxyl group. It is made by the aldol condensation of isobutyraldehyde, yielding a water-insoluble white solid. TMPD is used in producing linear unsaturated polyesters, and is particularly good for gel coats.

Trioctyl Phosphate. $[C_4H_9CH(C_2H_5)CH_2O]_3P{:}O$ (TOF, tri-2(2-ethylhexyl) phosphate) A plasticizer for PVC, imparting good low temperature flexibility, resistance to water extraction, flame and fungus restance, and minimum change in flexibility over a wide temperature range. It is also compatible with polyvinyl butyral, cellulose nitrate, ethyl cellulose, and CAB resins with a high butyral content.

Tri(n-Octyl n-Decyl) Trimellitate. (NODTM) A low-temperature plasticizer for vinyls and cellulosics, performing as well as polymeric plasticizers with respect to permanence. It is used in heavy-duty uses such as truck seating, window channeling, film and sheeting subjected to wide temperature change, and baby wear.

Trioctyl Trimellitate. $C_6H_3(COOC_8H_{17})_3$ (TOTM) A primary plasticizer for vinyls with good high-temperature aging characteristics. It combines the permanence of polymeric plasticizers with low temperature properties of monomerics. In vinyls, it is used for low-fogging auto interior parts and for wire insulation good for temperatures up to 105 °C. It is also used with cellulosics and acrylics.

Triol. A term sometimes used for a *trihydric alcohol,* that is an alcohol containing three hydroxyl (OH) radicals.

s-Trioxane. $CH_2OCH_2OCH_2O$ (triformol, trioxin, metaformaldehyde) The stable, trimer of formaldehyde; a colorless, crystalline solid. It is easily depolymerized in the presence of acids to monomeric formaldehyde, or may be further polymerized to form ACETAL RESINS, which see. s-Trioxane should not be confused with PARAFORMALDEHYDE, which see.

Triphenyl Phosphate. $(C_6H_5O)_3PO$ (TPP) A crystalline powder used as one of the original synthetic plasticizers for cellulose nitrate. It is also used as a flame retardant for vinyls, cellulosics, acrylics and polystyrene.

Tris(2-Chloroethyl) Phosphate. A plasticizer for polystyrene, cellulosics and vinyls. It is also effective as a flame retardant for unsaturated polyesters and urethane foams.

Tris(2,3-Dibromopropyl) Phosphate. $(CH_2BrCHBrCH_2O)_3PO$ A flame retardant for unsaturated polyesters, urethane foams and other plastics.

Tris(1,3-Dichloroisopropyl) Phosphate. A flame retardant for unsaturated polyesters.

Tris(2,3-Dichloropropyl) Phosphate. $(CH_2ClCHClCH_2O)_3PO$ A plasticizing flame retardant for many plastics including vinyls, cellulosics, acrylics, polyolefins, phenolics, polyesters and urethane foams.

Trisnonylphenyl Phosphite. (TNPP) An FDA-sanctioned heat stabilizer and antioxidant used in styrene-butadiene copolymers.

Tritactic Polymers. Isotactic or syndiotactic polymers which are also of the cis- or trans- form because the molecules are unsaturated and have double bonds.

Tritolyl Phosphate. See TRICRESYL PHOSPHATE.

Tri-Xylenyl Phosphate. (TXP) See TRI-DIMETHYLPHENYL PHOSPHATE.

Trommsdorff Effect. See AUTOACCELERATION.

TSC. A sometimes-used abbreviation for THERMAL STRESS CRACKING, which see.

TTB. See 2,2′,6,6′-TETRABROMO-3,3′,5,5′-TETRAMETHYL-4,4′-DIHYDROXYDI-PHENYL.

Tumbling. (tumble finishing, barreling) A finishing operation for small plastic articles by which gates, flash and fins are removed and/or surfaces are polished, by rotating them in a barrel together with wooden pegs, sawdust, and (sometimes) polishing compounds. The barrels are usually of octagonal shape with alternate open and closed panels, the open panels covered with screen to permit fragments of removed material to fall out. Blocks of dry ice may be added to the tumbling medium to embrittle the parts and thus facilitate cleaner flash rupture, but in this case the barrel must be closed to retard evaporation of the dry ice.

Tumbling Agitators. Cylindrical or cone-shaped vessels rotating about a horizontal or inclined axis, with internal ribs which lift the material and then let it tumble back into the charge. They are used mainly for dry blending operations such as adding color concentrates to molding powders.

Tunnel Gate. See SUBMARINE GATE.

Tunneling. A condition occurring in incompletely bonded laminates, characterized by release of longitudinal portions of the substrate and deformation of these portions to form tunnel-like structures.

Twaddell Scale. A method of stating specific gravity, designed to simplify measurements for unskilled persons. Its numbers are obtained from specific gravity numbers by dropping the decimal and the preceding numeral, reading the following numbers to two significant figures and doubling them. Thus, 1.35 sp. gr. becomes 70° Twad.

Twinning. A movement of planes of atoms in the lattice parallel to a specific (twinning) plane so that the lattice is divided into two symmetrical parts which are differently oriented. The amount of movement of each plane is proportional to its distance from the twinning plane.

Twin Screw Extruder. See EXTRUDER, TWIN SCREW.

Twin Shell Forming. A high speed thermoforming process for producing bottles and other hollow objects. Two thermoplastic sheets from separate roll unwind stations are conveyed through heating apparatus, then positioned between facing halves of vacuum-forming molds. After closing of the molds, vacuum is drawn on each half to simultaneously form and seal the edges of the articles. When molds are arranged on a pair of endless conveyors the process becomes continuous, sheets being fed into one end and a continuous web of formed products emerging from the other end ready for separation from the waste portion of the joined sheets.

Twist. A textile term referring to the number of turns per inch a multifilament yarn, staple yarn or other structure is turned or twisted around its longitudinal axis into a stable structure.

Two-Color Molding. See DOUBLE-SHOT MOLDING.

Two-Level Mold. A double-decked mold in which cavities are placed in two layers to reduce clamping force. It is used for parts with large areas.

Two-Shot Injection Molding. Confusingly, this term has been used in the literature for two processes that are distinctively different. One is described under DOUBLE SHOT MOLDING. The other process as described by engineers of the Western Electric Co. involved first injecting one material into a single-cavity die just until the polymer has commenced to chill against the cold wall of the mold, then immediately injecting a second polymer to force the first polymer to the cavity extremity. The second polymer, usually a reclaimed material, forms the interior of the molded article, the first virgin material completely forming the outside of the article.

Two-Shot Molding. See DOUBLE-SHOT MOLDING.

UF. Abbreviation for *urea-formaldehyde*. See AMINO RESINS.

UHMW Polyethylenes. Abbreviation for ultra-high molecular weight polyethylenes, those having molecular weights in the 1.5 to 3.0 million range. See also POLYETHYLENES.

Ultimate Elongation. In a tensile test, the elongation at rupture.

Ultimate Strength. Term used to describe the maximum unit stress a material will withstand when subjected to an applied load in a compression, tension or shear test.

Ultracentrifuge. A centrifuge capable of rotating from 20,000 to 60,000 RPM, creating forces up to several hundred thousand times gravity. Sedimentation studies of high polymers performed with an ultracentrifuge are used for determining weight-average molecular weights and molecular weight distributions.

Ultramarine Blue Pigments. Pigments comprising a complex of double silicates of sodium and aluminum in combination with sodium polysulphide. They produce bright, clean tones even in combination with white pigments, and are resistant to high temperatures employed in processing thermoplastics.

Ultrasonic Cleaning. A method used for thoroughly cleaning molded plastics for electrical components and mechanical parts. A transducer mounted on the side or bottom of a cleaning tank is excited by a frequency generator to produce high frequency vibrations in the cleaning medium. These vibrations dislodge contaminants from crevices and blind holes that normal cleaning methods would not affect.

Ultrasonic Degating. A degating method used for small plastic parts produced by a family mold. The molding machine operator removes the runner system and parts from the mold and loads the assembly into the degating machine. The runner is contacted by an ultrasonic horn. High frequency vibrations are transmitted through the runner to the thin gate areas which melt, dropping the parts through holes into a sorting system.

Ultrasonic Frequencies. Frequencies above the limit of human audibility, approximately 18,000 cycles per second.

Ultrasonic Inserting. A new method of incorporating metallic inserts in plastics articles by means of ultrasonic heating. A plain hole is molded in the plastic article by means of a core pin, the hole diameter being slightly less than that of the insert. Ultrasonic vibrations are applied as the metallic part is being inserted, which causes displaced plastic material to flow into threads, knurls, flutes or undercuts on the insert, mechanically locking it into place.

Ultrasonic Staking. The process of forming a head on a protruding portion of a plastic article for the purpose of holding a surrounding part in place, utilizing ultrasonic heating to melt the protrusion and pressure by a forming tool to form it into a head. The process is very fast, usually performed in a fraction of a second.

Ultrasonic Welding. (Ultrasonic sealing) A method of welding or sealing thermoplastics in which heating is accomplished with vibratory mechanical pressure at ultrasonic frequencies (20 to 40 kc.). Electrical energy is converted to ultrasonic vibrations by a transducer, directed to the area to be welded by means of a horn, and localized heat is generated by the friction of vibration at the surfaces to be joined. Other parts of the assembly are not heated. The process is most effective for rigid and semi-rigid plastics, since the energy is rapidly dissipated in soft, flexible materials.

Ultraviolet. The region of the electromagnetic spectrum between the violet end of visible light and the X-ray region, including wave lengths from 100 to 3900 A. Photons of radiations in the UV area have sufficient energy to initiate some chemical reactions and to degrade some plastics.

Ultraviolet Printing. (UV printing, UV-curing decoration) The process of printing or decorating with inks that cure rapidly by exposure to ultraviolet light. The inks are solvent-free, thus avoiding problems from air polution regulations, and contain a UV-sensitive catalyst that cures the ink in as little as one second. Mercury vapor lamps can be used as the source of UV light. For polymers that are subject to degradation by oxidation induced by the UV light, equipment has been designed to blanket the exposed area in a nitrogen atmosphere.

Ultraviolet Stabilizers. Exposure of many plastics to UV radiation, especially that in the near-violet region, can cause changes such as loss of gloss, crazing, chalking, discoloration, changes in electrical characteristics, embrittlement, and disintegration. Additives which protect plastics against such effects by preferentially absorbing the incident UV radiation and dissipating the associated energy in a harmless manner are sometimes called *ultraviolet absorbers* or *ultraviolet screening agents.* Additives which do not actually absorb UV radiation but protect the polymer in some other manner are called are called *ultraviolet stabilizers* or other names that indicate the mode of stabilization. For example, products that remove the energy absorbed by the polymer before photochemical degradation can take place are called *energy transfer agents* or *excited state quenchers.* Other modes of UV stabilization are singlet oxygen quenching, radical scavenging and hydroperoxide decomposition. Classes of UV stabilizers in commercial use are the benzophenones, benzotriazoles, substituted acrylates, aryl esters and compounds containing nickel or cobalt salts.

Ultraviolet Spectrophotometry. A method of analysis similar to INFRARED SPEC-TROPHOTOMETRY (which see) except that the spectrum is obtained with ultraviolet light. It is somewhat less sensitive than the IR method for polymer analysis, but is useful for detecting some plasticizers and antioxidants.

Undercure. A condition of inadequate physical properties in a thermosetting resin or elastomer resulting from too little time and/or temperature during the curing cycle.

Undercut. (n) An indentation or protuberance in a mold that tends to impede withdrawal of a molded part from the mold. Articles of soft materials such as flexible vinyls can be removed from molds with severe undercuts, but undercuts must be avoided in molds for rigid materials. Slight undercuts are sometimes deliberately formed in one half of a mold to cause the article to remain in a desired half until ejected.

Uniaxial Load. A condition whereby a material is stressed in only one direction.

Uniaxial Orientation. A method of ORIENTATION, which see, in which the orienting stress is applied only in one direction.

Unicellular Plastic. A term sometimes used for CLOSED CELL FOAMED PLASTIC, which see.

Unidirectional Laminate. A reinforced plastic laminate in which substantially all of the fibers are oriented in the same direction.

Unit Elongation. In a tensile test, the ratio of the elongation to the original length of the specimen; that is, the change in length per unit of original length.

Unit Mold. A mold designed for quick changing of interchangeable cavity parts.

Unloading Valve. A valve which limits the maximum pressure in a hydraulic line to a desired value by diverting the flow of fluid from a pump to a by-pass line.

Uns-. (unsym-) Abbreviation for unsymmetrical, a prefix denoting unsymmetrical disposition of substituents of organic compounds with respect to the carbon skeleton or a functional group.

Unsaturated Compounds. Compounds having more than one bond between two adjacent atoms, usually carbon atoms, and capable of adding other atoms at that point to reduce it to a single bond.

Unsaturated Polyester. See POLYESTER, UNSATURATED.

UP. Abbreviation used by British Standards Institution for unsaturated polyesters. See POLYESTER, UNSATURATED.

Upstroke Press. A hydraulic press in which the main ram is situated below the moving table, pressure being applied by an upward movement of the ram.

Urea. $CO(NH_2)_2$ (Carbamide) A white, crystalline powder derived from the decomposition of ammonium carbamate. It is used in the preparation of urea-formaldehyde resins. See also AMINO RESINS.

Urea Formaldehyde Foam. A foam produced by combining a urea formaldehyde resin with a detergent-type foaming agent under pressure. Upon release of the pressure a foam of about the consistency of a shaving cream emerges and cold-cures within 2 to 4 hours. The foam is of low density, is non-combustible, and dries within 1 to 2 days. The dried foam has some resiliency, good thermal insulation properties and is sound absorbent. Although not recommended for continuous exposure to temperatures in excess of 210 °F, the material does not decompose and release gases until heated to 1208 °F.

Urea-Formaldehyde Plastics. See AMINO RESINS.

Urea Plastics. See AMINO RESINS.

Urethane. $CO(NH_2)OC_2H_5$ (urethan, ethyl carbamate, ethyl urethane) A colorless crystalline substance used primarily in medicines, pesticides and fungicides. Strangely, urethane is not used in the production of urethane polymers or foams. The urethanes of the plastics industry are so named because the repeating units of their structures resemble the chemical urethane.

Urethane Coatings. The ASTM has designated five types of urethane coatings. Type I is a one-component system modified with a drying oil such as linseed or soya, which reacts with oxygen from the air to effect cure. It is used as a finish for wood. Type II is based on an isocyanate-terminated prepolymer in a solvent which dries by evaporation and cures by reaction with moisture in the air. It is used for coating wood, rubber and leather. Type III is based on a blocked isocyanate, a polyester, curing agent and suitable solvent. The applied coating is based to effect curing. It is used for wire coating and industrial finishes. Type IV is a two-component system, one being a prepolymer prepared from a diisocyanate and a polyol, the other being a catalyst such as a tertiary amine. It is used for heavy duty industrial finishes with good resistance to chemicals, abrasion and corrosion. Type V is a two-component system, one being a polyisocyanate (usually an adduct of diisocyanate and trimethylolpropane) and the other being a polyol, in solvents which evaporate after application of the coating. The reaction proceeds at ambient temperatures without the aid of a catalyst. It is used as a high-performance industrial coating.

Urethane Foams. (polyurethane foams, isocyanate foams) These foams differ from other members of the cellular plastics group in that the chemical reactions causing foaming occur simultaneously with the polymer-forming reactions. As in the case of polyurethane resins, the polymeric constituent of urethane foams is made by reacting a polyol with an isocyanate. When the isocyanate is in excess of the amount that will react with the polyol, and when water is present, the excess isocyanate will react with water to produce carbon dioxide which expands the mixture. The hardness of the foam is governed by the molecular weight of the polyol used. Low molecular weight polyols (approximately 700) produce rigid foams, and high molecular weight polyols (3000 to 4000) produce flexible foams. Polyols with molecular weights around 6000 are used for the so-called cold-cure or highly resilient foams. They are usually capped with ethylene oxide to provide terminal primary hydroxyl groups which increase their reactivity about threefold. Crosslinked foams are rigid or semi-rigid. Auxiliary blowing agents are sometimes used, especially in rigid foams. Other ingredients often incorporated in urethane foams are catalysts to control the speed of reaction, and a surfactant to stabilize the rising foam and control cell size. Three basic processes are used for making urethane foams: the prepolymer technique, the semi-prepolymer technique, and one-shot process. In the prepolymer technique, a polyol and an isocyanate are reacted to produce a compound which may be stored and subsequently mixed with water, catalyst and, in some cases, a foam stabilizer. In the semi-prepolymer process about 20% of the polyol is prereacted with all of the isocyanate, then this product is later reacted with a masterbatch containing the remainder of the ingredients. See also ONE-SHOT MOLDING, ISOCYANATES, POLYOL, POLYETHER FOAMS, RETICULATED URETHANE FOAMS, INTEGRAL SKIN MOLDING.

Urethane/Imide Modified Foam. A polyaryl polyisocyanate (PAPI) is reacted with 3,3'.4,4'-benzophenone tetracarboxylic dianhydride (BTDA) to form an isocyanate prepolymer. This prepolymer can be compounded with a polyol, a blowing agent, a catalyst and a cell stabilizer to form the modified foam. Such a foam containing 5% BTDA in the prepolymer has better thermal properties than conventional foams.

Urethane Plastics. See POLYURETHANES, URETHANE FOAMS.

UV Stabilizer. See ULTRAVIOLET STABILIZERS.

Vacuum Bag Molding. See BAG MOLDING.

Vacuum Casting. A method used for casting fluid thermosetting resins to avoid entrapment of gas bubbles. The mold is placed in a vacuum chamber, filled with the resin from an external hopper, then the vacuum is released.

Vacuum Deposition. See VACUUM METALLIZING.

Vacuum Forming. A method of forming plastic sheets or films into three-dimensional shapes, in which the plastic sheet is clamped in a frame suspended above a mold, heated until it becomes softened, drawn down into contact with the mold by means of vacuum, and cooled while in contact with the mold. See also THERMOFORMING, VACUUM SNAP-BACK FORMING, TWIN SHELL FORMING.

Vacuum Impregnating. The process of impregnating electrical components by subjecting the parts to a high vacuum, introducing the impregnant to cover the part, then releasing the vacuum. Epoxy, phenolic and polyester resins are often used in the process.

Vacuum Metallizing. A decorating process used to make plastics objects resemble shiny metals. The article to be coated is first thoroughly cleaned to remove grease, dirt or mold-release compound, then is coated with a lacquer by dipping or spraying. The prepared articles are placed on racks inside a chamber, a very high vacuum is created inside the chamber, then metal wire, e.g. aluminum, is vaporized by a heating unit and the metallic vapor condenses on the surfaces of the articles. A top coat of protective lacquer is applied to the metal coating. In a variation called the *Cathode sputtering process,* the metal is vaporized by an electrical discharge between electrodes in the high vacuum. This process gives a more uniform film. When the metallic coating is applied to the surface of the article which is normally viewed, the process is called *first surface vacuum metallizing.* Transparent plastics such as polystyrene and acrylics can be decorated on the reverse side, so that the serviceable surface is the plastic itself, in which case the process is called *second surface vacuum metallizing.*

Vacuum Snap-Back Forming. A variation of the vacuum forming process in which the heated plastic sheet is positioned above a vacuum box and below a male plug, first drawn into a concave shape by means of vacuum applied to the bottom of the vacuum box, then pulled upward or "snapped back" against the surface of the plug by means of vacuum drawn through the plug. The process is useful for very deep draws.

Vacuum Venting. The drawing of a vacuum on the mold interior or parts thereof in order to eliminate molding defects such as weld lines, voids, blisters, burned spots and short shots. The process is used in injection, transfer and compression molding. The vacuum may be drawn by means of tubes leading to vents in sharp corners, blind holes, etc. In a more recent modification of the process, the entire mold is encased in a vacuum-tight box with a parting line coincident with the parting line of the mold, its mating surfaces sealed by O-rings.

Valence. The property of an atom of an element which is measured by the number of atoms of hydrogen (or its equivalent) one atom of that element can hold in combination if negative, or can displace in a reaction if it is positive.

Valence Electrons. Electrons which are gained, lost or shared in a chemical reaction.

Valley Printing. (inlay printing) A printing process for flat plastic surfaces in which ink is applied to the raised portions of an embossing roll which simultaneously embosses the surface and deposits ink in the valleys of the embossed surface. Thus, the embossed areas and printed areas are always in perfect register. The plastic surface to be valley printed must be warm enough for deformation by the embossing roller, which can be achieved either by preheating with radiant heaters or hot rolls, or by taking the material while it is still warm from a calendering or extrusion operation. The process is similar to flexographic printing in that both print from raised portions of a cylinder. However, in flexographic printing the roll is flexible whereas in valley printing the roll is rigid.

Value. (color value) That attribute which describes the lightness of a color. A color may be classified as equivalent to some member of a series of shades ranging from black (the zero member) to white.

Vamac® . Du Pont's trademark for a family of ethylene-acrylic elastomers. Two types of the heat-, oil-, and weather-resistant materials are available, both in masterbatch form. It is claimed that only the more expensive silicones and fluoroelastomers offer better heat resistance, and that Vamac is superior to these materials in low temperature resistance.

Van Der Waals Forces. Forces that exist between molecules of a substance after all of the primary valences within covalent molecules are saturated. Also called *secondary valence forces* or *intermolecular forces*.

Vapor. As most frequently used, the term vapor means a substance which, although present in the gaseous phase, generally exists as a liquid or solid at room temperature. *Gas* is more frequently used for a substance which remains gaseous at room temperature.

Vapor Barrier. A layer of material through which water vapor will not pass readily or at all.

Vapor-Liquid Chromatography. See CHROMATAGRAPHY.

Vapor Transmission. See WATER VAPOR TRANSMISSION.

VCM. Abbreviation for vinyl chloride monomer. See VINYL CHLORIDE.

Veil. A thin mat of very fine, relatively long fibers used at the outermost layer of a composite in order to improve surface characteristics.

Vent. See AIR VENT.

Venturi Dispersion Plug. In injection molding, a plate having an orifice with a conical relief drilled therein which is fitted in the nozzle to aid in the dispersion of colorants in a resin.

Vermiculite. A mica-like mineral containing water, which causes it to expand from six to twenty times in volume when heated to a high temperature. It is sometimes used as a filler in plastics.

Vertical Extruder. An extruder arranged so that the barrel is vertical and extrusion is downward.

Vertical Flash Ring. The clearance between the force plug and the vertical wall of the cavity in a positive or semi-positive mold; also the ring of excess material which escapes from the cavity into this clearance space.

VF₂. Abbreviation for POLYVINYLIDENE FLUORIDE, which see.

Vibration Welding. A welding method developed in Switzerland by the Du Pont Plastics Technical Center. Heat is generated by moving one part with respect to the other within a frequency rnge of 90 to 120 Hz over displacements of 0.12 to 0.2 inches, at a speed of about 6 ft. per second.

Vibratory Feeders. Devices for conveying dry materials from storage hoppers to processing machines, comprising a tray or tube vibrated by mechanical or electrical pulses. The frequency and/or amplitude of the vibrations control the rate of flow.

Vibratory Mill. (Vibro-energy® mill) A grinding device consisting of a spring-suspended cylindrical container filled with a grinding media such as is used in a ball mill, driven by an oscillating device which causes the grinding media to vibrate throughout its mass. Its grinding action is more rapid and efficient than that of a ball mill.

Vicat Softening Point. The temperature at which a flat-ended needle of 1 sq. mm circular or square cross-section will penetrate a thermoplastic specimen to a depth of 1 mm under a specified load using a uniform rate of temperature rise. (ASTM D 1525-58T). This test is used for thermoplastics such as polyethylene, polystyrene, acrylics and cellulosics which have no definite melting point.

Vickers Hardness. A test similar to the BRINELL HARDNESS (which see) test, using an indentor in the form of a square-based diamond pyramid, with an angle of 136° between the opposite faces. The result is expressed as the load divided by the area of the impression.

Vinal. Generic name for a manufactured fiber in which the fiber-forming substance in any long chain synthetic polymer composed of at least 50% by weight of vinyl alcohol units, -CH₂CHOH-, and in which the total of the vinyl alcohol units and any one or more of the various acetal units is at least 85% by weight of the fiber. (Federal Trade Commission).

Vinyl. The unsaturated group CH₂:CH-, which is the basis for all vinyl plastics. See VINYL RESINS.

Vinyl Acetate. CH₃COOCH:CH₂   A colorless liquid obtained by the reaction of acetylene and acetic acid in the presence of a catalyst, e.g. mercuric oxide. It is the monomer for polyvinyl acetate, and a comonomer and intermediate for many members of the vinyl plastics family.

Vinyl Acetate Plastics. See POLYVINYL ACETATE.

Vinyl Alcohol. CH₂:CHOH (ethenol) For practical purposes, a non-existent material, since it is available only in the form of its esters and the polymer *polyvinyl alcohol,* which is not derived directly from vinyl alcohol.

Vinyl Alcohol Plastics. See POLYVINYL ALCOHOL.

Vinylation. The process of forming a vinyl derivative by reaction of alcohols, amines or phenols with acetylene. Such derivatives are used as polymerization intermediates.

Vinylbenzene. See STYRENE.

Vinyl Butyrate. CH_2:$CHOOCC_3H_7$ A volatile liquid monomer for polymers used in water-base paints.

N-Vinylcarbazole. $C_{12}H_8NCH$:CH_2 A monomer derived from acetylene and carbazole, used in the production of POLYVINYL CARBAZOLE, which see.

Vinyl Chloride. CH_2CHCl (chloroethylene, monochlorethylene) A colorless gas at normal temperatures and pressures, made by reacting ethylene with chlorine or hydrogen chloride to obtain ethylene dichloride, which is cracked to form vinyl chloride. When cooled to $-13.9\,°C$ the gas becomes a liquid, in which form it is usually handled. It is the monomer for POLYVINYL CHLORIDE, which see.

Vinyl Chloride-Ethylene Copolymers. Copolymers of minor amounts of ethylene with vinyl chloride possess superior heat stability and hot-strength, and require lesser amounts of impact modifiers to achieve impact strength than does straight PVC homopolymer to gain the same degree of impact strength. These copolymers are useful in producing films and bottles for packaging, since their better heat stability provides more latitude in selecting non-toxic stabilizers.

Vinyl Chloride Plastics. See POLYVINYL CHLORIDE.

Vinyl Cyanide. See ACRYLONITRILE.

Vinyl Ethers. See VINYL ETHYL ETHER, VINYL ISOBUTYL ETHER and VINYL METHYL ETHER.

Vinylethylene. See BUTADIENE.

Vinyl Ethyl Ether. CH_2:$CHOC_2H_5$ (EVE, ethyl vinyl ether) A colorless monomer which can be polymerized either in the liquid or the gaseous state. In plastics, it is used as a comonomer and intermediate.

Vinyl Fluoride. CH_2:CHF (fluoroethylene) A colorless gas, used as the monomer for POLYVINYL FLUORIDE, which see.

Vinyl Foams. Although cellular vinyls can be produced by many methods, including mechanical frothing and leaching out of soluble additives, the most widely used procedure is chemical blowing. From 1 to 2% of a blowing agent such as azobisformamide is incorporated in a vinyl compound or dispersion, remaining inert until it is decomposed by processing heat to release a gas. Such compounds are processed by conventional methods such as calendering, extrusion, injection molding, compression molding, slush casting and rotational casting. Vinyl foams are widely used in clothing, flooring, footwear, furniture and packaging.

Vinylformic Acid. See ACRYLIC ACID.

Vinylidene Chloride. CH_2:CCl_2 (1,1-dichloroethylene) A colorless, volatile liquid used as the monomer for POLYVINYLIDENE CHLORIDE, which see; and as a comonomer with vinyl chloride (see SARAN) and other monomers such as acrylonitrile.

Vinylidene Chloride Plastics. See SARAN, POLYVINYLIDENE CHLORIDE.

Vinylidene Fluoride. $CH_2:CF_2$ A colorless, nearly odorless gas prepared by the dehydrohalogenation of 1,1,1-chlorodifluoroethane, or by the dehalogenation of 1,2-dichloro-1,1-difluoroethane. It polymerizes readily in the presence of free-radical initiators to produce the homopolymer polyvinylidene fluoride, and is also copolymerized with olefins and other fluorocarbon monomers to make many fluorocarbon elastomers. The homopolymer is a tough linear thermoplastic with high impact strength, abrasion resistance, chemical resistance, and good electrical properties. Conventional molding processes may be employed.

Vinylidene Group. The group $CH_2 = C =$.

Vinyl Isobutyl Ether. $CH_2:CHOCH_2CH(CH_3)_2$ (IVE, isobutyl vinyl ether) A colorless flammable liquid used for polymers and copolymers suitable for use in coatings, adhesives and lacquers; modifiers for alkyd resins and polystyrene; and plasticizer for cellulose nitrate.

Vinyl Methyl Ether. $CH_2:CHOCH_3$ A colorless gas or liquid depending on pressure and temperature, polymerizable to POLYVINYL METHYL ETHER, which see. It is also used as a modifier for alkyd resins and polystyrene, and a plasticizer for cellulose nitrate.

Vinyl Plastics. See VINYL RESINS.

Vinyl Propionate. $CH_2:CHOOCC_2H_5$ A volatile liquid used for emulsion paint polymers.

N-Vinyl-2-Pyrrolidone. $CH_2CH:NCH_2CH_2CH_2CO$ A monomer derived from acetylene and formaldehyde. See POLYVINYL PYRROLIDONE.

Vinyl Resins. According to strict chemical nomenclature this term includes all resins and polymers made from monomers containing the vinyl group $CH_2:CH-$. Thus, in the chemical literature, polystyrene, polymethyl methacrylate and many other styrene and acrylic polymers and copolymers are classified as vinyls. However, in the plastics literature the above materials are considered worthy of separate classification, and the term vinyl plastics refers primarily to POLYVINYL CHLORIDE (which see) and secondarily to the following:
  polyvinyl acetal
  polyvinyl acetate
  polyvinyl alcohol
  polyvinyl butyral
  polyvinyl carbazole
  polyvinyl dichloride
  polyvinyl formal
  polyvinylidene chloride
  polyvinyl isobutyl ether
  polyvinyl pyrrolidone

Vinyl Stearate. $CH_3(CH_2)_{16}COOCH:CH_2$ A white, waxy solid, used as an internal plasticizer by means of copolymerization.

Vinylstyrene. See DIVINYL BENZENE.

Vinyl Toluene. $CH_2CHC_6H_4CH_3$ A colorless liquid, commercial forms comprising a 60-40% mixture of the meta- and para- forms, used as a solvent and as a polymerizable monomer in place of styrene in the production of polyester resins.

Vinyltrichlorosilane. $CH_2:CHSiCl_3$ A silane coupling agent used in reinforced polyesters.

Vinyltriethoxysilane. $CH_2:CHSi(OC_2H_5)_3$ A silane coupling agent used in reinforced polyesters, polyethylene and polypropylene.

Vinyl-Tris(beta-Methoxyethoxy)Silane. $CH_2:CHSi(OCH_2CH_2OCH_3)_3$ A silane coupling agent used in reinforced polyester and epoxy resin structures.

Vinyon. Generic name for a manufactured fiber in which the fiber-forming substance is any long chain synthetic polymer composed of at least 85% by weight of vinyl chloride units, $-CH_2CHCl-$. (Federal Trade Commission).

Virgin Material. Any plastic compound or resin that has not been subjected to use or processing other than that required for its original manufacture.

Viscoelasticity. The tendency of plastics to respond to stress as if they were a combination of elastic solids and viscous fluids. This property, possessed by all plastics to some degree, dictates that while plastics have solid-like characteristics such as elasticity, strength and form-stability, they also have liquid-like characteristics such as flow depending on time, temperature, rate and amount of loading.

Viscometer. (viscosimeter) An instrument used for measuring the viscosity and flow properties of fluids. A commonly used type (Brookfield) measures the force required to rotate a disc or hollow cup immersed in the specimen fluid at a predetermined speed. Of the many other types, some employ rising bubbles, falling or rolling balls, and cups with orifices through which the fluid flows by gravity. Instruments for measuring flow properties of highly viscous fluids and molten polymers are more often called *plastometers* or *rheometers*. For other examples of viscometers see AIR BUBBLE VISCOMETER, CAPILLARY VISCOMETER, CAPILLARY RHEOMETER, EXTRUSION RHEOMETER, BRABENDER PLASTOGRAPH, EXTENSIOMETER, FORD VISCOSITY CUPS, SHELL CUP VISCOMETER, STORMER VISCOMETER, ZAHN VISCOSITY CUPS.

Viscose. A solution of xanthated cellulose in dilute sodium hydroxide from which rayon fibers are formed. The xanthated cellulose is produced by treating alkali cellulose (e.g. wood fibers treated with sodium hydroxide) with oxygen and carbon disulfide. Rayon produced by this method is known as *viscose rayon.*

Viscosity. A measure of the resistance of flow due to internal friction when one layer of fluid is caused to move in relationship to another layer. The c.g.s. units of measure are the Poise, centipoise, Stoke and centistoke. The Poise represents *absolute viscosity,* the tangential force per unit area of either of two horizontal planes at unit distance apart, the space between being filled with the substance. A liquid with an absolute viscosity of one Poise requires a force of one dyne to maintain a velocity differential of one centimeter per second over a surface one centimeter square. The centipoise is one one-hundredth of a Poise. The Stoke is the measure of *kinematic* (kinetic) *viscosity,* the ratio of absolute viscosity to the density. The centistoke is one one-hundredth of a Stoke. The following table of approximate viscosities in centipoises at room temperature is useful for rough comparisons.

water	1
kerosene	10
motor oil, SAE 10	100
castor oil, glycerine	1,000
corn syrup	10,000
molasses	100,000

When the ratio of shearing stress to the rate of shear is constant, as is the case with water and thin motor oils, the fluid is called a *Newtonian fluid*. In the case of non-Newtonian fluids, the ratio varies with the shearing stress, and viscosities of such fluids are called *apparent viscosities*. In the new SI system, it is proposed that values for the Poise be stated as Pascal seconds, the conversion factor being 1 Poise equal to 1×10^{-1} Pa.s. The centipoise would become 1×10^{-3} Pa.s. However, the SI also equates the *stoke* (unit of kinematic viscosity) with square meters per second, one stoke times 1×10^{-4} being equal to viscosity in m^2 per second. See also the following terms which are related to viscosity:

ABSOLUTE VISCOSITY	NEWTONIAN FLOW
BINGHAM BODY	POISEUILLE'S EQUATION
CAPILLARY VISCOMETER	PSEUDOPLASTIC FLUID
CAPILLARY RHEOMETER	REDUCED VISCOSITY
CONSISTENCY	RELATIVE VISCOSITY
CUP FLOW TEST	RHEOLOGY
DILATANCY	SAYBOLT VISCOSITY
EXTRUSION RHEOMETER	SPECIFIC VISCOSITY
INHERENT VISCOSITY	THIXOTROPY
INITIAL VISCOSITY	VISCOELASTICITY
INTRINISIC VISCOSITY	VISCOMETER
KINEMATIC VISCOSITY	VISCOSITY COEFFICIENT
K-VALUE	VISCOSITY NUMBER
KREBS UNIT	VISCOSITY RATIO
LAMINAR FLOW	VISCOUS FLOW
MELT INDEX	YIELD VALUE

Viscosity Coefficient. The shearing stress necessary to induce a unit velocity flow gradient in a material. Note: In actual measurement, the viscosity coefficient of a material is obtained from the ratio of shearing stress to shearing rate. This assumes the ratio to be constant and independent of the shearing stress, a condition which is satisfied only by Newtonian fluids. Consequently, in all other cases, values obtained are apparent and represent one point on the flow curve. In the metric system, the viscosity coefficient is expressed in poises, units being dyne sec. per sq. cm. In the SI system the viscosity coefficient is expressed in Newton-seconds per square meter. See also VISCOSITY. (ASTM D 883-65T).

Viscosity Depressant. A substance which when added in relatively minor amount to a liquid lowers its viscosity. Such materials, for example ethoxylated fatty acids, are often incorporated in vinyl plastisols to lower their viscosities without increasing plasticizer levels. The viscosity depressant may be incorporated in the plastisol during the original mixing or may be added later to reduce excessive viscosity after prolonged storage.

Viscosity Number. The IUPAC term for REDUCED VISCOSITY, which see.

Viscosity Ratio. The IUPAC term for RELATIVE VISCOSITY, which see.

Viscous. A term used to loosely denote that a material is thick and sluggish in flow rather than thin and free flowing.

Viscous Elasticity. A degree of elasticity in which the time necessary to recover initial dimensions is longer than a stated time by 5%.

Viscous Flow. A type of fluid movement in which all particles of the fluid flow in a straight line parallel to the axis of a containing pipe or channel, with little or no mixing or turbidity.

Viton. Du Pont's trademark for a series of fluoroelastomers with a wide range of properties. Principal uses are gaskets, O-rings, oil seals, diaphragms, pump and valve linings, hose, tubing and coated fabrics. High temperature resistance is the most outstanding valuable property in grades used for O-rings and seals. Other grades possess resistance to steam and hot water, chemicals, and below-zero temperature.

Void. (n) (1) In a solid plastic, an unfilled space of such size that it scatters radiant energy such as light. (2) A cavity unintentionally formed in a cellular material and substantially larger than the characteristic individual cells. (ASTM D 883-75a). (3) An empty space in any material or medium.

Volatile. Capable of being driven off as a vapor at room or slightly elevated temperatures.

Volatile Content. The percent of volatiles which are driven off as vapor from a plastic or impregnated reinforcement.

Volatile Loss. The loss in weight of a substance caused by vaporization of a constituent.

Volt. The unit of electromotive force, equal to the difference in potential required to make a current of one ampere flow through a substance with a resistance of one ohm. Volts are calculated by dividing watts by amperes in a circuit. The term volt and its abbreviation "V" are to be retained in the new "Metric" system advocated by all standards setting organizations.

Voltage Breakdown. See DIELECTRIC STRENGTH.

Volume. The numerous units of volume in use throughout the world are to be replaced in the new SI by cubic meter (m^3). Thus, one teaspoon would be expressed as 4.928 922 E-06 m^3. The old cgs unit of one liter would become 1.000 000*E-03 m^3.

Volume Resistance. The ratio of the direct voltage applied to two electrodes which are in contact with, or embedded in, a specimen to that portion of the current between them that is distributed through the volume of the specimen.

Volume Resistivity. (specific insulation resistance) The ratio of the potential gradient parallel to the current in a material, to the current density. In the metric system, volume resistivity is numerically equal to the direct current resistance between opposite faces of a one centimeter cube of the material, expressed in ohm-centimeters.

Vulcanization. The chemical reaction which induces extensive changes in the physical properties of a rubber or elastomer, brought about by reacting the material with sulfur and/or other suitable agents. The changes include decreased plastic flow, reduced surface tackiness, increased elasticity, much greater tensile strength, and considerably less solubility. More recently, certain thermoplastics, e.g. polyethylene, have been modified to be vulcanizable. Cross-linking is encouraged, thereby giving resistance to deformation or flow above the melting point.

Vulcanized Fiber. Cellulosic material which has been partially gelatinized by action of a chemical (usually zinc chloride), then heavily compressed or rolled to required thickness, leached free from the gelatinizing agent, and dried. Used for electrical insulation and in packaging.

Wall Stress. In a filament wound part, usually a pressure vessel, the stress calculated using the load and the entire laminate cross-sectional area.

Warm Forging. The process of forming thermoplastic sheets or billets into desired shapes by pressing them between dies in a press, when either the material has been preheated or hot dies are used. The blanks may be billets formed by extrusion, or may be die cut from sheets. Dies may be inexpensive castings of the "matched metal" type. The process is usually employed for relatively thick parts of polyolefins, including glass-reinforced compounds. One method of performing the process employs two shaped metal punches with a common die ring located between the upper and lower punches, the ring containing a preheated plastic billet. Another method uses a rubber pad on one platen to force the plastic sheet or billet into conformity with a metal die on the other platen. Warm forgings accurately reproduce mold detail, and exhibit low mold shrinkage, improved thermal dimensional stability, impact strength and creep resistance. See also COLD FORMING, SOLID PHASE FORMING.

Warp. In the fabric industry, those threads in a woven fabric which are parallel to the selvedge, that is, placed lengthwise in the loom.

Warpage. Distortion caused by nonuniform change of internal stresses. See also DISHED. (ASTM D 883-75a).

Wash. An area where the reinforcement has moved during closing of the mold, resulting in a resin-rich area.

Water Absorption. The amount of water absorbed by a plastic article when immersed in water for a stipulated period of time. All plastics will absorb moisture to some extent, varying from almost zero in the case of PTFE to complete solubility for some types of PVA and polyethylene oxide. Water absorption can cause swelling, dissolving, leaching, plasticizing and/or hydrolyzing, which events can result in discoloration, embrittlement, loss of mechanical and electrical properties, lower resistance to heat and weathering, and stress cracking. However, the amount of water absorbed by any particular polymer does not necessarily indicate the extent of harmful effects that might result. The ASTM test for water absorption of plastics is D 570-63, found in Part 35 of the 1976 Book of ASTM Standards.

Water-Extended Polyesters. (WEP) Unsaturated polyester resins that have been extended with water rather than conventional fillers. In the process usually employed, a liquid polyester resin containing a promoter is mixed with water containing MEK peroxide as a catalyst in a blending machine that produces a low-viscosity emulsion that can be poured into open molds, and which cures without additional heat by its own exothermic reaction. When the process conditions are properly controlled, the water except that at the surface of the casting is retained indefinitely as microscopic droplets in the cured casting. The resultant structure is similar to that of wood, and WEP castings can be nailed, stapled and finished in much the same way as wood. However, variations in physical properties such as shrinkage and warpage upon long-term loss of water, are greater than those of wood. Within these limitations, the working properties of WEP combined with their economy and ease of processing have stimulated interest in their use as a replacement for wood, plaster and other materials in some furniture parts, and decorative applications such as frames, lamp bases and statuary.

Water-Soluble Resins. Synthetic water-soluble resins are produced by polymerization reactions in which the chain growth results from breaking of ring structures or double bonds of the monomers. Examples of such resins are alkyl- and hydroxyalkyl cellulose derivatives, carboxymethyl cellulose, polyvinyl alcohol, polyvinyl pyrrolidone, polyacrylic acid, polyacrylamide, polyethylene oxide, and polyethylene-imide.

Water Vapor Transmission. The amount of water vapor passing through a given area of a plastic sheet or film in a given time, when the sheet or film is maintained at a constant temperature and when its faces are exposed to certain different relative humidities. The result is usually expressed as grams per 24 hours per square meter.

Watt. The unit of power expressing the time rate at which work is done. Equivalent to the IUPAC-approved Joule, one watt is equivalent to ten million ergs per second. In the case of alternating current, the watt is computed as the product of volts times amperes and the phase angle between the current and the impressed electromotive force.

Weathering. A broad term encompassing exposure of plastics to solar or ultraviolet light, temperature, oxygen, humidity, rain, snow, wind, and air-borne biological and chemical agents.

Weathering, Artificial. See ARTIFICIAL WEATHERING.

Weatherometer. An instrument which is utilized to subject articles to accelerated weathering conditions, e.g. rich UV source and water spray.

Web. (1) A thin sheet in process in a machine. In extrusion coating, the *molten web* is that which issues from the die, and the *substrate web* is the material being coated. (2) A continuous length of sheet material handled in roll form as contrasted with the same material cut into sheets.

Web Coating. See EXTRUSION COATING, SPREAD COATING.

Weber. The new SI unit of magnetic flux, defined as the magnetic flux which, linking a circuit of one turn, produces in it an electromotive force of one volt as it is reduced to zero at a uniform rate in one second.

Web Gate. See DIAPHRAGM GATE.

Weft. The transverse threads or fibers in a woven fabric; those fibers running perpendicular to the warp. Also called *filler* or *filling yarn.*

Weight. The force with which a body is attracted to the earth. The Cgs unit of weight is the dyne, the force which will produce an acceleration of one centimeter per second per second in a gram mass. In the new SI system, it is proposed to replace the dyne with the newton, expressed as $Kg.m/s^2$. When an object is weighed on a balance with arms, the numerical values of dynes, weight, mass and newtons are identical because the force of gravity is cancelled out. In using a spring scale, the force of gravity causes a slight difference between weight and mass.

Weight-Average Molecular Weight. (M_w) The first moment of a plot of the weight of polymer in each molecular weight range against molecular weight. The value of M_w can be estimated by light scattering or sedimentation equilibrium measurements. See also MOLECULAR WEIGHT, MOLECULAR WEIGHT DISTRIBUTION.

Weissenberg Effect. A phenomenon sometimes encountered in rotational viscometric studies at high speeds, particularly when the alignment between the cups is not perfect, characterized by a tendency of the polymer solution to climb the wall of the cup or cylinder which is rotating.

Weissenberg Rheogoniometer. A cone-and-plate rheometer capable of measuring both the viscous and the elastic response of polymer melts. The tangential force caused by resistance to flow when the cone turns with respect to the plate is a measure of the viscous flow. The normal force, that is perpendicular to the plane of rotation and tending to separate the cone and plate, represents elasticity.

Weldbonding. A process developed in the USSR and introduced in the U.S. by the Air Force Materials Laboratory. The process as developed for the aerospace industry combined spot welding and adhesive bonding of aluminum structures. It has been found to provide an economical and efficient method of laying-up and oven curing large epoxy-bonded assemblies.

Welding. The joining of two or more pieces of plastic by fusion at adjoining or nearby areas, either with or without the addition of plastic from another source. The term includes HEAT SEALING, with which it is synonymous in some countries, including Britain; but heat sealing is more often used in connection with films and sheeting in the U.S. See also BUTT-FUSION, ELECTROMECHANICAL VIBRATION WELDING, DIELECTRIC HEAT SEALING, EXTRUDED BEAD SEALING, HEAT SEALING, FRICTION WELDING, HIGH FRE-QUENCY WELDING, HOT GAS WELDING, HOT PLATE WELDING, INDUCTION WELDING, IMPULSE SEALING, JIG WELDING, RADIO FREQUENCY WELDING, SOLVENT WELDING, SPIN WELDING, STITCHING, THERMOBANDE WELDING, ULTRASONIC WELDING, VIBRATION WELDING.

Weld Mark. (weld line, flow line) A flaw on a molded plastic article caused by the meeting of two flow fronts during the molding or extrusion operation.

Werner Complex. See COORDINATION COMPOUND.

Wet Flexural Strength. The flexural strength measured after boiling the specimen in water.

Wet Layup. In the reinforced plastics industry, the process of forming an article by first apply-ing a liquid resin (sometimes called gel coat) to the surface of a mold, then applying a reinforc-ing backing layer.

Wet-Out. The condition of an impregnated reinforcement wherein substantially all voids between the sized strands and filaments are filled with resin.

Wet-Out Rate. The time required for a plastic to fill the interstices of a reinforcement material and wet the surfaces of the fibers, usually determined by optical or light transmission means.

Wet Spinning. See SPINNING.

Wet Strength. The strength of paper when saturated with water, especially used in discussions of processes whereby the strength of paper is increased by treatment with plastic resins. Also, the strength of an adhesive joint determined immediately after removal from a liquid in which it has beeɪ. immersed under specified conditions of time, temperature and pressure.

Wetting Agents. Compounds that cause a liquid to penetrate more easily into, or to spread over the surface of, another material. Common wetting agents are soaps, detergents and surface-active agents. They are widely used in polymerization reactions and in preparing emulsions of plastics.

Wet Winding. The filament winding process wherein the strand is impregnated with resin just prior to contact with the mandrel.

Whiskers. A colloquial term used for nearly perfect, single-crystal fibers produced synthetically under controlled conditions from minerals such as aluminum oxide, aluminum nitride, beryllium oxide, boron carbide, graphite, magnesium oxide, silicon carbide and silicon nitride. They range in diameter from 1 to 30 microns, and in length from 1 to several thousand microns. Whiskers are available as loose fibers, mats or felts. Having a fiber tensile strength from 5 to 10 times that of glass, they impart extremely high strength to reinforced plastics.

White Bole. (bolus alba) See KAOLIN.

Whiting. A finely divided form of CALCIUM CARBONATE, which see, obtained by milling high-calcium limestone, marble, shell or chemically precipitated calcium carbonate.

Window. A globule of incompletely plasticated material in a thermoplastic film, sheet or molding which is visible when viewed by transmitted light. It is equivalent to *fish eye,* except that the term window is usually employed when the material is colored or opaque.

Wire Train. The entire assembly which is utilized to produce a resin-coated wire, normally consisting of an extruder, a crosshead and a die, cooling means, and feed and take-up spools for the wire.

Wollastonite. See CALCIUM SILICATE.

Wood Alcohol. An alcohol obtained by the destructive distillation of wood, containing METHYL ALCOHOL (which see) as its principal ingredient. It is highly toxic, capable of causing blindness or death when inhaled or ingested.

Wood Flour. (wood meal) Finely pulverized dried wood, used as a filler in thermosetting molding compounds such as phenolics and ureas. The woods used are resin-free softwoods such as pine, fir and spruce, and, to a lesser extent, hardwoods. Wood shredded to fibrous form is also used as a reinforcement rather than a filler.

Woodgraining. A group of processes used to impart wood-like appearance to sheets or shaped articles. The substrates may be of plastic, wood, steel or any other material. Among the processes used are conventional laminating techniques, multiple-coat painting methods, hot stamping, and introduction of several colors into the melt during molding.

Work. When a force acts against resistance to produce motion in a body the force is said to do work. The old cgs unit of work is the erg, or force of one dyne acting through a distance of one centimeter. In the new SI, this unit is to be replaced by the newton meter, 1×10^{-7} ergs.

Working Life. (pot life) The period during which a compound, after mixing with a catalyst, solvent or other compounding ingredients, remains suitable for its intended use.

Wrinkle. (n) An imperfection in reinforced plastics that has the appearance of a wave molded into one or more plies of fabric or other reinforcing material. (ASTM D 883-75a).

WVT. Abbreviation for WATER VAPOR TRANSMISSION, which see.

Xenon Arc Light Ageing. A test for evaluating the light stability of plastics, employing a xenon gas discharge lamp of special design which emits radiation duplicating the spectra of natural sunlight better than most artificial sources.

X-ray Microscopy. The technique of examining x-rays by means of a microscope. In a variation called Point Projection X-ray Microscopy, an enlarged image is obtained from x-rays emitted from a pinhole point source. The technique is useful for studying the structure of materials such as foamed plastics, laminates, fibers and filaments.

X-Rays. Electromagnetic waves with wavelengths ranging from 0.01 to 50 Ångstrom units, produced by the bombardment of a target with cathode rays.

Xylene. $C_6H_4(CH_3)_2$ A mixture of three isomers, ortho-, meta-, and para-xylene, used as a solvent for alkyd resins.

ortho-Xylene. $1,2\text{-}C_6H_4(CH_3)_2$ (1,2-dimethylbenzene) A colorless liquid used as a feedstock in the production of phthalic anhydride. It can be extracted from Xylene by distillation, and can be isomerized to para-Xylene.

para-Xylene. (1,4-Dimethylbenzene) $1,4\text{-}C_6H_4(CH_3)_2$ A colorless liquid used in the synthesis of terephthalic acid and dimethylterephthalate, both of which are intermediates for polyester fibers and films.

para-Xylene-alpha,alpha´-Diol. $C_6H_4(CH_2OH)_2$ A white crystalline solid, used as a crosslinking agent in polyurethanes, and in the production of polyesters and polycarbonates.

Xylenol Resin. A phenolic-type resin produced by condensing xylenol (a coal tar derivative) with an aldehyde.

Xylox Resins® . A family of heat-resistant resins made by the condensation of aralkyl ethers and phenols, resulting in hydroxy-phenylene-p-xylene prepolymers which can be cured to hard, intractable resins by reaction with hexamethylenetetramine or epoxy compounds. These thermosetting resins have all the processing advantages, chemical resistance and machining qualities of phenolic and epoxy resins, with superior mechanical strength and electrical properties at elevated temperatures. Xylox is the trade name of Albright & Wilson Ltd. for the resins marketed in the U.S. by Ciba-Geigy.

Yarn. Generic term for strands of fibers or filaments in a form suitable for weaving or otherwise intertwining to form a fabric.

Yellowness Index. A measure of the tendency of plastics to turn yellow upon long-term exposure to light.

Yield Point. In tensile testing, yield point is the first point on the stress-strain curve at which an increase in strain occurs without an increase in stress. This is the point at which permanent deformation of the stressed specimen begins to take place.

Yield Strength. The stress at which a material exhibits a specified limiting deviation from the proportionality of stress to strain. Unless otherwise specified, this stress will be the stress at the yield point. See also OFFSET YIELD STRENGTH. (ASTM D 638-60T).

Yield Value. In viscosity measurements, yield value is the force that must be applied to a fluid layer before any movement is produced. A good example of a fluid with a yield value is ketchup.

Young's Modulus. The ratio of tensile stress to tensile strain below the proportional limit. See also MODULUS OF ELASTICITY.

Zahn Viscosity Cup. A small U-shaped cup suspended form a looped wire, with an orifice of any one of five sizes at the base. The entire cup is submerged in the test sample then withdrawn. The time in seconds from the moment the top of the cup emerges from the sample until the stream from the orifice first breaks is the measure of viscosity. This instrument provides a rapid, although somewhat less precise, method of measuring kinematic viscosity of liquids.

Z-Calender. A calender with four rolls arranged so that the material passes through them in the form of the letter Z.

Zein. A naturally occuring, high molecular weight copolymer of amino acids linked by peptide bonds, derived from corn. It is considered a member of the protein family of plastics, the main member of which is casein plastic. Zein plastics, rarely used today, have been used for fibers, films and coatings.

Ziegler Catalysts. A large group of catalysts made by reacting a compound of a transition metal chosen from groups IV through VIII of the periodic table with an alkyl, hydride or other compound of a metal from groups I through III. A typical example is the reaction product of an aluminum alkyl with titanium tetrachloride or titanium trichloride. These catalysts were first developed by the German scientist Karl Ziegler for the polymerization of ethylene. Subsequent work by G. Natta showed that these and similar catalysts are useful for preparing stereoregular polyolefins; thus, the family of catalysts is sometimes called Ziegler-Natta catalysts.

Zinc Baryta White. See LITHOPONE.

Zinc Borates. White, amorphous powders of indefinite composition, containing various amounts of zinc oxide and boric oxide. They are used as flame retardants in PVC, polyvinylidene chloride, polyesters and polyolefins, quite often in combination with antimony trioxide.

Zinc Oxide. ZnO (Chinese white, zinc white, flowers of zinc) An amorphous white or yellowish powder, used as a pigment in plastics. It is said to have the greatest ultra-violet light absorbing power of all commercially available pigments. It has been found that mixtures of zinc oxide and secondary organic additives such as butyl zimate have a strong synergistic effect and increase the protection against UV light as much as 66 times that of ZnO alone.

Zinc Palmitate. $Zn(C_{16}H_{31}O_2)_2$ An amorphous white powder used as a lubricant in plastics.

Zinc Ricinoleate. A white powder used as a stabilizer in vinyl plastics.

Zinc Stearate. $Zn(C_{18}C_{35}O_2)_2$ A white powder used as a lubricant.

Zirconium Oxide. ZrO_2 (zirconia, zirconium dioxide) A white amorphous powder used as a pigment when good electrical properties are required.

ZMBT. Abbreviation for the zinc salt of 2-mercaptobenzothiazole, a light cream colored powder used as an accelerator in natural rubber and butadiene-styrene copolymers.